OXFORD APPLIED MATHEMATICS
AND COMPUTING SCIENCE SERIES

Nonlinear Ordinary Differential Equations
(Second edition)

D. W. JORDAN AND P. SMITH

**Oxford Applied Mathematics
and Computing Science Series**

General Editors
J. N. Buxton, R. F. Churchhouse, and
A. B. Tayler

OXFORD APPLIED MATHEMATICS AND COMPUTING SCIENCE SERIES

D. W. JORDAN
AND ι

P. SMITH
University of Keele

Nonlinear Ordinary Differential Equations

SECOND EDITION

CLARENDON PRESS · OXFORD

Oxford University Press, Walton Street, Oxford OX2 6DP

Oxford New York Toronto
Delhi Bombay Calcutta Madras Karachi
Petaling Jaya Singapore Hong Kong Tokyo
Nairobi Dar es Salaam Cape Town
Melbourne Auckland

and associated companies in
Berlin Ibadan

Oxford is a trade mark of Oxford University Press

Published in the United States
by Oxford University Press, New York

First published 1977
Second edition 1987
Reprinted 1988 (twice) (with corrections), 1989, 1990

British Library Cataloguing in Publication Data
Jordan, D. W.
Nonlinear ordinary differential equations.
2nd ed.—(Oxford applied mathematics
and computing science series)
1. Differential equations, Nonlinear
I. Title II. Smith, P. (Peter). 1935–
515.3'52 QA372
ISBN 0–19–859657–X
ISBN 0–19–859656–1 Pbk

Library of Congress Cataloging in Publication Data
Jordan, D. W. (Dominic William)
Nonlinear ordinary differential equations.
(Oxford applied mathematics and computing science series)
Bibliography: p.
Includes index.
1. Differential equations, Nonlinear
I. Smith, Peter, 1935– .II. Title. III. Series.
QA372.J648 1987 515.3'52 86–18000
ISBN 0–19–859657–X
ISBN 0–19–859656–1 (pbk.)

Printed in Great Britain
by Courier International Ltd
Tiptree, Essex

Preface

SINCE THE first printing of this book in 1977 the literature on nonlinear systems has grown very greatly. The wide applicability of the subject to the physical, biological, and social sciences generates a continuous supply of new problems of practical and theoretical interest. Accordingly this new edition has been thoroughly revised, several chapters have been completely rewritten, and it includes a completely new chapter on bifurcations, structural stability, and the chaotic behaviour of oscillators treated from an elementary point of view.

The book grew from courses in nonlinear differential equations given over many years at the University of Keele. It is an introduction to dynamical systems in the context of systems of differential equations, intended for students of mathematics and science who are mainly interested in the more direct applications of the subject. The level is about that of a second- or third-year undergraduate course in Britain. It has been found that much of the material in Chapters 1 to 4 and 8, 10, and 11 can be covered in a one-term course by students having a background of elementary differential equations and some linear algebra. The book is designed to accommodate courses of varying emphasis, the chapters forming fairly self-contained groups from which a coherent selection can be made without large parts of the argument being lost.

We have given the subject a qualitative slant throughout, emphasizing how the presence of a nonlinear element may introduce totally novel phenomena. In order that these new features should not seem too shadowy, many of the diagrams are drawn from computed data and are not merely sketches. In Chapter 3 we advise the reader on how to do his own computing using the simplest means, a practice which adds a dimension of conviction to the whole subject.

Chapters 1 to 4 treat mainly two-dimensional systems and second-order equations in the phase plane. The treatment is kept at an intuitive and elementary level, but we try to encourage the reader to feel that almost immediately he has available useful new investigative techniques. The main features of the phase plane: equilibrium points, periodic solutions, limit-cycles, domains of attraction; and the associated stability ideas are treated informally. Quantitative estimates for periodic solutions are obtained by energy balance and harmonic balance methods.

Chapters 5 to 7 deal with expansions in terms of a small parameter, singular perturbations, and the phenomena associated with forced oscillations: amplitude limitation, harmonic and subharmonic response, and jump phenomena and entrainment. Stability and instability of the solutions is determined by analysis of the van der Pol plane for these cases. Chapters 8 to 10 deal formally with the question of stability. Chapter 8 includes what is necessary of the elementary theory of linear systems, and Chapter 9 includes the Floquet theory and the theory of Mathieu's equation required for determining the stability of paths by perturbation when the paths are given only approximately. In Chapter 11 criteria for the existence of periodic solutions in some representative cases are presented. Chapter 12 illustrates some of the basic ideas and vocabulary connected with bifurcations, structural stability, chaotic oscillators, and strange attractors.

There are over 400 examples, some worked in full in the text and others left as exercises, often with solutions and hints. The exercises provide a way of indicating developments for which there is no room in the text, and sometimes to present more specialized illustrative material. On the whole we have tried to keep the text itself free from scientific technicality and to present equations in a simple reduced form, believing that students have enough to do to follow the underlying arguments.

We are grateful to several correspondents for suggestions and observations on the earlier editions, and to the University of Keele for computing and other facilities.

Keele, 1986   D.W.J.
                                                                                          P.S.

Contents

1 Second-order differential equations in the phase plane

IT IS NOT in general possible to obtain analytic solutions to an arbitrary differential equation. This is not simply because ingenuity fails, but because the repertory of standard functions (polynomials, exp, sin, and so on) in terms of which solutions may be expressed is too limited to accommodate the variety of differential equations encountered in practice. Even if an analytic solution can be found, the 'formula' is often too complicated to display clearly the principal features of the solution; this is particularly true of implicit solutions and of solutions which are in the form of integrals or infinite series.

The qualitative study of differential equations is concerned with how to deduce important characteristics of the solutions of differential equations without actually solving them. Since these considerations apply equally to linear equations and to the much greater variety of nonlinear equations, this book is mainly about nonlinear equations. In this chapter we introduce a geometrical device, the phase plane, which is used extensively for obtaining directly from the differential equation such properties as equilibrium, periodicity, unlimited growth, stability, and so on. The classical pendulum problem shows how the phase plane may be used to reveal all the main features of the solutions of a particular differential equation.

1.1. Phase diagram for the pendulum equation

The simple pendulum consists of a bob of mass m suspended from a fixed point O by a light string or rod of length a, which is allowed to swing in a vertical plane. If no friction is present then the equation of motion is

$$ma^2\ddot{\theta} + mga\sin\theta = 0, \qquad (1.1)$$

where θ is the inclination of the string to the downward vertical (Fig. 1.1). Since $\ddot{\theta} = \dot{\theta}(\mathrm{d}\dot{\theta}/\mathrm{d}\theta)$, eqn (1.1) becomes

$$ma^2\dot{\theta}\frac{\mathrm{d}\dot{\theta}}{\mathrm{d}\theta} + mga\sin\theta = 0.$$

This equation relates $\dot{\theta}$ and θ instead of θ and t. By integrating with

FIG. 1.1. The simple pendulum

respect to θ we obtain

$$ma^2 \int \dot\theta \, d\dot\theta + mga \int \sin\theta \, d\theta = C$$

where C is a constant. Therefore

$$\tfrac{1}{2}ma^2\dot\theta^2 - mga\cos\theta = C. \tag{1.2}$$

Equation (1.2) expresses conservation of energy, the terms on the left being, in order, the kinetic energy and potential energy of the pendulum. The value of C for a particular motion can be settled by using the initial conditions: the values of θ and $\dot\theta$ at $t = 0$. Equation (1.2) then gives the relation between $\dot\theta$ and θ for the motion corresponding to these initial conditions. By choosing different values for C we can obtain this relation for any possible motion.

Equation (1.2) is also a differential equation for θ in terms of t, but it cannot be solved in terms of elementary functions (see McLachlan 1956). It is therefore not easy to obtain a useful representation of θ as a function of time. We shall show how it is possible, by working directly with eqn (1.2), to reveal the main characteristics of the solutions.

The relation (1.2) can be represented in a diagram. Set up a cartesian *phase plane* having θ and $\dot\theta$ as its axes (Fig. 1.2) and plot the one-parameter family of curves generated by (1.2) for different values of C.

A given pair of values $(\theta, \dot\theta)$ is called a *state* of the 'system' (in this case a pendulum), and the diagram shows how any state evolves as time progresses. We know that a given state determines all subsequent states,

FIG. 1.2. Phase diagram for the simple pendulum

since it serves as initial conditions for the subsequent motion.

The curves depicted in Fig. 1.2 are known as the *phase paths*, *trajectories*, or *integral curves* corresponding to eqn (1.1), and the complete figure is called the *phase diagram* or *phase portrait* for the system. Each phase path corresponds to a particular possible motion of the system. Associated with each path is a *direction*, indicated by an arrow in Fig. 1.2, showing how the state of the system changes as time *increases*; the direction of the arrows is settled by observing that when $\dot{\theta}$ is positive, θ must be increasing with time and when $\dot{\theta}$ is negative, θ must be decreasing with time.

We shall see that despite the non-appearance of the time variable in the phase plane display, we can deduce several physical features of the pendulum's possible motions from Fig. 1.2. Consider first the possible states of physical equilibrium of the pendulum. The obvious one is when the pendulum hangs without swinging; then $\theta = 0$, $\dot{\theta} = 0$, which corresponds to the origin in Fig. 1.2. The corresponding function $\theta(t)$ $= 0$ is a perfectly legitimate *constant solution* of (1.1): the phase path degenerates to a *single point*. If the suspension consists of a light rod there is a second position of equilibrium, where it is balanced vertically on end. This is the state $\theta = \pi$, $\dot{\theta} = 0$, another constant solution, represented by the point A on the phase diagram. The same physical condition is described by $\theta = -\pi$, $\dot{\theta} = 0$, represented by the point B, and indeed the state $\theta = n\pi$, $\dot{\theta} = 0$, where n is any integer, corresponds to one of these two equilibrium conditions. In fact we have displayed in Fig. 1.2 only part of the phase diagram, whose pattern repeats periodically: there is not in this case a one-to-one relationship between the physical condition of the pendulum and points on its phase diagram.

Since the points O, A, B represent states of physical equilibrium, they are called *equilibrium points* or *critical points* on the phase diagram.

Now consider the family of closed curves immediately surrounding the origin in Fig. 1.2. These indicate periodic motions, in which the pendulum swings to and fro about the vertical. The amplitude of the swing is the maximum value of θ encountered on the curve. For small enough amplitudes, the curves represent the usual 'small amplitude' solutions of the pendulum equation in which eqn (1.1) is simplified by writing $\sin \theta \approx \theta$. The phase paths are nearly ellipses in the small amplitude region.

The wavy lines at the top and bottom of Fig. 1.2, on which $\dot{\theta}$ is of constant sign and θ continuously increases or decreases, correspond to whirling motions of the pendulum. The fluctuations in $\dot{\theta}$ are due to the gravitational influence, and for such phase paths on which $\dot{\theta}$ is very large these fluctuations become imperceptible: the phase paths become nearly straight lines parallel to the θ axis.

We can discuss also the *stability* of the two typical equilibrium points O and A. If the initial state is displaced slightly from O, it goes on to one of the nearby closed curves and the pendulum oscillates with small amplitude about O. We describe the equilibrium point at O as being *stable*. If the initial state is slightly displaced from A (the vertically upward equilibrium position) however, it will normally fall on a phase path which carries the state far from the equilibrium state A into a large oscillation or a whirling condition (see Fig. 1.3). This equilibrium point is therefore described as *unstable*.

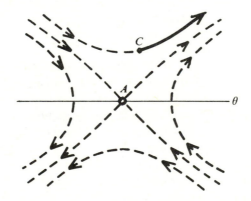

Fig. 1.3. Unstable equilibrium point for the pendulum; displaced state C

1.2. Autonomous equations in the phase plane

The second-order differential equation of general type

$$\ddot{x} = f(x, \dot{x}, t) \qquad (1.3)$$

can be interpreted as an equation of motion for a mechanical system, in which x represents displacement of a particle of unit mass, $\dot{x}$ its velocity, $\ddot{x}$ its acceleration, and f the applied force, so that (1.3) expresses Newton's law of motion for the particle:

$$\text{acceleration} = \text{force per unit mass}.$$

A mechanical system is in equilibrium if its state does not change with time. This implies that *an equilibrium state corresponds to a constant solution* of (1.3), and conversely. A constant solution implies in particular that $\dot{x}$ and $\ddot{x}$ must be simultaneously zero. Note that $\dot{x} = 0$ is not alone sufficient for equilibrium: a swinging pendulum is instantaneously at rest at its maximum amplitude, but this is obviously not a state of equilibrium. Such constant solutions are therefore the constant solutions (if any) of

$$f(x, 0, t) = 0. \qquad (1.4)$$

We distinguish between two types of equation:
(i) the *autonomous* type in which f does not depend explicitly on t;
(ii) the *non-autonomous* or *forced* equation where t appears explicitly in the function f.

A typical non-autonomous equation is the linear oscillator with a harmonic forcing term

$$\ddot{x} + k\dot{x} + \omega_0^2 x = F \cos \omega t,$$

in which $f(x, \dot{x}, t) = -k\dot{x} - \omega_0^2 x + F \cos \omega t$. There are no equilibrium states. Equilibrium states are not usually associated with non-autonomous equations although they can occur as, for example, in the equation (Mathieu's equation, Chapter 9)

$$\ddot{x} + (\alpha + \beta \cos t)x = 0,$$

which has an equilibrium state at $x = 0$.

In the present chapter we shall consider only autonomous equations; in this case t is absent and we shall write, instead of (1.3),

$$\ddot{x} = f(x, \dot{x}). \qquad (1.5)$$

The constant solutions (which imply equilibrium states) are the solutions (if any) of

$$f(x, 0) = 0. \tag{1.6}$$

For example, the equation $\ddot{x} = (1 - x^2) + x\dot{x}$ has the constant solutions $x = 1$ and $x = -1$.

Now consider the representation of eqn (1.5) on a phase plane. The state of the system at a time $t = t_0$ consists of the pair of numbers $(x(t_0), \dot{x}(t_0))$. This state can be regarded as *a pair of initial conditions*, and therefore determines subsequent (and earlier) states. The succession of states can be depicted on a phase plane, as in Section 1.1 for the pendulum, in which the axes are for x and $\dot{x}$. We will relabel the $\dot{x}$ axis as y, thus defining y by

$$\dot{x} = y; \tag{1.7}$$

then $\ddot{x} = \dot{y}$ and (1.5) becomes

$$\dot{y} = f(x, y). \tag{1.8}$$

Equations (1.7) and (1.8) can be looked on as two simultaneous first-order equations for the functions $x(t)$, $y(t)$ (or $\dot{x}(t)$). To obtain the relation between y and x which will give the phase paths, eliminate t by dividing (1.8) by (1.7): $\dot{y}/\dot{x} = f(x, y)/y$, or

$$\frac{dy}{dx} = \frac{f(x, y)}{y}. \tag{1.9}$$

The solutions of this equation are the phase paths.

Example 1.1   *Find the phase paths for the equation $\ddot{x} + \alpha \sin x = 0$.*
This is essentially eqn (1.1). We define

$$\dot{x} = y$$

and the equation gives

$$\dot{y} = -\alpha \sin x.$$

The equation for the phase paths is

$$\frac{dy}{dx} = -\frac{\alpha \sin x}{y},$$

which separates to give

$$\int y \, dy = -\alpha \int \sin x \, dx + C,$$

or

$$\tfrac{1}{2}y^2 - \alpha \cos x = C,$$

where C is the parameter of the phase paths. The diagram in the plane is as Fig. 1.2.

The process of obtaining constant solutions of eqn (1.5) amounts to putting $\dot{x} = 0$, $\ddot{x} = 0$, or to putting $\dot{x} = 0$, $\dot{y} = 0$ in (1.7) and (1.8). An equilibrium point (x_0, y_0) on the phase diagram is therefore any solution of the pair

$$y = 0, f(x, y) = 0. \tag{1.10}$$

Note that

(a) *equilibrium points are always situated on the x axis*;

(b) *except at equilibrium points the phase paths cut the x axis at right angles* (from (1.7), (1.8) and (1.10));

(c) *closed paths represent periodic solutions, since on completing a circuit the original state is returned to, and the motion simply repeats itself indefinitely.*

Now consider the time taken between two points on a phase path. Figure 1.4 shows a segment $\mathscr{C}$ of the phase path joining two points A and B. P represents any intermediate state: we call it a *representative point* of the path. P moves along $\mathscr{C}$ with velocity $(\dot{x}, \dot{y})$, or $(\dot{x}, \ddot{x})$ by (1.7). The time taken, T_{AB}, is therefore given by

$$T_{AB} = \int_{\mathscr{C}} \mathrm{d}t \equiv \int_{\mathscr{C}} \left(\frac{\mathrm{d}x}{\mathrm{d}t}\right)^{-1} \left(\frac{\mathrm{d}x}{\mathrm{d}t}\right) \mathrm{d}t = \int_{\mathscr{C}} \frac{\mathrm{d}x}{y}, \tag{1.11}$$

which is calculable when $\mathscr{C}$ is given. The phase diagram therefore merely suppresses and does not completely lose track of the time variable since (1.11) depends only on the geometry of the phase path.

FIG. 1.4.

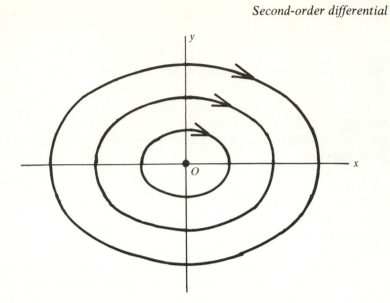

Fig. 1.5. Centre for the simple harmonic oscillator

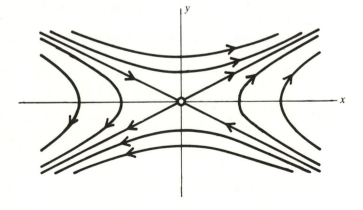

Fig. 1.6. Saddle point

Example 1.2 *Construct the phase diagram for the simple harmonic oscillator equation* $\ddot{x} + \omega^2 x = 0$.

This approximates to the pendulum equation for small-amplitude swings. Corresponding to eqns (1.7) and (1.8) we have

$$\dot{x} = y, \quad \dot{y} = -\omega^2 x.$$

There is one equilibrium point at

$$x = 0, \quad y = 0.$$

The phase paths are the solutions of

$$\frac{dy}{dx} = -\omega^2 \frac{x}{y},$$

which is separable, leading to

$$y^2 + \omega^2 x^2 = C.$$

The phase portrait therefore consists of a family of ellipses concentric with the origin (Fig. 1.5). All solutions are therefore periodic. The equilibrium point is stable.

An equilibrium point surrounded in its immediate neighbourhood (not necessarily over the whole plane) by closed paths is called a *centre*. A centre is a *stable* equilibrium point.

Example 1.3 *Construct the phase diagram for the equation* $\ddot{x} - \omega^2 x = 0$.

The equivalent first-order pair ((1.7) and (1.8)) are

$$\dot{x} = y, \quad \dot{y} = \omega^2 x.$$

There is a single equilibrium point at

$$x = 0, \quad y = 0.$$

The phase paths are the solutions of

$$\frac{dy}{dx} = \omega^2 \frac{x}{y};$$

therefore their equations are

$$y^2 - \omega^2 x^2 = C.$$

These paths are hyperbolas with asymptotes $y = \pm \omega x$ as shown in Fig. 1.6.

An equilibrium point with paths of this type in its neighbourhood is called a *saddle point*. Such a point is *unstable* since a small displacement from the equilibrium state will generally involve a solution which goes far from this state. (The question of stability is discussed precisely in Chapter 8.) In the figures, stable equilibrium points are usually indicated by a full dot ●, and unstable ones by an 'open' dot ○.

1.3. Conservative systems

In mechanical terms, consider a system with one degree of freedom and let x be a generalized coordinate (position, angle, etc.). Let $\mathscr{T}$ and $\mathscr{V}$ be the kinetic and potential energy functions, and assume that they take the form

$$\mathscr{T} = \tfrac{1}{2}m(x)\dot{x}^2, \quad \mathscr{V} = \mathscr{V}(x) \tag{1.12}$$

where m is another function, $m(x) > 0$. If the system is *conservative*, then the total energy, $\mathscr{E}$, is constant during motion:

$$\tfrac{1}{2}m(x)\dot{x}^2 + \mathscr{V}(x) = \mathscr{E}, \quad \text{constant,} \tag{1.13}$$

which gives the phase paths. The type of equation of motion which leads to (1.13) can be obtained by taking the time derivative of (1.13):

$$m(x)\ddot{x} + \tfrac{1}{2}m'(x)\dot{x}^2 + \mathscr{V}'(x) = 0. \tag{1.14}$$

Equation (1.14) can be simplified by introducing a new variable, u, in place of x (in effect another generalized coordinate) by

$$u = \int \sqrt{\{m(x)\}}\, \mathrm{d}x.$$

Equation (1.14) becomes

$$\ddot{u} + \mathscr{L}'(u) = 0, \tag{1.15}$$

where $\mathscr{L}'(u) = \mathscr{V}'(x)m^{-1/2}(x)$, and the corresponding energy-type equation for the phase paths is

$$\tfrac{1}{2}\dot{u}^2 + \mathscr{L}(u) = C. \tag{1.16}$$

We shall need only the form (1.15), but it should be remembered that equations of the form (1.14) also represent conservative systems.

Translating (1.15) and (1.16) back into the usual notation, we take the *standard equation for a conservative system* to be

$$\ddot{x} = f(x), \tag{1.17}$$

where the 'force per unit mass' f, is independent of $\dot{x}$. Suppose that f is continuous. Define for the new problem a function $\mathscr{V}$ by

$$\mathscr{V}(x) = -\int f(x)\,dx, \quad \text{or} \quad \mathscr{V}'(x) = -f(x). \tag{1.18}$$

Then $\mathscr{V}$ may be spoken of as the potential energy function (or just the 'potential') for (1.17). The equilibrium points of (1.17) are then given by

$$-f(x) = \mathscr{V}'(x) = 0 \tag{1.19}$$

(that is, the stationary points of the potential energy, as expected). By writing $\dot{x} = y$ as in (1.7), the phase paths are given by the 'energy equation'

$$\tfrac{1}{2}y^2 + \mathscr{V}(x) = C \tag{1.20}$$

for various values of C, or by

$$y = \pm\sqrt{(2C - 2\mathscr{V}(x))}. \tag{1.21}$$

Figure 1.7 shows the types of equilibrium point arising from the three types of turning point of $\mathscr{V}$: a local minimum always leads to a centre, a maximum to a saddle point, and a point of inflection to a cusp.

The construction can be thought of in the following way. For a fixed C, $2C - 2\mathscr{V}(x)$ can be read off from the top frame for the range of x for which this is non-negative, y calculated from (1.21) for each x, and the symmetrically-placed pair of points inserted in the lower frame.

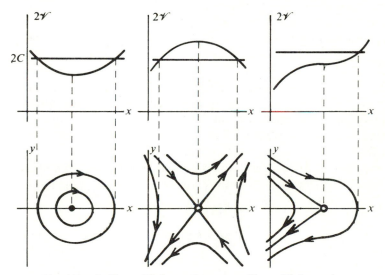

FIG. 1.7. ● Stable equilibrium point, ○ unstable equilibrium point

Alternatively we can view the matter in terms of $f(x)$, regarded as the 'restoring force' for a particle on a nonlinear spring. If $f(x)$ changes sign from positive to negative on passing through the equilibrium point ($\mathscr{V}$ having a minimum), the system behaves like an ordinary linear spring and oscillates, implying a centre. If it changes from negative to positive, the particle is repelled from the equilibrium point and there is a saddle.

Example 1.4 *Sketch the phase diagram for the equation* $\ddot{x} = x^3 - x$.

This represents a conservative system (in fact the pendulum equation (1.1) for moderate amplitudes, writing $\sin\theta \approx \theta - \frac{1}{6}\theta^3$, leads to an equation reducible to this one). We have $\mathscr{V}''(x) = x - x^3$ and $\mathscr{V}(x) = -\frac{1}{4}x^4 + \frac{1}{2}x^2$ (to an additive constant). Figure 1.8 shows the construction of the phase diagram.

There are three equilibrium points: at $(0,0)$, a centre; at $(1,0)$, a saddle; and at $(-1,0)$, a saddle. The reconciliation between the types of phase path originating round these points is achieved by special paths called the *separatrices* (shown as broken lines). These correspond to values of C of 0 and $\frac{1}{4}$, the ordinates of the maxima and minimum of $\mathscr{V}$. They start or end at equilibrium points—they must not be mistaken for closed paths.

1.4. The damped linear oscillator

Generally speaking (apart from (1.14)) equations of the form

$$\ddot{x} = f(x, \dot{x}) \tag{1.22}$$

do not arise from conservative systems, and can be expected to show new phenomena. The simplest such system is a linear oscillator with linear damping, having the equation

$$\ddot{x} + k\dot{x} + cx = 0 \tag{1.23}$$

where $c > 0$, $k > 0$. An equation of this form describes, for example, a spring-mass system with a dashpot or a circuit containing inductance, capacitance, and resistance, and serves as a model for many other oscillating systems. We shall show how the familiar features of damped oscillations show up on the phase plane.

Equation (1.23) is a standard type of linear equation. The nature of its solutions depend on whether the roots of the auxiliary equation

$$m^2 + km + c = 0$$

are real and different, complex, or coincident and real. The roots are given by

$$\begin{matrix} m_1 \\ m_2 \end{matrix} = \frac{1}{2}\{-k \pm \sqrt{(k^2 - 4c)}\},$$

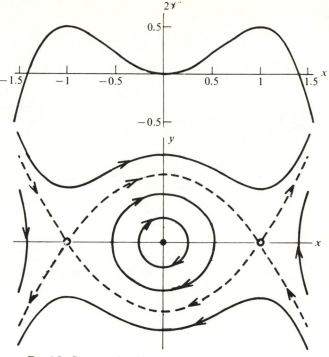

Fig. 1.8. Centre and saddle points. _ _ _ _ _ separatrices

and the discriminant, Δ,

$$\Delta = k^2 - 4c \tag{1.24}$$

is therefore the parameter which determines the general type of motion.

Strong damping ($\Delta > 0$)

The solutions are given by

$$x(t) = A\, e^{m_1 t} + B\, e^{m_2 t}, \tag{1.25}$$

where m_1 and m_2 are real and negative, A and B being any constants. Figure 1.9 shows two typical solutions. There is no oscillation and the t axis is cut at most once.

To construct the phase paths we could write as usual

$$\dot{x} = y, \quad \dot{y} = -cx - ky. \tag{1.26}$$

There is a single equilibrium point at $x = 0, y = 0$. The equation for the phase paths is

Fig. 1.9. Solution curves for a heavily-damped simple harmonic oscillator

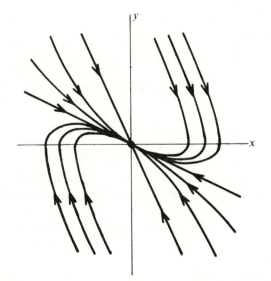

Fig. 1.10. Stable node

$$\frac{dy}{dx} = -c\frac{x}{y} - k. \qquad (1.27)$$

A general approach to linear systems such as (1.26) will be described in Chapter 2. For the present we remark that the solutions of (1.27) are too complicated for simple interpretation. We therefore proceed in the following way. From (1.25)

$$x = A\,e^{m_1 t} + B\,e^{m_2 t}, \quad \text{so} \quad y = \dot{x} = Am_1\,e^{m_1 t} + Bm_2\,e^{m_2 t}, \qquad (1.28)$$

and for fixed A and B, (1.28) constitutes a parametric representation of a phase path. The phase paths in Fig. 1.10 are plotted in this way for certain values of k and c.

This shows a new type of equilibrium point, called a *node*. For all slight displacements from $x = 0$, $\dot{x} = 0$, the state returns to the equilibrium point. It is therefore a *stable node*. That all the phase paths terminate at the origin can be seen by letting $t \to \infty$ in (1.28).

Weak damping $(\Delta < 0)$

The exponents are complex with negative real part, and the solutions are

$$x(t) = A\exp(-\tfrac{1}{2}kt)\cos\{\tfrac{1}{2}\sqrt{(-\Delta)}t + \alpha\} \qquad (1.29)$$

where A, α are arbitrary constants. A typical solution is shown in Fig. 1.11(a): it represents an oscillation with exponentially decreasing amplitude, decaying more rapidly for larger k. Its image on the phase plane, plotted parametrically as before, is shown in Fig. 1.11(b).

The equilibrium point at the origin is called a *stable spiral* or a *stable focus*.

Critical damping $(\Delta = 0)$

In this case $m_1 = m_2 = -\tfrac{1}{2}k$ and the solutions are

$$x(t) = (A + Bt)\exp(-\tfrac{1}{2}kt).$$

Fig. 1.11. Stable spiral

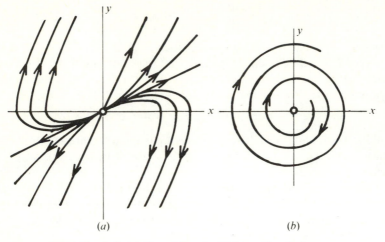

FIG. 1.12. (*a*) Unstable node; (*b*) Unstable spiral

The solutions resemble those for strong damping and the phase diagram shows a stable node.

Permutations of signs of the parameters k and c are possible.

(i) $c < 0, k \neq 0$. m_1 and m_2 are real but have different signs. The phase diagram shows a saddle point.

(ii) $c > 0, k < 0$. This case is called *negative damping*. Instead of energy being lost by the equivalent of a resistance or friction, energy is generated constantly in the system. The node or spiral is now *unstable* as shown in Fig. 1.12, since a slight disturbance from equilibrium leads to the system being carried far from the equilibrium state.

1.5. Nonlinear damping

Returning to the system

$$\ddot{x} = f(x, \dot{x}),$$

assume that f takes the form

$$f(x, \dot{x}) = -h(x, \dot{x}) - g(x), \tag{1.30}$$

where h does not contain an additive function of x. Then

$$\ddot{x} + h(x, \dot{x}) + g(x) = 0. \tag{1.31}$$

In mechanical terms the equation describes the displacement x of a particle of unit mass under a force system containing a conservative element, $g(x)$, and a dissipative or energy-generating component, $h(x, \dot{x})$.

In the case of a spring, $-g(x)$ is the restoring force. If g is of the appropriate type we should expect a tendency to oscillate, modified by the presence of the term $h(x, \dot{x})$.

The kinetic energy $\mathscr{T}$ is equal to $\frac{1}{2}\dot{x}^2$. We define a potential energy function (or, simply, a *potential*) for (1.31) by

$$\mathscr{V}(x) = \int g(x)\,dx, \quad \text{so that} \quad g(x) = \mathscr{V}'(x). \tag{1.32}$$

The total energy $\mathscr{E}$ defined by

$$\mathscr{E} = \mathscr{T} + \mathscr{V} = \tfrac{1}{2}\dot{x}^2 + \int g(x)\,dx \tag{1.33}$$

is not in general constant. Consider how $\mathscr{E}$ changes as a particular motion progresses: that is, along a phase path.

$$\frac{d\mathscr{E}}{dt} = \frac{d\mathscr{T}}{dt} + \frac{d\mathscr{V}}{dt} = \dot{x}\ddot{x} + \mathscr{V}'(x)\dot{x}$$

$$= \dot{x}\{-f(x, \dot{x}) + g(x)\} \quad \text{(on the path)}$$

$$= -\dot{x}h(x, \dot{x}) \tag{1.34}$$

by (1.31). Integrate (1.34) with respect to t from $t = \tau_0$ to $t = \tau$; then

$$\mathscr{E}(\tau) - \mathscr{E}(\tau_0) = -\int_{\tau_0}^{\tau} \dot{x}h(x, \dot{x})\,dt \tag{1.35}$$

(where $x(t)$ is the solution of (1.30) on the chosen phase path).

It may be possible to say, for a phase path lying in some region of the phase plane, either that
(i) $\dot{x}h(x, \dot{x}) = yh(x, y) > 0$, in which case $\mathscr{E}(\tau) < \mathscr{E}(\tau_0)$: that is, the energy decreases and h has a damping effect, contributing to a general decrease in amplitude; or that
(ii) $\dot{x}h(x, \dot{x}) = yh(x, y) < 0$, in which case $\mathscr{E}(\tau) > \mathscr{E}(\tau_0)$: the effect is of an internal source injecting energy into the system.

A system may contain both characteristics. For example, a pendulum clock has energy stored in a weight which is supplied to the pendulum, and a balance is achieved between the rate of energy supplied and the loss through friction to maintain a steady oscillation.

Example 1.5 *Examine the equation $\ddot{x} + |\dot{x}|\dot{x} + x = 0$ for damping effects.*

$$h(x, \dot{x}) = |\dot{x}|\dot{x},$$

and

$$yh(x, y) = |y|y^2 > 0, \quad y \neq 0.$$

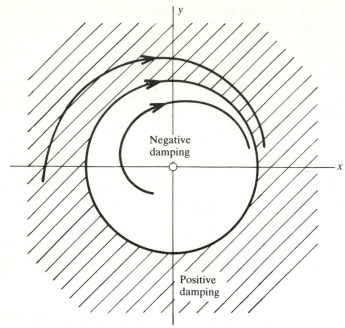

FIG. 1.13. Approach of two phase paths to a stable limit cycle

Except for the equilibrium point $(0,0)$ there is a loss of energy along every phase path no matter where it goes in the phase plane. We should therefore not be surprised if, from any initial state, the corresponding phase path eventually entered the origin, and motion ceased.

Example 1.6 *Examine the equation* $\ddot{x} + (x^2 + \dot{x}^2 - 1)\dot{x} + x = 0$ *for damping or energy input effects.*

Putting $\dot{x} = y$,

$$yh(x, y) = (x^2 + y^2 - 1)y^2.$$

When $x^2 + y^2 < 1$, $yh(x, y) < 0$ and when $x^2 + y^2 > 1$, $yh(x, y) > 0$.

The regions of energy loss and energy input on the phase plane are shown in Fig. 1.13. It can be verified that $x = \cos t$ is a solution of the differential equation, and this is represented by the circle $x^2 + y^2 = 1$ on the phase plane. The phase diagram consists of this circle, together with paths approaching it from the outside, paths approaching it from the inside, and the equilibrium point at the origin. The *isolated* closed path $x^2 + y^2 = 1$; isolated in the sense that there is no other closed path in its neighbourhood; is called a *limit cycle*. All paths approach the circle as $t \to \infty$. A limit cycle is therefore one of the most important features of a physical system.

A useful way to approach this and similar equations is to use polar coordinates (r, θ) where $x = r\cos\theta$, $y = r\sin\theta$. It follows that, since $r^2 = x^2 + y^2$ and $\tan\theta = y/x$,

$$\dot{r} = (x\dot{x} + y\dot{y})/r, \quad \dot{\theta} = (x\dot{y} - \dot{x}y)/r^2.$$

Thus

$$\dot{r} = -r(r^2 - 1)\sin^2\theta, \tag{1.36}$$

$$\dot{\theta} = -1 - (r^2 - 1)\sin\theta\cos\theta. \tag{1.37}$$

A particular solution of (1.36) is given by $r = 1$, the unit circle previously obtained. Also from (1.36) $\dot{r} > 0$ for $r < 1$ ($\sin\theta \neq 0$), and $\dot{r} < 0$ for $r > 1$ ($\sin\theta \neq 0$). We infer that outside the limit cycle r decreases and inside r increases with time, implying that the limit cycle is stable, the paths approaching it from both inside and outside.

A more decisive way of seeing what is happening in such cases, not involving reference to energy, is shown by the following example.

Example 1.7 *Show that all solutions of the equation* $\ddot{x} + |\dot{x}|\dot{x} + x^3 = 0$ *tend to zero as* $t \to \infty$.

It is easy to confirm the following important identity

$$\ddot{x} = \dot{x}\frac{d\dot{x}}{dx} = \frac{1}{2}\frac{d(\dot{x}^2)}{dx}.$$

The equation may then be written

$$\frac{1}{2}\frac{d(\dot{x}^2)}{dx} + x^3 = -|\dot{x}|\dot{x},$$

or, in terms of the phase plane variables,

$$\frac{1}{2}\frac{d(y^2)}{dx} + x^3 = -|y|y. \tag{1.38}$$

Consider a segment of a path starting at the point A, where $x = a$, and ending at B, where $x = b$. By integrating (1.38) with respect to x from a to b we find that

$$[\tfrac{1}{2}y^2 + \tfrac{1}{4}x^4]_A^B = -\int_a^b |y|y \, dx = -\int_{t_A}^{t_B} |y|y^2 \, dt \tag{1.39}$$

(since $dx = \dot{x} \, dt = y \, dt$). The right-hand side is negative everywhere, so (1.39) implies that, as the representative point moves along any path, the value of $\tfrac{1}{2}y^2 + \tfrac{1}{4}x^4$ (corresponding to the potential energy for this example) steadily decreases. But the family of curves defined by

$$\tfrac{1}{2}y^2 + \tfrac{1}{4}x^4 = constant$$

is a family of ovals closing in on the origin as the constant gets smaller. The above argument shows that a representative point crosses these successively in an inward direction, and so the path approaches the origin.

This method may be used in Examples 1.5 and 1.6, in which the analogous family consists of the circles $\frac{1}{2}y^2 + \frac{1}{2}x^2 = constant$.

In the system of Example 1.6 the regions of positive and negative damping are separated by the circle $x^2 + y^2 = 1$. In the important equation (van der Pol's equation)

$$\ddot{x} + e(x^2 - 1)\dot{x} + x = 0, \quad e > 0,$$

negative damping occurs in the strip $|x| < 1$ and positive damping in the half-planes $|x| > 1$. We shall see later that this type of damping also leads to a limit cycle. Van der Pol's equation originally arose as an idealization of a spontaneously oscillating, or self-excited, valve circuit (Minorsky 1962). It will provide for us the simplest form of equation having this pattern of positive and negative damping which gives rise to a limit cycle.

A closely-connected equation is Rayleigh's equation

$$\ddot{x} + e(\tfrac{1}{3}\dot{x}^2 - 1)\dot{x} + x = 0, \tag{1.40}$$

which originated in connection with a theory of the oscillation of a violin string. It has a similar distribution of positive and negative damping regions in the phase plane, and in fact reduces to van der Pol's equation by differentiating and putting $\dot{x} = z$.

1.6. Some applications

(i) *Dry friction*

Dry (or Coulomb) friction occurs when the surfaces of two solids are in contact and in relative motion without lubrication. The system shown in Fig. 1.14 illustrates dry friction. A continuous belt is driven by rollers at a constant speed v_0. A block of mass m connected to a fixed support by a spring of stiffness c rests on the belt. If F is the frictional force between

FIG. 1.14.

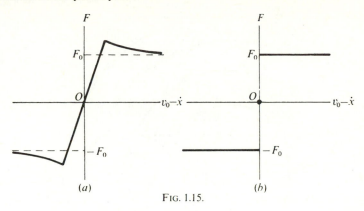

FIG. 1.15.

the block and the belt and x is the extension of the spring, then the equation of motion is

$$m\ddot{x} + cx = F.$$

Assume that F depends on the slip velocity, $v_0 - \dot{x}$; a typical relation is shown in Fig. 1.15(a). We will replace this function by a simpler one having a discontinuity at the origin (Fig. 1.15(b)):

$$F = F_0 \operatorname{sgn}(v_0 - \dot{x})$$

where F_0 is a positive constant (see Fig. 1.15(b)) and the sgn (signum) function is defined by

$$\operatorname{sgn}(u) = \begin{cases} 1, & u > 0, \\ 0, & u = 0, \\ -1, & u < 0. \end{cases}$$

The equation of motion becomes

$$m\ddot{x} + cx = F_0 \operatorname{sgn}(v_0 - \dot{x}).$$

The term on the right is equal to F_0 when $v_0 > \dot{x}$, and $-F_0$ when $v_0 < \dot{x}$, and we obtain the following solutions for the phase paths in these regions:

$$y = \dot{x} > v_0: \quad my^2 + (x + F_0/c)^2 c = \text{constant},$$
$$y = \dot{x} < v_0: \quad my^2 + (x - F_0/c)^2 c = \text{constant}.$$

These are families of ellipses, the first having its centre at $(-F_0/c, 0)$ and the second at $(F_0/c, 0)$. Figure 1.16 shows the corresponding phase diagram, plotted as $m^{\frac{1}{2}}y$ against $c^{\frac{1}{2}}y$ to give circular paths.

There is a single equilibrium point, at $(F_0/c, 0)$, which is a centre. Points on y (or $\dot{x}$) $= v_0$ are not covered by the differential equation since this is where F is discontinuous, so the behaviour must be deduced from other physical arguments. On encountering the state $\dot{x} = v_0$ for

FIG. 1.16.

$|x| < F_0/c$, the block will move with the belt until the maximum available friction, F_0, is insufficient to resist the increasing spring tension. This is when $x = F_0/c$; the block then goes into an oscillation represented by the closed path through $(F_0/c, v_0)$. In fact, for any initial conditions lying outside this ellipse, the system ultimately settles into this oscillation. A computed phase diagram corresponding to a frictional force as in Fig. 1.15(*a*) is displayed in Exercise 50, Chapter 3.

(ii) *The brake*

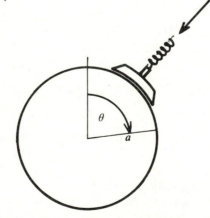

FIG. 1.17. A brake

Consider a simple brake shoe applied to the hub of a wheel as shown in Fig. 1.17. The friction force will depend on the pressure and the angular velocity of the wheel, $\dot{\theta}$. We assume again a simplified dry-friction relation corresponding to constant pressure

$$F = -F_0 \operatorname{sgn}(\dot{\theta})$$

so if the wheel is otherwise freely spinning its equation of motion is

$$I\ddot{\theta} = -F_0 a \operatorname{sgn}(\dot{\theta}),$$

where I is the moment of inertia of the wheel and a the radius of the brake drum. The phase paths are found by rewriting the differential equation

$$I\dot{\theta}\frac{d\dot{\theta}}{d\theta} = -F_0 a \operatorname{sgn}(\dot{\theta}),$$

whence for $\dot{\theta} > 0$

$$\tfrac{1}{2}I\dot{\theta}^2 = -F_0 a\theta + C$$

and for $\dot{\theta} < 0$

$$\tfrac{1}{2}I\dot{\theta}^2 = F_0 a\theta + C.$$

These represent two families of parabolas as shown in Fig. 1.18. $(\theta, 0)$ is an equilibrium point for every θ.

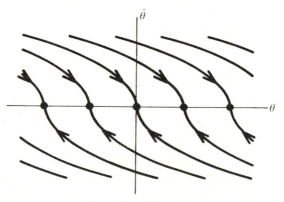

FIG. 1.18.

(iii) *The pendulum clock: a limit cycle*

Figure 1.19 shows the main features of the pendulum clock. The 'escape wheel' is a toothed wheel, which drives the hands of the clock through a succession of gears. It has a spindle around which is wound a wire with a weight at its free end. The escape wheel is arrested by the 'anchor' which has two teeth. The anchor is attached to the shaft of the

FIG. 1.19. A weight-driven clock mechanism

pendulum and rocks with it, controlling the rotation of the escape wheel. Its teeth are so designed that as the pendulum reaches its maximum amplitude on one side one tooth of the escape wheel is released, and the escape wheel is then stopped again by the other tooth on the anchor. Every time this happens the anchor receives a small impulse or pressure. It is this impulse which maintains the oscillation of the pendulum, which would otherwise die away. The loss of potential energy due to the weight's descent is therefore fed periodically into the pendulum via the anchor mechanism.

It can be shown that the system will settle into steady oscillations of fixed amplitude independently of sporadic disturbance and of initial conditions. If the pendulum is swinging with too great an amplitude, its loss of energy per cycle due to friction is large, and the impulse supplied by the escapement is insufficient to offset this. The amplitude consequently decreases. If the amplitude is too small, the frictional loss is small; the impulses will over-compensate and the amplitude will build up. A balanced state is therefore approached, which appears in the $\theta, \dot{\theta}$ plane (Fig. 1.20) as an isolated closed curve $\mathscr{C}$. Such an isolated periodic oscillation, or *limit cycle* (see Example 1.6) can occur only in systems described by nonlinear equations, and the following simple model shows

FIG. 1.20. Phase diagram for a clock, impulse applied near A

where the nonlinearity is located. The motion can be approximated by the equation

$$I\ddot{\theta} + k\dot{\theta} + c\theta = f(\theta), \qquad (1.41)$$

where I is the moment of inertia of the pendulum, k is a small damping constant, c is another constant determined by gravity, θ is the angular displacement, and $f(\theta)$ is the moment, supplied once per cycle by the escapement mechanism. Since f is periodic in θ, it must be a nonlinear function of θ.

Such an oscillation, generated by an energy source whose input is not regulated externally, but which *automatically* synchronizes with the existing oscillation, is called *self-excited*. Here the build-up is limited by the friction.

1.7. Parameter-dependent conservative systems

Suppose $x(t)$ satisfies

$$\ddot{x} = f(x, \lambda)$$

where λ is a parameter. The equilibrium points of the system are given by $f(x, \lambda) = 0$, and in general their location will depend on the parameter λ. In mechanical terms, for a particle of unit mass with displacement x, $f(x, \lambda)$ represents the force experienced by the particle. Suppose there exists a function $\mathcal{V}(x, \lambda)$ such that $f(x, \lambda) = -\partial\mathcal{V}/\partial x$ for each value of λ; then $\mathcal{V}(x, \lambda)$ is the potential energy of the system and equilibrium points correspond to stationary values of the potential energy. As indicated in Section 1.3, we expect a minimum of potential energy to correspond to a stable equilibrium point, and other stationary values (the maximum and point of inflexion) to be unstable. In fact, $\mathscr{V}$ is a minimum at $x = x_1$ if $\partial \mathscr{V}/\partial x$ changes from negative to positive on passing through x_1; this implies that $f(x, \lambda)$ changes sign from positive to negative as x increases through $x = x_1$.

There exists a simple method of displaying the stability of equilibrium points for parameter-dependent systems in which both the number and stability of equilibrium points may vary with λ. We assume $f(x, \lambda)$ to be continuous in both x and λ. Plot the curve $f(x, \lambda) = 0$ in the λ, x plane; this curve represents the equilibrium points. Shade the domains in which $f(x, \lambda) > 0$ as shown in Fig. 1.21. If a segment of the curve has shading below it, the corresponding equilibrium points are stable, since for fixed λ, f changes from positive to negative as x increases.

FIG. 1.21. Stability curves for the equilibrium points of $\ddot{x} = f(x, \lambda)$

For example, the solid line between A and B corresponds to stable equilibrium points. A and B are unstable: C is also unstable since f is positive on both sides of C. The nature of equilibrium points can easily be read from the figure; when $\lambda = \lambda_0$ as shown, the system has three equilibrium points, two of which are stable. A, B, and C are known as *bifurcation points*. As λ varies through such points the equilibrium point may split into two or more, or several equilibrium points may appear or merge into a single one. More information on bifurcation can be found in Chapter 12.

Example 1.7 *A bead slides on a smooth circular wire of radius a which is*

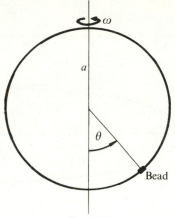

FIG. 1.22.

constrained to rotate about a vertical diameter with constant angular velocity ω. Analyse the stability of the bead.

The bead has a velocity component $a\dot{\theta}$ tangential to the wire and a component $a\omega \sin \theta$ perpendicular to the wire, where θ is the inclination of the radius to the bead to the downward vertical as shown in Fig. 1.22. The kinetic energy $\mathscr{T}$ and potential energy $\mathscr{V}$ are given by

$$\mathscr{T} = \tfrac{1}{2}ma^2(\dot{\theta}^2 + \omega^2 \sin^2 \theta), \quad \mathscr{V} = -mga \cos \theta.$$

Since the system is subject to a moving constraint (that is, the angular velocity of the wire is imposed), the usual energy equation does not hold. Lagrange's equation for the system is

$$\frac{\mathrm{d}}{\mathrm{d}t}\left(\frac{\partial \mathscr{T}}{\partial \dot{\theta}}\right) - \frac{\partial \mathscr{T}}{\partial \theta} = -\frac{\partial \mathscr{V}}{\partial \theta},$$

which gives

$$a\ddot{\theta} = a\omega^2 \sin \theta \cos \theta - g \sin \theta.$$

Set $a\omega^2/g = \lambda$. Then

$$a\ddot{\theta} = a\dot{\theta}\frac{\mathrm{d}\dot{\theta}}{\mathrm{d}\theta} = g \sin \theta(\lambda \cos \theta - 1)$$

which, after integration, becomes

$$\tfrac{1}{2}a\dot{\theta}^2 = g(1 - \tfrac{1}{2}\lambda \cos \theta)\cos \theta + C,$$

the equation of the phase paths.

In the previous theory

$$f(\theta, \lambda) = g \sin \theta(\lambda \cos \theta - 1)/a$$

and the equilibrium points are given by $f(\theta, \lambda) = 0$, which is satisfied when

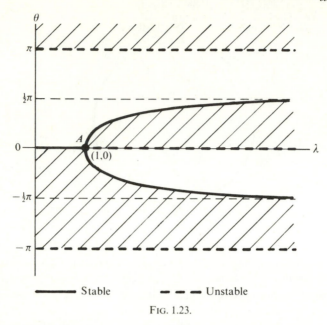

FIG. 1.23.

$\sin \theta = 0$ or $\cos = \lambda^{-1}$. From the periodicity of the problem, $\theta = \pi$ and $\theta = -\pi$ correspond to the same state of the system.

The regions where $f < 0$ and $f > 0$ are separated by curves where $f = 0$, and can be located, therefore, by checking the sign at particular points; for example, $f(\frac{1}{2}\pi, 1) = -g/a < 0$. Figure 1.23 shows the stable and unstable equilibrium positions of the bead. A is a stable bifurcation point.

Phase diagrams for the system may be constructed as in Section 1.3 for fixed values of λ. Two possibilities are shown in Fig. 1.24. Note that they confirm the stability predictions of Fig. 1.23.

FIG. 1.24.

Exercises

1. Locate the equilibrium points and sketch the phase diagrams in their neighbourhood for the following equations:

   (i) $\ddot{x} - k\dot{x} = 0$.

  (ii) $\ddot{x} - 8x\dot{x} = 0$.

 (iii) $\ddot{x} = k\,(|x| > 1)$, $\ddot{x} = 0\,(|x| < 1)$.

 (iv) $\ddot{x} + 3\dot{x} + 2x = 0$.

  (v) $\ddot{x} - 4\dot{x} + 40x = 0$.

 (vi) $\ddot{x} + 3|\dot{x}| + 2x = 0$.

(vii) $\ddot{x} + k\,\mathrm{sgn}\,(\dot{x}) + c\,\mathrm{sgn}\,(x) = 0$, $c > k$. Show that the path starting at $(x_0, 0)$ reaches $((c-k)^2 x_0/(c+k)^2, 0)$ after one circuit of the origin. Deduce that the origin is a spiral point.

(viii) $\ddot{x} + x\,\mathrm{sgn}\,(x) = 0$.

2. Sketch the phase diagram for the equation $\ddot{x} = -x - \alpha x^3$, considering all values of α. Check the stability of the equilibrium points by the method of Section 1.7.

3. A certain dynamical system is governed by the equation $\ddot{x} + \dot{x}^2 + x = 0$. Show that the origin is a centre in the phase plane, and that the open and closed paths are separated by the path $2y^2 = 1 - 2x$.

4. Sketch the phase diagrams for the equation $\ddot{x} + e^x = a$, for $a < 0, a = 0$, and $a > 0$.

5. Sketch the phase diagram for the equation $\ddot{x} - e^x = a$, for $a < 0, a = 0$, and $a > 0$.

6. The potential energy $\mathscr{V}(x)$ of a conservative system is continuous, and is strictly increasing for $x < -1$, zero for $|x| \leqslant 1$, and strictly decreasing for $x > 1$. Locate the equilibrium points and sketch the phase diagram for the system.

7. Figure 1.25 shows a pendulum striking an inclined wall. Sketch the phase diagram, for α positive and α negative, when (i) there is no loss of energy at impact, (ii) the magnitude of the velocity is halved on impact.

Fig. 1.25.

8. Show that the time elapsed, T, along a phase path $\mathscr{C}$ of the system $\dot{x} = y$, $\dot{y} = f(x, y)$ is given, in a form alternative to (1.11), by

$$T = \int_{\mathscr{C}} (y^2 + f^2)^{-1/2} \, ds,$$

where ds is an element of distance along $\mathscr{C}$.

By writing $\delta s \simeq (y^2 + f^2)^{1/2} \delta t$, indicate, very roughly, equal time intervals along the phase paths of the system $\dot{x} = y, \dot{y} = 2x$.

9. On the phase diagram for the equation $\ddot{x} + x = 0$, the phase paths are circles. Use (1.11) in the form $\delta t \simeq \delta x / y$ to indicate, roughly, equal time steps along several phase paths.

10. Repeat Exercise 9 for the equation $\ddot{x} + 9x = 0$, in which the phase paths are ellipses.

11. The pendulum equation, $\ddot{x} + \omega^2 \sin x = 0$, can be approximated for moderate amplitudes by the equation $\ddot{x} + \omega^2(x - \tfrac{1}{6}x^3) = 0$. Sketch the phase diagram for the latter equation, and explain the differences between it and Fig. 1.2.

12. The displacement, x, of a spring-mounted mass under the action of Coulomb dry friction is assumed to satisfy

$$m\ddot{x} + cx = -F_0 \operatorname{sgn}(\dot{x}),$$

where m, c, and F_0 are positive constants (Section 1.6). The motion starts at $t = 0$, with $x = x_0 > 3F_0/c$ and $\dot{x} = 0$. Subsequently, whenever $x = -\alpha$, where $2F_0/c - x_0 < -\alpha < 0$ and $\dot{x} > 0$, a trigger operates, to increase suddenly the forward velocity so that the kinetic energy increases by a constant amount E. Show that if $E > 8F_0^2/c$, a periodic motion is approached, and show that the largest value of x in the periodic motion is equal to $F_0/c + E/4F_0$.

13. In Exercise 12, suppose that the energy is increased by E at $x = -\alpha$ for both $\dot{x} < 0$ and $\dot{x} > 0$; that is, there are two injections of energy per cycle. Show that periodic motion is possible if $E > 6F_0^2/c$, and find the amplitude of the oscillation.

14. The 'friction pendulum' consists of a pendulum attached to a sleeve, which embraces a close-fitting cylinder (Fig. 1.26). The cylinder is turned at a constant rate Ω. The sleeve is subject to Coulomb dry friction through the couple $G = -F_0 \operatorname{sgn}(\dot{\theta} - \Omega)$. Write down the equation of motion, find the equilibrium states, and sketch the phase diagram.

15. By plotting the 'potential energy' of the nonlinear conservative system $\ddot{x} = x^4 - x^2$, construct the phase diagram of the system. A particular path has the initial conditions $x = \tfrac{1}{2}, \dot{x} = 0$ at $t = 0$. Is the subsequent motion periodic?

FIG. 1.26.

16. The system $\ddot{x} + x = -F_0 \operatorname{sgn}(\dot{x})$, $F_0 > 0$, has the initial conditions $x = x_0 > 0$, $\dot{x} = 0$. Show that the phase path will spiral exactly n times before entering equilibrium (Section 1.6) if $(4n-1)F_0 < x_0 < (4n+1)F_0$.

17. A pendulum of length a has a bob of mass m which is subject to a horizontal force $m\omega^2 a \sin\theta$, where θ is the inclination to the downward vertical. Show that the equation of motion is $\ddot{\theta} = \omega^2(\cos\theta - \lambda)\sin\theta$, where $\lambda = g/\omega^2 a$. Investigate the stability of the equilibrium states by the method of Section 1.7 for parameter-dependent systems. Sketch the phase diagrams for various λ.

18. Investigate the stability of the equilibrium points of the parameter-dependent system $\ddot{x} = (x - \lambda)(x^2 - \lambda)$.

19. If a bead slides on a smooth parabolic wire rotating with constant angular velocity ω about a vertical axis, then the distance x of the particle from the axis of rotation satisfies $(1 + x^2)\ddot{x} + (g - \omega^2 + \dot{x}^2)x = 0$. Analyse the motion of the bead in the phase plane.

20. A particle is attached to a fixed point 0 on a smooth horizontal plane by an elastic string. When unstretched, the length of the string is a. The equation of motion of the particle, which is constrained to move on a straight line through 0, is

$$\ddot{x} = -x + a \operatorname{sgn}(x), \quad |x| > a \quad \text{(when the string is stretched)},$$

$$\ddot{x} = 0, \quad |x| < a \quad \text{(when the string is slack)},$$

x being the displacement from 0. Find the equilibrium points and the equations of the phase paths, and sketch the phase diagram.

21. The equation of motion of a conservative system is $\ddot{x} + g(x) = 0$, where $g(0) = 0$; $g(x) < 0$ for $x < 0$ and $g(x) > 0$ for $x > 0$; and

$$\int_0^x g(u)\,du \to \infty \quad \text{as} \quad x \to \pm\infty. \tag{a}$$

Show that the motion is always periodic.

By considering $g(x) = xe^{-x^2}$, show that if (a) does not hold, the motions are not all necessarily periodic.

22. The wave function $u(x, t)$ satisfies the partial differential equation

$$\frac{\partial^2 u}{\partial x^2} + \alpha \frac{\partial u}{\partial x} + \beta u^3 + \gamma \frac{\partial u}{\partial t} = 0,$$

where α, β, and γ are positive constants. Show that there exist permanent wave solutions of the form $u(x, t) = U(x - ct)$ for any c, where $U(\zeta)$ satisfies

$$\frac{d^2 U}{d\zeta^2} + (\alpha - \gamma c) \frac{dU}{d\zeta} + \beta U^3 = 0.$$

Using Exercise 21, show that when $c = \alpha/\gamma$, all such waves are periodic.

23. The linear oscillator $\ddot{x} + \dot{x} + x = 0$ is set in motion with initial conditions $x = 0$, $\dot{x} = v$, at $t = 0$. After the first and each subsequent cycle the kinetic energy is instantaneously increased by a constant, E, in such a manner as to increase $\dot{x}$. Show that if $E = \frac{1}{2}v^2(1 - e^{4\pi/\sqrt{3}})$, a periodic motion occurs. Find the maximum value of x in a cycle.

24. Show how solutions of Exercise 23 having arbitrary initial conditions spiral on to the periodic solution (which is a limit cycle). Sketch the phase diagram.

25. The kinetic energy, $\mathcal{T}$, and the potential energy, $\mathcal{V}$, of a system with one degree of freedom are given by

$$\mathcal{T} = T_0(x) + \dot{x}T_1(x) + \dot{x}^2 T_2(x), \quad \mathcal{V} = \mathcal{V}(x).$$

Use Lagrange's equation

$$\frac{d}{dt}\left(\frac{\partial \mathcal{T}}{\partial \dot{x}}\right) - \frac{\partial \mathcal{T}}{\partial x} = -\frac{\partial \mathcal{V}}{\partial x}$$

to obtain the equation of motion of the system. Show that the equilibrium points are stationary points of $T_0(x) - \mathcal{V}(x)$, and that the phase paths are given by the energy equation

$$T_2(x)\dot{x}^2 - T_0(x) + \mathcal{V}(x) = \text{constant}.$$

26. Sketch the phase diagram for the equation $\ddot{x} = -f(x + \dot{x})$, where

$$f(u) = \begin{cases} f_0, & u \geqslant c, \\ f_0 u/c, & |u| \leqslant c, \\ -f_0, & u \leqslant -c, \end{cases}$$

where f_0, c are constants, $f_0 > 0$, and $c > 0$. How does the system behave as $c \to 0$?

27. Sketch the phase diagram for the equation $\ddot{x} = u$, where

$$u = -\operatorname{sgn}(\sqrt{2}|x|^{1/2}\operatorname{sgn}(x) + \dot{x}).$$

(u is an elementary control variable which can switch between $+1$ and -1. The curve $\sqrt{2}|x|^{1/2}\operatorname{sgn}(x) + y = 0$ is called the switching curve.)

28. The relativistic equation for an oscillator is

$$\frac{d}{dt}\left\{\frac{m_0\dot{x}}{\sqrt{[1-(\dot{x}/c)^2]}}\right\} + kx = 0$$

where m_0, c, and k are positive constants. Show that the phase paths are given by

$$\frac{m_0 c^2}{\sqrt{[1-(y/c)^2]}} + \tfrac{1}{2}kx^2 = \text{constant}.$$

If $y = 0$ when $x = a$, show that the period, T, of an oscillation is given by

$$T = \frac{4}{c\sqrt{\varepsilon}}\int_0^a \frac{[1+\varepsilon(a^2-x^2)]\,dx}{\sqrt{(a^2-x^2)}\sqrt{[2+\varepsilon(a^2-x^2)]}}, \quad \varepsilon = k/2m_0 c^2.$$

The constant ε is small; by expanding the integrand in powers of ε show that

$$T \approx 2\pi\sqrt{(m_0/k)}\,(1 + \tfrac{3}{8}\varepsilon a^2).$$

29. A mass m is attached to the mid-point of an elastic string of length $2a$ and stiffness λ. There is no gravity acting, and the tension is zero in the equilibrium

FIG. 1.27.

position. Obtain the equation of motion for transverse oscillations and sketch the phase paths.

30. Analyse, from the point of view of Section 1.5, the equation of motion for the dry-friction problem of Section 1.6, and reconcile this with the phase diagram Fig. 1.16.

31. Show that, in the phase diagram for a pendulum clock (Section 1.6), there is a net loss of energy along paths far enough from the origin, and a net gain along paths close enough to the origin. (Put eqn (1.41) into the form (1.31) and use (1.35).)

32. A pendulum with a magnetic bob oscillates in a vertical plane over a magnet, which repels the bob according to the inverse square law, so that the equation of motion is (Fig. 1.28)

$$ma^2\ddot{\theta} = -mga\sin\theta + Fh\sin\phi,$$

FIG. 1.28.

where $h > a$ and $F = c/(a^2 + h^2 - 2ah\cos\theta)$ and c is a constant. Find the equilibrium positions of the bob, and classify them as centres and saddle-points according to the parameters of the problem. Describe the motion of the pendulum.

33. A pendulum with equation $\ddot{x} + \sin x = 0$ oscillates with amplitude a. Show that its period, T, is equal to $4K(\beta)$, where $\beta = \sin^2 \frac{1}{2}a$ and

$$K(\beta) = \int_0^{\pi/2} \frac{d\phi}{\sqrt{(1 - \beta \sin^2 \phi)}}.$$

$K(\beta)$ has the power series representation

$$K(\beta) = \frac{1}{2}\pi[1 + (\tfrac{1}{2})^2\beta + (\tfrac{1 \cdot 3}{2 \cdot 4})^2\beta^2 + \ldots], \quad |\beta| < 1.$$

Deduce that, for small amplitudes,

$$T = 2\pi(1 + \tfrac{1}{16}a^2 + \tfrac{11}{3072}a^4) + O(a^6).$$

34. Repeat Exercise 33 with the equation $\ddot{x} + x - \varepsilon x^3 = 0$, and show that

$$T = \frac{4\sqrt{2}}{\sqrt{(2 - \varepsilon a^2)}} K(\beta), \quad \beta = \frac{\varepsilon a^2}{2 - \varepsilon a^2},$$

and that

$$T = 2\pi(1 + \tfrac{3}{8}\varepsilon a^2 + \tfrac{57}{256}\varepsilon^2 a^4) + O(a^6), \quad \varepsilon a^2 < 2.$$

35. Show that equations of the form $\ddot{x} + g(x)\dot{x}^2 + h(x) = 0$ are effectively conservative. (Find a transformation of x which puts the equation into the usual conservative form. Cf. eqn (1.14).)

36. Compare the phase diagrams of the following systems and the second-order equations in x to which they give rise:
(i) $\dot{x} = 1, \dot{y} = 0$, (ii) $\dot{x} = y, \dot{y} = 0$, (iii) $\dot{x} = xy, \dot{y} = 0$.

37. Show that the phase plane for the equation

$$\ddot{x} - \varepsilon x \dot{x} + x = 0$$

has a centre at the origin, by first calculating the phase paths.

38. Show that the equation $\ddot{x} + x + \varepsilon x^3 = 0$ with $x(0) = a, \dot{x}(0) = 0$ has phase paths given by

$$\dot{x}^2 + x^2 + \tfrac{1}{2}\varepsilon x^4 = (1 + \tfrac{1}{2}\varepsilon a^2)a^2.$$

Show that the origin is a centre. Are all phase paths closed, and hence all solutions periodic?

39. Locate the equilibrium points of the equation

$$\ddot{x} + \lambda + x^3 - x = 0,$$

in the x, λ plane. Show that the phase paths are given by

$$\tfrac{1}{2}\dot{x}^2 + \lambda x + \tfrac{1}{4}x^4 - \tfrac{1}{2}x^2 = \text{constant}.$$

Sketch, in the $x, \dot{x}, \lambda$ space, the surface which represents the set of phase diagrams for which the constant (the energy) is zero. Suppose the system is oscillating with zero energy and, by means of two controllers, λ is slowly reduced and the energy is kept zero. What happens as λ passes through the value $-\tfrac{1}{3}\sqrt{(\tfrac{2}{3})}$?

40. Burgers' equation

$$\frac{\partial \phi}{\partial t} + \phi \frac{\partial \phi}{\partial x} = c \frac{\partial^2 \phi}{\partial x^2}$$

shows diffusion and nonlinear effects. Find the equation for permanent waves by putting $\phi = U(x - ct)$, where c is the constant wave speed. Find the equilibrium points and the phase paths for the resulting equation and interpret the phase diagram.

2 First-order systems in two variables and linearization

CHAPTER 1 describes the application of phase-plane methods to the equation $\ddot{x} = f(x, \dot{x})$ through the equivalent first-order system $\dot{x} = y$, $\dot{y} = f(x, y)$. This approach permits a useful line of argument based on a mechanical interpretation of the original equation. Frequently, however, the appropriate formulation of mechanical, biological, and geometrical problems is not through a second-order equation at all, but directly as a more general type of first-order system of the form $\dot{x} = X(x, y)$, $\dot{y} = Y(x, y)$. The appearance of these equations is an invitation to construct a phase plane with x, y coordinates in which solutions are represented by curves $(x(t), y(t))$ where $x(t), y(t)$ are the solutions. The constant solutions are represented by equilibrium points obtained by solving the equations $X(x, y) = 0$, $Y(x, y) = 0$, and these may now occur anywhere in the plane. Near the equilibrium points we may make a linear approximation to $X(x, y), Y(x, y)$, solve the simpler equations obtained, and so determine the local character of the paths. This enables the stability of the equilibrium states to be settled and is a starting-point for global investigations of the solutions.

2.1. The general phase plane

Consider the general autonomous first-order system

$$\dot{x} = X(x, y), \quad \dot{y} = Y(x, y) \tag{2.1}$$

of which the type considered in Chapter 1,

$$\dot{x} = y, \quad \dot{y} = f(x, y), \tag{2.2}$$

is a special case. As in Section 1.2, a system is called *autonomous* when the time variable does not appear in the right-hand side of (2.1). We shall give examples later of how such systems arise.

The solutions $x(t), y(t)$ of (2.1) may be represented on a plane with cartesian axes x, y. Then as t increases $(x(t), y(t))$ traces out a directed curve in the plane called a *phase path*.

The appropriate form for the initial conditions of (2.1) is

$$x = x_0, \quad y = y_0 \quad \text{at} \quad t = t_0$$

where x_0 and y_0 are the *initial values* at time t_0; by the existence and uniqueness theorem (Appendix A) there is one and only one solution satisfying this condition when (x_0, y_0) is an 'ordinary point'. This does not at once mean that there is one and only one phase path through the point (x_0, y_0) on the phase diagram, because this same point could serve as the initial conditions for other starting times. Therefore it might seem that other phase paths through the same point could result: the phase diagram would then be a tangle of criss-crossed curves. We shall see that this will not be so by forming the differential equation for the phase paths. Since $\dot{y}/\dot{x} = \mathrm{d}y/\mathrm{d}x$ on a path the required equation is

$$\frac{\mathrm{d}y}{\mathrm{d}x} = \frac{Y(x, y)}{X(x, y)}. \tag{2.3}$$

There are two classes of exceptional points where solution curves of (2.3) (the phase paths) may meet. First, there may be points where Y/X has some manifest singularity so that the uniqueness theorem fails: these are *singular points*. Of more immediate interest, there are usually points where

$$X(x, y) = 0, \quad Y(x, y) = 0, \tag{2.4}$$

which are the *equilibrium points*. If a solution of (2.4) is x_1, y_1, then $x(t) = x_1, y(t) = y_1$ are *constant solutions* of (2.1), and are degenerate phase paths. The terms *fixed point* and *critical point* are also used. There is one and only one solution curve passing through any ordinary point of (2.3), which may have singular points where (2.1) does not. Therefore there is one and only one phase path containing the ordinary point (x_0, y_0), no matter at what time the initial condition (x_0, y_0) is prescribed. Thus, infinitely many solutions of (2.1), differing only by time displacements, map onto a single phase path.

Equation (2.3) does not give any indication of the direction to be associated with a phase path, and this must be settled by reference to (2.1): the signs of X and Y at a point determine the direction through the point, and generally the directions at all other points can be settled by continuity.

The diagram depicting the phase paths is called the *phase diagram*. The point (x, y) is called a *state* of the system, and the phase diagram depicts the evolution of the states of the system from arbitrary initial states.

Example 2.1   *Compare the phase diagrams of the systems*

(i) $\dot{x} = y, \dot{y} = -x$;   (ii) $\dot{x} = xy, \dot{y} = -x^2$.

The equation for the paths is the same for both, namely

$$\frac{dy}{dx} = -\frac{x}{y}$$

(strictly, for $y \neq 0$, and for $x \neq 0$ in the second case), giving a family of circles in both cases. However, in case (i) there is an equilibrium point only at the origin, but in case (ii) every point on the y axis is an equilibrium point. The directions, too, are different. By considering the signs of $\dot{x}, \dot{y}$ in the various quadrants the phase diagram of Fig. 2.1 is produced.

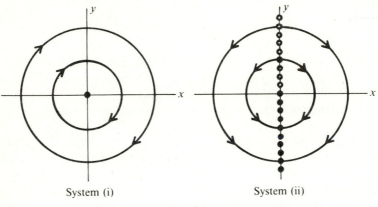

System (i)   System (ii)

Fᵢɢ. 2.1.

A second-order differential equation can be reduced to the general form (2.2) in an arbitrary number of ways and this occasionally has advantages akin to changing the variable to simplify a differential equation. For example the reduction (2.1) applied to $x\ddot{x} - \dot{x}^2 - x^3 = 0$ leads to the system

$$\dot{x} = y, \quad \dot{y} = y^2/x + x^2. \tag{2.5}$$

Suppose we define, instead of y, another variable y_1 by $y_1(t) = \dot{x}(t)/x(t)$, that is, by

$$\dot{x} = xy_1. \tag{2.6}$$

Then from (2.6), $\ddot{x} = x\dot{y}_1 + \dot{x}y_1 = x\dot{y}_1 + xy_1^2$ (using (2.6) again). But from the differential equation, $\ddot{x} = \dot{x}^2/x + x^2 = xy_1^2 + x^2$. Therefore

$$\dot{y}_1 = x. \tag{2.7}$$

The pair of equations (2.6) and (2.7) afford a representation alternative to (2.5). The phase diagram x, y_1 will, of course, be different in appearance from the x, y diagram.

The time T elapsing along a segment $\mathscr{C}$ of a phase path connecting two states (see Fig. 1.4) is given by

$$T = \int_{\mathscr{C}} \mathrm{d}t = \int_{\mathscr{C}} \left(\frac{\mathrm{d}x}{\mathrm{d}t}\right)^{-1} \left(\frac{\mathrm{d}x}{\mathrm{d}t}\right) \mathrm{d}t = \int_{\mathscr{C}} \frac{\mathrm{d}x}{X(x,y)}. \qquad (2.8)$$

Alternatively, let $\mathrm{d}s$ be a length element of $\mathscr{C}$. Then $\mathrm{d}s^2 = \mathrm{d}x^2 + \mathrm{d}y^2$ on the path, and

$$T = \int_{\mathscr{C}} \left(\frac{\mathrm{d}s}{\mathrm{d}t}\right)^{-1} \left(\frac{\mathrm{d}s}{\mathrm{d}t}\right) \mathrm{d}t = \int_{\mathscr{C}} \frac{\mathrm{d}s}{(X^2 + Y^2)^{1/2}}. \qquad (2.9)$$

The integrals above depend only on X and Y and the geometry of the phase path; therefore the time scale is implicit in the phase diagram.

2.2. Some population models

In the following examples systems of the type (2.1) arise naturally. Further examples from biology can be found in Pielou (1969) and Rosen (1973).

Example 2.2 *A predator–prey problem* (Volterra's model)

In a lake there are two species of fish: A, which lives on plants of which there is a plentiful supply, and B (the predator) which subsists by eating A (the prey). We shall construct a crude model for the interaction of A and B.

Let $x(t)$ be the population of A and $y(t)$ that of B. We assume that A is relatively long-lived and rapidly breeding if left alone. Then in time δt there is a population increase given by

$$ax\delta t, \quad a > 0$$

due to births and 'natural' deaths, and a 'negative increase'

$$-cxy\delta t, \quad c > 0$$

owing to A's being eaten by B (the number being eaten in this time being assumed proportional to the number of encounters between types A and B). The net population increase of A, δx, is given by

$$\delta x = ax\delta t - cxy\delta t,$$

so that in the limit $\delta t \to 0$

$$\dot{x} = ax - cxy. \qquad (2.10)$$

We now assume that, in the absence of prey, the starvation rate of B predominates over the birth rate, but that the compensating growth of B is again proportional to the number of encounters with A. This gives

$$\dot{y} = -by + dxy \tag{2.11}$$

with $b > 0$, $d > 0$. Equations (2.10) and (2.11) are a pair of simultaneous nonlinear equations of the form (2.2).

We now plot the phase diagram in the x, y plane. Only the quadrant

$$x \geqslant 0, \quad y \geqslant 0$$

is of interest. The equilibrium points are where

$$X(x, y) \equiv ax - cxy = 0, \quad Y(x, y) \equiv -by + dxy = 0;$$

that is at $(0, 0)$ and $(b/d, a/c)$. The phase paths are given by $dy/dx = Y/X$, or

$$\frac{dy}{dx} = \frac{(-b + dx)y}{(a - cy)x},$$

which is a separable equation leading to

$$\int \frac{(a - cy)}{y} dy = \int \frac{(-b + dx)}{x} dx,$$

or

$$a \log_e y + b \log_e x - cy - dx = C \tag{2.12}$$

where C is an arbitrary constant, the parameter of the family. Writing (2.12) in the form $(a \log_e y - cy) + (b \log_e x - dx) = C$, the result of Exercise 25 shows that this is a system of closed curves centred on the equilibrium point $(b/d, a/c)$.

Figure 2.2 shows a detailed plot of the phase paths for a particular case. The direction on the paths, indicating the change of state as t increases, can be obtained by, say, finding the sign of $\dot{x}$ on $x = b/d$. In fact, the direction at a single point, even on $x = 0$ or $y = 0$, determines the directions at all points by continuity.

Since the paths are closed, the fluctuations of $x(t)$ and $y(t)$, starting from any initial population, are predicted as periodic, the maximum population of A being about a quarter of a period behind the maximum population of B. As A gets eaten, causing B to thrive, the population x of A is reduced, causing eventually a drop in that of B. The shortage of predators then leads to a resurgence of A and the cycle starts again. A sudden change in state due to external causes, such as a bad season for the plants, puts the state on to another closed curve, but no tendency to an equilibrium population, nor for the population to disappear, is predicted. If we expect such a tendency, then we must construct a different model (see Exercises 12 and 13).

In Section 2.4 we shall show how the nature of equilibrium points, and hence the general character of the solutions (periodic, tending to equilibrium, unstable, etc.), can be determined without solving the equations explicitly.

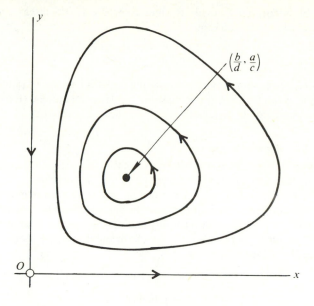

FIG. 2.2. Typical phase diagram for the predator–prey problem

Example 2.3 *A general epidemic model*

Consider the spread of a non-fatal disease in a population which is assumed to have constant size over the period of the epidemic. At time t suppose the population consists of

$x(t)$ susceptibles: those so far uninfected and therefore liable to infection;

$y(t)$ infectives: those who have the disease and are still at large;

$z(t)$ who are isolated, or who have recovered and are therefore immune.

Assume there is a steady contact rate between susceptibles and infectives and that a constant proportion of these contacts result in transmission. Then in time δt, δx of the susceptibles become infective, where

$$\delta x = -\beta x y \delta t,$$

and β is constant.

If γ is the rate at which current infectives become isolated, then

$$\delta y = \beta x y \delta t - \gamma y \delta t.$$

The number of new isolates δz is given by

$$\delta z = \gamma y \delta t.$$

Now let $\delta t \to 0$. Then the system

$$\dot{x} = -\beta x y, \quad \dot{y} = \beta x y - \gamma y, \quad \dot{z} = \gamma y, \tag{2.13}$$

with suitable initial conditions, determines the progress of the disease. Note that the result of adding the equations is

$$\frac{d}{dt}(x+y+z) = 0;$$

that is to say, the assumption of a constant population is built in to the model. The analysis of this problem in the phase plane is left as an exercise (Exercise 29). We shall instead look in detail at a more complicated situation:

Example 2.4 *Recurrent epidemic*

Suppose that the problem is as before, except that the stock of susceptibles $x(t)$ is being added to at a constant rate μ per unit time. This condition could be the result of fresh births in the presence of a childhood disease such as measles in the absence of vaccination. In order to balance the population in the simplest way we shall assume that deaths occur naturally and only among the immune, that is, among the $z(t)$ older people most of whom have had the disease. For a constant population the equations become

$$\dot{x} = -\beta xy + \mu, \tag{2.14}$$

$$\dot{y} = \beta xy - \gamma y, \tag{2.15}$$

$$\dot{z} = \gamma y - \mu \tag{2.16}$$

(note that $(d/dt)(x+y+z) = 0$: the population is steady).

Consider the variation of x and y, the active participants, represented on the x, y phase plane. We need only (2.14) and (2.15), which show an equilibrium point $(\gamma/\beta, \mu/\gamma)$.

Instead of trying to solve the equation for the phase paths we shall try to get an idea of what the phase diagram is like by forming linear approximations to the right-hand sides of (2.14), (2.15) in the neighbourhood of the equilibrium point. Near the equilibrium point we write

$$x = \gamma/\beta + \xi, \quad y = \mu/\gamma + \eta \tag{2.17}$$

(ξ, η small) so that $\dot{x} = \dot{\xi}$ and $\dot{y} = \dot{\eta}$. Retaining only the linear terms in the expansion of the right sides of (2.14), (2.15), we obtain

$$\dot{\xi} = -\frac{\beta\mu}{\gamma}\xi - \gamma\eta, \tag{2.18}$$

$$\dot{\eta} = \frac{\beta\mu}{\gamma}\xi. \tag{2.19}$$

We are said to have *linearized* (2.14) and (2.15) near the equilibrium point. Elimination of ξ gives

$$\gamma\ddot{\eta} + (\beta\mu)\dot{\eta} + (\beta\mu\gamma)\eta = 0. \tag{2.20}$$

This is the equation for the damped linear oscillator (Section 1.4), and we may compare (2.20) with eqn (1.23) (or the system (1.26)) of Chapter 1, but it is necessary to remember that it only holds as an approximation close to the equilibrium point of (2.14) and (2.15). When the 'damping' is light ($\beta\mu/\gamma^2 < 4$) the phase path is a spiral. Figure 2.3 shows some phase paths for a particular case. All starting conditions lead to the stable equilibrium point E: this point is called the 'endemic state' for the disease.

FIG. 2.3. Typical phase diagram for the recurrent epidemic

2.3. Linear approximation at equilibrium points

Approximation to a nonlinear system by linearizing it at an equilibrium point, as in the last example, is a most important and generally useful technique. If the geometrical nature of the equilibrium points can be settled in this way the general character of the phase diagram is often clear. Consider the system

$$\dot{x} = X(x, y), \quad \dot{y} = Y(x, y). \tag{2.21}$$

Suppose that the equilibrium point to be studied has been moved to the origin by a translation of axes, if necessary, so that $X(0, 0) = Y(0, 0) = 0$. We can therefore write, by a Taylor expansion,

$$X(x, y) = ax + by + P(x, y), \quad Y(x, y) = cx + dy + Q(x, y),$$

where $P(x, y) = O(r^2)$ and $Q(x, y) = O(r^2)$ as $r = \sqrt{(x^2 + y^2)} \to 0$, and

$$a = \frac{\partial X}{\partial x}(0, 0), \quad b = \frac{\partial X}{\partial y}(0, 0), \quad c = \frac{\partial Y}{\partial x}(0, 0), \quad d = \frac{\partial Y}{\partial y}(0, 0).$$

(2.22)

The *linear approximation* to (2.21) in the neighbourhood of the origin is defined as the system

$$\dot{x} = ax + by, \quad \dot{y} = cx + dy.$$ (2.23)

We expect that the solutions of (2.23) will be geometrically similar to those of (2.21) near the origin, an expectation fulfilled in most cases (but see Exercise 7: a centre may be an exception).

2.4. The general solution of a linear system

It is known that there are non-trivial solutions of (2.23) of the form

$$x = re^{\lambda t}, \quad y = se^{\lambda t}$$ (2.24)

where r and s are related constants, and λ is another constant. All these constants may be complex. To find values for λ, put (2.24) into (2.23); we obtain

$$(a - \lambda)r + bs = 0,$$
$$cr + (d - \lambda)s = 0.$$ (2.25)

Non-trivial solutions exist if and only if

$$\begin{vmatrix} a - \lambda & b \\ c & d - \lambda \end{vmatrix} = 0,$$

or

$$\lambda^2 - (a + d)\lambda + (ad - bc) = 0,$$ (2.26)

which is called the characteristic equation. When this equation has two different roots, λ_1, λ_2, two linearly independent families of solutions are generated by (2.24), corresponding to $\lambda = \lambda_1$ and $\lambda = \lambda_2$ respectively. We shall consider only this case here.

By substituting first λ_1 and then λ_2 into the equations (2.25) we find the permissible corresponding non-zero pairs r, s. Since the equations are homogeneous, if r, s is one solution, all solutions are given by Cr, Cs, where C is any constant. Let $r = r_1$, $s = s_1$ be any one fixed solution corresponding to $\lambda = \lambda_1$, and similarly for λ_2. Then, because of the linearity of the system (2.23) the general solution is

$$x(t) = C_1 r_1 e^{\lambda_1 t} + C_2 r_2 e^{\lambda_2 t},$$
$$y(t) = C_1 s_1 e^{\lambda_1 t} + C_2 s_2 e^{\lambda_2 t}, \tag{2.27}$$

where C_1 and C_2 are arbitrary constants. Note that when λ_1, λ_2 are complex, r_1, s_1, r_2, s_2 are complex; therefore, if the solutions x, y are to be real, we must allow C_1, C_2 to be complex in general.

Example 2.5 *Find the general solution of the system*

$$\dot{x} = x + y, \quad \dot{y} = -5x - 3y.$$

The characteristic equation (2.26) is

$$\lambda^2 + 2\lambda + 2 = 0$$

so that

$$\lambda_1 = -1 + i, \quad \lambda_2 = -1 - i.$$

Equations (2.25) are, for λ_1

$$(2-i)r_1 + s_1 = 0, \quad -5r_1 - (2+i)s_1 = 0.$$

(Note that these equations are equivalent.) A particular solution is

$$r_1 = 1, \quad s_1 = -2 + i.$$

Since $\lambda_2 = \bar{\lambda}_1$ a solution of the equations corresponding to λ_2 is $r_2 = \bar{r}_1, s_2 = \bar{s}_1$ or

$$r_2 = 1, \quad s_2 = -2 - i.$$

In the form (2.27) the general solution is therefore

$$x(t) = C_1 e^{(-1+i)t} + C_2 e^{(-1-i)t},$$
$$y(t) = C_1(-2+i)e^{(-1+i)t} + C_2(-2-i)e^{(-1-i)t}.$$

If we choose $C_2 = \bar{C}_1$ with C_1 arbitrary, all the real solutions are obtained. Put $C_1 = \frac{1}{2}c_1 + \frac{1}{2}ic_2$ where c_1, c_2 are arbitrary. Then this simplifies to

$$x(t) = e^{-t}(c_1 \cos t - c_2 \sin t),$$
$$y(t) = -e^{-t}\{(2c_1 + c_2)\cos t + (c_1 - 2c_2)\sin t\}.$$

2.5. Classifying equilibrium points

We are interested not so much in the solutions as in the corresponding pattern of phase paths. Write the linearized system (2.23) in the form

$$\dot{x} = Ax \tag{2.28}$$

where

$$A = \begin{pmatrix} a & b \\ c & d \end{pmatrix}, \quad \mathbf{x} = \begin{pmatrix} x \\ y \end{pmatrix}, \quad \dot{\mathbf{x}} = \begin{pmatrix} \dot{x} \\ \dot{y} \end{pmatrix}. \tag{2.29}$$

Look for two linearly independent solutions of the form

$$\mathbf{x} = \mathbf{s}\,\mathrm{e}^{\lambda t}, \tag{2.30}$$

where

$$\mathbf{s} = \begin{pmatrix} r \\ s \end{pmatrix} \neq \mathbf{0}. \tag{2.31}$$

Then $\dot{\mathbf{x}} = \lambda \mathbf{s}\,\mathrm{e}^{\lambda t}$, and (2.28) gives

$$(A - \lambda I)\mathbf{s} = \mathbf{0} \tag{2.32}$$

where I is the identity matrix. Non-zero $\mathbf{s}$ satisfies (2.31) if and only if

$$\det(A - \lambda I) = 0,$$

or

$$\begin{vmatrix} a - \lambda & b \\ c & d - \lambda \end{vmatrix} = 0. \tag{2.33}$$

(2.33) is (2.26) again, and (2.32) is eqn (2.25) again.

Therefore λ_1, λ_2, the solutions of (2.33), are the *eigenvalues* of A; and $\mathbf{s} = \mathbf{s}_1$ and $\mathbf{s} = \mathbf{s}_2$, the solutions of (2.32), are a pair of *eigenvectors* corresponding respectively to λ_1 and λ_2. Provided $\lambda_1 \neq \lambda_2$, the general solution of (2.28) is

$$\mathbf{x} = C_1 \mathbf{s}_1\,\mathrm{e}^{\lambda_1 t} + C_2 \mathbf{s}_2\,\mathrm{e}^{\lambda_2 t}, \tag{2.34}$$

which is eqn (2.27).

Now reconsider the problem of finding the phase paths. In general their equation is

$$\frac{\mathrm{d}y}{\mathrm{d}x} = \frac{cx + dy}{ax + by}. \tag{2.35}$$

This can always be solved but the reader can soon convince himself that, except in special cases, the results are too complicated to give a simple discussion of the geometrical nature of the solutions.

We can obtain linear transformations which reduce eqn (2.35) to manageable form. A nonsingular linear transformation from x, y to u, v:

$$\mathbf{u} = S\mathbf{x}, \quad \mathbf{u} = \begin{pmatrix} u \\ v \end{pmatrix} \tag{2.36}$$

does not change the nature of the singular points; for example spirals in the x, y plane remain spirals in the u, v plane, and so on. Writing

$$\mathbf{x} = S^{-1}\mathbf{u}, \quad \dot{\mathbf{x}} = S^{-1}\dot{\mathbf{u}},$$

(2.28) becomes $S^{-1}\dot{\mathbf{u}} = AS^{-1}\mathbf{u}$, or

$$\dot{\mathbf{u}} = SAS^{-1}\mathbf{u} = B\mathbf{u}, \tag{2.37}$$

say. It is known from algebraic theory (Ayres 1962) that S can be chosen so that B takes one of several canonical forms which in general are simpler than A, the particular form depending on the nature of the eigenvalues of A. In the new coordinates u, v, the equations are therefore simpler although the topological character of the transformed equilibrium point at the origin is not affected. We do not need to calculate the S for this purpose: we only need the fact that the relevant S exists.

The principal cases, neglecting certain degenerate ones, are the following:

(i) λ_1, λ_2 *real, different, and non-zero*
We can choose S so that

$$\dot{u} = \lambda_1 u, \quad \dot{v} = \lambda_2 v,$$

where also $\lambda_2 > \lambda_1$. Then the equation for the phase paths is

$$\frac{\mathrm{d}v}{\mathrm{d}u} = \frac{\lambda_2}{\lambda_1} \frac{v}{u}$$

whose solutions are given by

$$v = C|u|^{\lambda_2/\lambda_1} \tag{2.38}$$

where C is arbitrary. $u = 0$ is also a phase path.

Since we arranged the eigenvalues so that $\lambda_2 > \lambda_1$ only two possible patterns emerge, depending on whether λ_1 and λ_2 have the same or opposite signs (Fig. 2.4). In the former case, of a *node*, its stability and hence the direction of the 'arrows' are determined from (2.34): if λ_1, $\lambda_2 > 0$, then all solutions are exponentially increasing and the node is

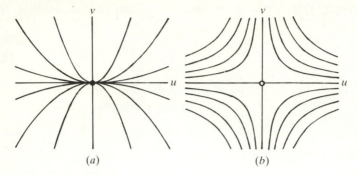

FIG. 2.4. λ_1, λ_2 real, non-zero, $\lambda_2 > \lambda_1$: (*a*) λ_1, λ_2 have the same sign (node); (*b*) λ_1, λ_2 have opposite signs (saddle)

unstable, all arrows pointing away from the origin. If $\lambda_1, \lambda_2 < 0$, then the node is *stable*.

The detailed analysis of a particular case proceeds as in the following example.

Example 2.6   *Sketch the phase paths of the system*

$$\dot{x} = x - 2y, \quad \dot{y} = 3x - 4y. \tag{2.39}$$

The eigenvalues of the system are given by

$$0 = \begin{vmatrix} 1 - \lambda & -2 \\ 3 & -4 - \lambda \end{vmatrix} \equiv \lambda^2 + 3\lambda + 2$$

so that $\lambda_1 = -2$, $\lambda_2 = -1$. Since they are real and have the same sign a node is implied in the u, v plane.

The paths in the x, y plane must be obtainable from Fig. 2.4(*a*) by a linear transformation from (u, v) to (x, y). There may be changes of scale, shears, rotations, and reflections, but straight lines remain straight and parallel lines remain parallel. The stability property is unaffected, and since $\lambda_1, \lambda_2 < 0$, the node is stable.

To obtain a more precise idea of the shape of the node, we look for those solutions of (2.39) which correspond to the straight lines $u = 0$, $v = 0$ in Fig. 2.4(*a*). Since the transformation preserves them as straight lines they take the form $y = kx$, and substitution into (2.35) gives

$$k = \frac{3 - 4k}{1 - 2k}$$

from which $k = 1$ and 3/2. (These directions are those of the eigenvectors of A, since $\dot{x} = A\mathbf{x} = \lambda\mathbf{x}$ along the eigenvectors.)

FIG. 2.5.

The second feature of Fig. 2.4(*a*) preserved by the linear transformation is that as $t \to +\infty$ (that is, approaching the origin) *all the curves become tangential to one of the solutions just found, and as $t \to -\infty$ they become parallel to the other.* The associated correspondence between Figs. 2.4(*a*) and 2.5 can be established by noting that, from (2.39), the paths cross the line $y = \frac{3}{4}x$ horizontally. The only possibility (Fig. 2.5) is that all paths are tangential to $y = x$ at the origin.

The *saddle* (Fig. 2.4(*b*)), for which λ_1, λ_2 have opposite signs, is simpler to construct, as the following example illustrates.

Example 2.7 *Sketch the phase paths of the system*

$$\dot{x} = 3x + 2y, \quad \dot{y} = -2x - 2y. \tag{2.40}$$

The eigenvalues of the system are given by

$$0 = \begin{vmatrix} 3 - \lambda & 2 \\ -2 & -2 - \lambda \end{vmatrix} \equiv \lambda^2 - \lambda - 2$$

so that $\lambda_1 = -1$, $\lambda_2 = 2$. Since these are real and of opposite sign the equilibrium point at the origin is a saddle.

In Fig. 2.4(*a*), $u = 0$ and $v = 0$ transform into straight-line paths in the (x, y)-plane. To find them, substitute $y = kx$ into the present version of (2.35) to give

$$k = \frac{-2 - 2k}{3 + 2k}$$

with solutions $k = -1/2$ *and* -2 (Fig. 2.6). (As in the last Example, these are the directions of the eigenvectors of A.)

The directions of the arrows are settled by putting, for example, $y = 0$, $x > 0$ into (2.40), showing that $\dot{y} < 0$ on the positive x-axis.

FIG. 2.6.

(ii) *Complex eigenvalues*: $\lambda_1 = \bar{\lambda}_2 = \alpha + i\beta$, $\beta \neq 0$

The transformation matrix S of (2.36) can be chosen so that the reduced eqn (2.37) takes the form

$$\dot{u} = \alpha u - \beta v, \quad \dot{v} = \beta u + \alpha v. \tag{2.41}$$

Put $z = u + iv$: then, by (2.41),

$$\dot{z} = (\alpha + i\beta)z,$$

and by writing $z = r(t) e^{i\theta(t)}$ where $r = |z|$ we obtain the polar equations

$$\dot{r} = \alpha r, \quad \dot{\theta} = \beta.$$

The solutions of these equations are $r = r_0 e^{\alpha t}$ and $\theta = \beta t + \theta_0$ where $r(0) = r_0$ and $\theta(0) = \theta_0$. The origin is therefore a *stable spiral* if $\alpha < 0$; an *unstable spiral* if $\alpha > 0$; and a *centre* if $\alpha = 0$.

Example 2.8   *Sketch the phase paths of the system*

$$\dot{x} = -x - 5y, \quad \dot{y} = x + 3y. \tag{2.42}$$

The eigenvalues are given by

$$0 = \begin{vmatrix} -1 - \lambda & -5 \\ 1 & 3 - \lambda \end{vmatrix} \equiv \lambda^2 - 2\lambda + 2$$

so that λ_1, $\lambda_2 = 1 \pm i$. The phase diagram is therefore an unstable spiral. The spirals 'unwind' with increasing time, but may be directed *clockwise* or *anticlockwise* in general; to determine the direction in this case, put, for example $y = 0$, $x > 0$ in (2.42). The second equation implies that $\dot{y} > 0$ on the positive x-axis, so that the spirals unwind in the anticlockwise direction.

(iii) *Degenerate cases*

These occur when there is a single repeated eigenvalue, and when an eigenvalue is zero. These cases will not be discussed in detail.

2.6. Constructing a phase diagram

Suppose that the given system

$$\dot{x} = X(x, y), \quad \dot{y} = Y(x, y) \tag{2.43}$$

has an equilibrium point at (x_0, y_0):

$$X(x_0, y_0) = 0, \quad Y(x_0, y_0) = 0. \tag{2.44}$$

As we remarked in Section 2.3, the pattern of phase paths close to (x_0, y_0) may be decided by linearizing the equations at this point by retaining only linear terms of the Taylor series for X and Y there. It is simplest to use the method leading up to eqn (2.22) to obtain the coefficients. If local coordinates are defined by

$$\xi = x - x_0, \quad \eta = y - y_0,$$

then, approximately,

$$\begin{pmatrix} \dot{\xi} \\ \dot{\eta} \end{pmatrix} = \begin{pmatrix} a & b \\ b & d \end{pmatrix} \begin{pmatrix} \xi \\ \eta \end{pmatrix} \tag{2.45}$$

where the coefficients are given by

$$\begin{pmatrix} a & b \\ c & d \end{pmatrix} = \begin{pmatrix} X_x(x_0, y_0) & X_y(x_0, y_0) \\ Y_x(x_0, y_0) & Y_y(x_0, y_0) \end{pmatrix}. \tag{2.46}$$

The equilibrium point is then classified using the methods of the last section. This is done for each equilibrium point in turn, and it is then possible to make a fair guess at the complete pattern of the phase paths, as in the following example.

Example 2.9 *Sketch the phase diagram for the nonlinear system*

$$\dot{x} = x - y, \quad \dot{y} = 1 - xy. \tag{2.47}$$

The equilibrium points are at $(-1, -1)$ and $(1, 1)$. The matrix for linearization, to be evaluated at each equilibrium point in turn, is

$$\begin{pmatrix} X_x & X_y \\ Y_x & Y_y \end{pmatrix} = \begin{pmatrix} 1 & -1 \\ -y & -x \end{pmatrix}. \tag{2.48}$$

At $(-1, -1)$ eqns (2.46) and (2.48) give

$$\begin{pmatrix} \dot{\xi} \\ \dot{\eta} \end{pmatrix} = \begin{pmatrix} 1 & -1 \\ 1 & 1 \end{pmatrix} \begin{pmatrix} \xi \\ \eta \end{pmatrix}, \tag{2.49}$$

where $\xi = x+1$, $\eta = y+1$. The eigenvalues of the coefficient matrix are found to be $\lambda_1, \lambda_2 = 1 \pm i$ implying an unstable spiral. To obtain the direction of rotation, it is sufficient to use the linear equations (2.53) (or the original equations may be used): putting $\eta = 0$, $\xi > 0$ we find $\dot{\eta} = \xi > 0$, indicating that the rotation is *anticlockwise*.

At $(1, 1)$, we find from (2.46) and (2.48) that

$$\begin{pmatrix} \dot{\xi} \\ \dot{\eta} \end{pmatrix} = \begin{pmatrix} 1 & -1 \\ -1 & -1 \end{pmatrix} \begin{pmatrix} \xi \\ \eta \end{pmatrix}, \tag{2.50}$$

where $\xi = x-1$, $\eta = y-1$. The eigenvalues are given by $\lambda_1, \lambda_2 = \pm\sqrt{2}$, which implies a saddle. The directions of the 'straight-line' paths from the saddle (which become curved separatrices when away from the equilibrium point), are resolved by the technique of Example 2.7: from (2.50)

$$\frac{\dot{\eta}}{\dot{\xi}} = \frac{d\eta}{d\xi} = \frac{-\xi - \eta}{\xi - \eta}. \tag{2.51}$$

We know that two solutions of this equation have the form $\eta = k\xi$ for some values of k. By substituting into (2.51) we obtain $k^2 - 2k - 1 = 0$, so that $k = 1 \pm \sqrt{2}$.

Finally the phase diagram is put together as in Fig. 2.7, where the phase paths in the neighbourhoods of the equilibrium points are now known. The process can be assisted by sketching in the direction fields on the lines $x = 0$, $x = 1$, etc., and also the curve $1 - xy = 0$ on which the phase paths have zero slopes and the line $y = x$ on which the paths have infinite slopes.

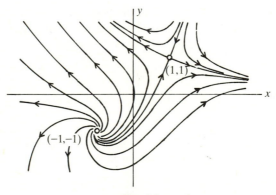

FIG. 2.7.

2.7. Transitions between types of equilibrium point

The types of equilibrium point discussed above can be classified in terms of two parameters of the system as follows. The determinantal equation giving the eigenvalues, (2.26), can be written as

$$\lambda^2 - p\lambda + q = 0$$

where

$$p = a+d, \quad q = ad-bc.$$

Hence

$$\lambda_1, \lambda_2 = \tfrac{1}{2}p \pm \tfrac{1}{2}\sqrt{\Delta}$$

where

$$\Delta = p^2 - 4q.$$

The various cases are displayed in Fig. 2.8. The degenerate cases occur on $\Delta = 0$ and $q = 0$. Note that the centre constitutes a transition between stable and unstable spirals: the existence of a centre depends on particular exact relations between the coefficients of the system and is therefore a rather fragile feature. This has the consequence that if the linear approximation of a nonlinear system predicts a centre it cannot be reliably concluded that the equilibrium point is truly a centre—it could be a stable or unstable spiral (see, for example, Exercise 7). Further reference to this matter in connection with bifurcations is made in Chapter 12.

If there exists a neighbourhood of an equilibrium point such that phase paths which start from all points in the neighbourhood

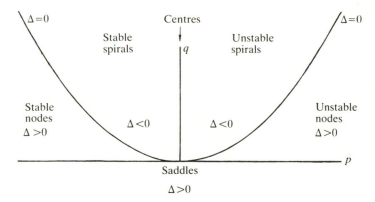

FIG. 2.8. General classification for the linear system: $\dot{x} = ax + by$, $\dot{y} = cx + dy$ with $p = a+d$, $q = ad-bc$, and $\Delta = p^2-4q$

ultimately approach the equilibrium point, the equilibrium point is known as an *attractor*. Both the stable node and stable spiral are examples of attractors. An attractor with all path directions reversed is known as a *repellor*. The unstable node and unstable spiral are both repellors (but the saddle is not).

If the eigenvalues of the linearized equation have non-zero real parts at the equilibrium point then the point is said to be *hyperbolic*. As we shall see in Chapter 10, for hyperbolic points the phase diagrams of the nonlinear equation and its linearization are locally qualitatively the same. Spirals, nodes, and saddles are hyperbolic but the centre is not.

Exercises

1. Sketch phase diagrams for the following linear systems:
   - (i) $\dot{x} = x - 5y,$ $\dot{y} = x - y;$
   - (ii) $\dot{x} = x + y,$ $\dot{y} = x - 2y;$
   - (iii) $\dot{x} = -4x + 2y,$ $\dot{y} = 3x - 2y;$
   - (iv) $\dot{x} = x + 2y,$ $\dot{y} = 2x + 2y;$
   - (v) $\dot{x} = 4x - 2y,$ $\dot{y} = 3x - y;$
   - (vi) $\dot{x} = 2x + y,$ $\dot{y} = -x + y.$

2. The following systems either generate a single eigenvalue, or a zero eigenvalue, or in other ways vary from the types illustrated in Section 2.4. Sketch their phase diagrams.
   - (i) $\dot{x} = 3x - y,$ $\dot{y} = x + y;$
   - (ii) $\dot{x} = x - y,$ $\dot{y} = 2x - 2y;$
   - (iii) $\dot{x} = x,$ $\dot{y} = 2x - 3y;$
   - (iv) $\dot{x} = x,$ $\dot{y} = x + 3y;$
   - (v) $\dot{x} = -y,$ $\dot{y} = 2x - 4y;$
   - (vi) $\dot{x} = x,$ $\dot{y} = y;$
   - (vii) $\dot{x} = 0,$ $\dot{y} = x.$

3. Locate and classify the equilibrium points of the following systems. Sketch the phase diagrams: it will often be helpful to obtain isoclines and path directions at other points in the field.
   - (i) $\dot{x} = x - y,$ $\dot{y} = x + y - 2xy;$
   - (ii) $\dot{x} = y\,e^{y},$ $\dot{y} = 1 - x^2;$
   - (iii) $\dot{x} = 1 - xy,$ $\dot{y} = (x - 1)y;$
   - (iv) $\dot{x} = (1 + x - 2y)x,$ $\dot{y} = (x - 1)y;$
   - (v) $\dot{x} = x - y,$ $\dot{y} = x^2 - 1;$
   - (vi) $\dot{x} = -6y + 2xy - 8,$ $\dot{y} = y^2 - x^2;$
   - (vii) $\dot{x} = 4 - 4x^2 - y^2,$ $\dot{y} = 3xy;$
   - (viii) $\dot{x} = -y\sqrt{(1 - x^2)},$ $\dot{y} = x\sqrt{(1 - x^2)}$ for $|x| \leqslant 1;$
   - (ix) $\dot{x} = \sin y,$ $\dot{y} = -\sin x.$

4. Construct phase diagrams for the following differential equations, using the phase plane in which $y = \dot{x}$.

(i) $\ddot{x} + x - x^3 = 0$; (ii) $\ddot{x} + x + x^3 = 0$;
(iii) $\ddot{x} + \dot{x} + x - x^3 = 0$; (iv) $\ddot{x} + \dot{x} + x + x^3 = 0$;
(v) $\ddot{x} = \sin x (2 \cos x - 1)$.

5. Confirm that the system $\dot{x} = x - 5y$, $\dot{y} = x - y$ consists of a centre. By substitution into the equation for the paths or otherwise show that the family of ellipses given by

$$x^2 - 2xy + 5y^2 = constant$$

describes the paths. Show that the axes are inclined at about $13.3°$ (the major axis) and $-76.7°$ (the minor axis) to the x direction, and that the ratio of major to minor axis length is about 8.33.

6. The family of curves which are orthogonal to the family described by the equation $(dy/dx) = f(x, y)$ is given by the solution of $(dy/dx) = -[1/f(x, y)]$. (These are called the orthogonal trajectories of the first family.) Prove that the family which is orthogonal to a centre which is associated with a linear system is a node.

7. Show that the origin is a spiral point of the system $\dot{x} = -y - x\sqrt{(x^2 + y^2)}$, $\dot{y} = x - y\sqrt{(x^2 + y^2)}$, but a centre for its linear approximation.

8. Show that the systems $\dot{x} = y$, $\dot{y} = -x - y^2$, and $\dot{x} = x + y_1$, $\dot{y}_1 = -2x - y_1 - (x + y_1)^2$, both represent the equation $\ddot{x} + \dot{x}^2 + x = 0$ in different phase planes.

9. Use eqn (2.9) in the form $\delta s \simeq \delta t \sqrt{(X^2 + Y^2)}$ to mark off approximately equal time steps on some of the phase paths of $\dot{x} = xy$, $\dot{y} = xy - y^2$.

10. Obtain approximations to the phase paths described by (2.12) in the neighbourhood of the equilibrium point $x = b/d$, $y = a/c$. (Write $x = b/d + \xi$, $y = a/c + \eta$, and expand the logarithms to second-order terms in ξ and η.)

11. For the system $\dot{x} = ax + by$, $\dot{y} = cx + dy$, where $ad - bc = 0$, show that all points on the line $cx + dy = 0$ are equilibrium points.

12. The interaction between two species is governed by the deterministic model $\dot{H} = (a_1 - b_1 H - c_1 P)H$, $\dot{P} = (-a_2 + c_2 H)P$, where H is the population of the host (or prey), and P is that of the parasite (or predator), all constants being positive. (Compare Example 2.2: the term $-b_1 H^2$ represents interference with the host population when it gets too large.) Find the equilibrium states for the populations, and find how they vary with time from various initial populations.

13. With the same terminology as in Exercise 12, analyse the system $\dot{H} = (a_1 - b_1 H - c_1 P)H$, $\dot{P} = (a_2 - b_2 P + c_2 H)P$, all the constants being positive. (In this model the parasite can survive on alternative food supplies, although the

prevalence of the host encourages growth in the population.) Find the equilibrium states. Confirm that the parasite population can persist even if the host dies out.

14. Consider the host–parasite population model $\dot{H} = (a_1 - c_1 P)H$, $\dot{P} = (a_2 - c_2 P/H)P$, where the constants are positive. Analyse the system in the H, P plane.

15. In the population model $\dot{F} = -\alpha F + \beta\mu(M)F$, $\dot{M} = -\alpha M + \gamma\mu(M)F$, where $\alpha > 0, \beta > 0, \gamma > 0$, F and M are the female and male populations. In both cases the death-rates are α. The birth-rate is governed by the coefficient $\mu(M) = 1 - e^{-kM}$, $k > 0$, so that for large M the birth-rate of females is βF and that for males is γF, the rates being unequal in general. Show that if $\beta > \alpha$ then there are two equilibrium states, at $(0, 0)$ and at $([-\beta/\gamma k]\log[(\beta-\alpha)/\beta], [-1/k]\log[(\beta-\alpha)/\beta])$.

Show that the origin is stable and that the other equilibrium point is a saddle-point, according to their linear approximations. Verify that $M = \gamma F/\beta$ is a particular solution. Sketch the phase diagram and discuss the stability of the populations.

16. A rumour spreads through a closed population of constant size $N + 1$. At time t the total population can be classified into three categories:

x persons who are ignorant of the rumour;

y persons who are actively spreading the rumour;

z persons who have heard the rumour but have stopped spreading it: if two persons who are spreading the rumour meet then they stop spreading it.

The contact rate between any two categories is constant, μ.

Show that the equations

$$\dot{x} = -\mu xy, \quad \dot{y} = \mu[xy - y(y-1) - yz]$$

give a deterministic model of the problem. Find the equations of the phase paths and sketch the phase diagram.

Show that, when initially $y = 1$ and $x = N$, the number of people who ultimately never hear the rumour is x_1, where

$$2N + 1 - 2x_1 + N\log(x_1/N) = 0.$$

17. The one-dimensional steady flow of a gas with viscosity and heat conduction satisfies the equations

$$\frac{\mu_0}{\rho c_1}\frac{dv}{dx} = \sqrt{(2v)}\left[2v - \sqrt{(2v)} + \theta\right]$$

$$\frac{k}{gR\rho c_1}\frac{d\theta}{dx} = \sqrt{(2v)}\left[\frac{\theta}{\gamma - 1} - v + \sqrt{(2v)} - c\right]$$

where $v = u^2/2c_1^2$, $c = c_2^2/c_1^2$ and $\theta = gRT/c_1^2 = p/\rho c_1^2$. In this notation, x is measured in the direction of flow, u is the velocity, T is the temperature, ρ is the density, p the pressure, R the gas constant, k the coefficient of thermal conductivity, μ_0 the coefficient of viscosity, γ the ratio of the specific heats, and c_1, c_2 are arbitrary constants. Find the equilibrium states of the system.

18. A particle moves under a central attractive force γ/r^2 per unit mass, where r, θ are the polar coordinates of the particle in its plane of motion. Show that

$$\frac{d^2u}{d\theta^2} + u = \frac{\gamma}{h^2} u^{\alpha-2},$$

where $u = r^{-1}$, h is the angular momentum about the origin per unit mass of the particle, and γ is a constant. Find the non-trivial equilibrium point in the $u, du/d\theta$ plane and classify it according to its linear approximation. What can you say about the stability of the circular orbit under this central force?

19. The relativistic equation for the central orbit of a planet is

$$\frac{d^2u}{d\theta^2} + u = \alpha + \varepsilon u^2$$

where $u = 1/r$, and r, θ are the polar coordinates of the planet in the plane of its motion. The term εu is the 'Einstein correction', and α and ε are positive constants, with ε very small. Find the equilibrium point which corresponds to a perturbation of the Newtonian orbit. Show that the equilibrium point is a centre in the $u, du/d\theta$ plane according to the linear approximation. Confirm this by using the method of Section 1.3.

20. A top is set spinning at an axial rate n about its pivotal point, which is fixed in space. The equations for its motion, in terms of the angles θ and μ are (see Fig. 2.9)

$$A\ddot{\theta} - A(\Omega + \dot{\mu})^2 \sin\theta\cos\theta + Cn(\Omega + \dot{\mu})\sin\theta - Mgh\sin\theta = 0,$$

$$A\dot{\theta}^2 + A(\Omega + \dot{\mu})^2 \sin^2\theta + 2Mgh\cos\theta = E;$$

where (A, A, C) are the principal moments of inertia about O, M is the mass of the top, h is the distance between the mass-centre and the pivot, and E is a constant. Show that one equilibrium state is given by $\theta = \alpha$, provided E and μ satisfy

$$A\Omega^2 \cos\alpha - Cn\Omega + Mgh = 0, \quad A\Omega^2 \sin^2\alpha + 2Mgh\cos\alpha = E.$$

Discuss the motion of the top in this state.

Suppose that $E = 2Mgh$, so that $\theta = 0$ is an equilibrium state. Show that, close to this state, θ satisfies

$$A\ddot{\theta} + (Cn\Omega - A\Omega^2 - Mgh)\theta = 0.$$

For what condition on Ω is the motion stable?

Fig. 2.9.

21. Figure 2.10 shows an 'up-and-over' garage door AB, of height $2h$, mass m, and moment of inertia I about G. The force P is applied by springs and weights, and it can be designed to have suitable characteristics. The unknown reactions S and R, and the frictional force F are as shown.

(i) If the runner B is smooth, show that it is possible to design the door so that it is in equilibrium at every angle θ, when P is a suitable constant. Sketch the phase diagram.

(ii) Suppose now that a frictional force F, with $|F/R| = \mu$, is introduced at B. What values should $P(\theta)$ have at $\theta = 0$ and $\theta = \frac{1}{2}\pi$ in order that the open and closed positions should be in equilibrium?

Fig. 2.10. Garage door

(iii) A larger design problem can be attempted. Construct a function $P(\theta)$ which will ensure that the open and closed positions are stable nodes (why nodes?) according to the linear approximations. Do you expect the door to have a further equilibrium position? (This problem does not, of course, have a unique solution.)

22. A disc of radius a is freely pivoted at its centre A so that it can turn in a vertical plane. An elastic string, of natural length $2a$ and stiffness λ connects a point B on the circumference of the disc to a fixed point O, distance $2a$ above A. Show that θ satisfies

$$I\ddot{\theta} = -Ta\sin\phi, \quad T = \lambda a[(5 - 4\cos\theta)^{1/2} - 2],$$

where I is the moment of inertia of the disc about A, $O\hat{A}B = \theta$ and $A\hat{B}O = \phi$. Find the equilibrium states of the disc.

23. A man rows a boat across a river of width a occupying the strip $0 \leqslant x \leqslant a$ in the x, y plane, always rowing towards a fixed point on one bank, say $(0, 0)$. He rows at a constant speed u relative to the water, and the river flows at a constant speed v. Show that

$$\dot{x} = -ux/\surd(x^2 + y^2), \quad \dot{y} = v - uy/\surd(x^2 + y^2),$$

where (x, y) are the coordinates of the boat. Show that the phase paths are given by $y + \surd(x^2 + y^2) = Ax^{1-\alpha}$, where $\alpha = v/u$. Sketch the phase diagram for $\alpha < 1$ and interpret it. What kind of point is the origin? What happens to the boat if $\alpha > 1$?

24. In a simple model of a national economy, $\dot{I} = I - \alpha C$, $\dot{C} = \beta(I - C - G)$, where I is the national income, C is the rate of consumer spending, and G the rate of government expenditure; the constants α and β satisfy $1 < \alpha < \infty$, $1 \leqslant \beta < \infty$. Show that if the rate of government expenditure G_0 is constant there is an equilibrium state. Classify the equilibrium state and show that the economy oscillates when $\beta = 1$.

Consider the situation when government expenditure is related to the national income by the rule $G = G_0 + kI$, where $k > 0$. Show that there is no equilibrium state if $k \geqslant (\alpha - 1)/\alpha$. How does the economy then behave?

Discuss an economy in which $G = G_0 + kI^2$, and show that there are two equilibrium states.

25. Let $f(x)$ and $g(y)$ have local minima at $x = a$ and $y = b$ respectively. Show that $f(x) + g(y)$ has a minimum at (a, b). Deduce that there exists a neighbourhood of (a, b) in which all solutions of the family of equations

$$f(x) + g(y) = \text{constant}$$

represent closed curves surrounding (a, b).

Show that $(0,0)$ is a centre for the system $\dot{x} = y^5$, $\dot{y} = -x^3$, and that all paths are closed curves.

26. For the predator–prey problem in Section 2.2, show by using Exercise 25 that all solutions in $H > 0$, $P > 0$ are periodic.

27. Show that the phase paths of the Hamiltonian system $\dot{x} = -\partial H/\partial y$, $\dot{y} = \partial H/\partial x$ are given by $H(x, y) = $ constant. (This is a generalization of the conservative system of Section 1.3.) Equilibrium points occur at the stationary points of $H(x, y)$. If (x_0, y_0) is an equilibrium point, show that (x_0, y_0) is stable according to the linear approximation if $H(x, y)$ has a maximum or a minimum at the point. (Assume that all the second derivatives of H are non-zero at x_0, y_0.)

28. The equilibrium points of the nonlinear parameter-dependent system $\dot{x} = y$, $\dot{y} = -f(x, y, \lambda)$ lie on the curve $f(x, 0, \lambda) = 0$ in the x, λ plane. Show that an equilibrium point x_1, λ_1 is stable and that all neighbouring solutions tend to this point (according to the linear approximation) if $f_x(x_1, 0, \lambda_1) < 0$ and $f_y(x_1, 0, \lambda_1) < 0$.

Interpret stability in terms of the surface $f(x, y, \lambda) = 0$ in the x, y, λ space and the regions in which $f(x, y, \lambda) > 0$.

29. Find the equations for the phase paths for the general epidemic described (Section 2.2) by the system

$$\dot{x} = -\beta xy, \quad \dot{y} = \beta xy - \gamma y, \quad \dot{z} = \gamma y.$$

Sketch the phase diagram in the x, y plane. Confirm that the number of infectives reaches its maximum when $x = \gamma/\beta$.

30. Two species x and y are competing for a common limited food supply. Their growth equations are given by

$$\dot{x} = xP(x, y), \quad \dot{y} = yQ(x, y).$$

Express in mathematical terms the following constraints on the populations:
(i) if either species increases, the growth rate of the other goes down;
(ii) if one species is absent, the other shows limited growth characteristics;
(iii) if either population is very large, then neither species can multiply.

Assuming that P and Q have the necessary properties, sketch a first-quadrant phase diagram in which also $P = 0$ and $Q = 0$ do not intersect, and $P = 0$ is 'below' $Q = 0$. Indicate the signs of $\dot{x}$ and $\dot{y}$ in the regions into which the quadrant is now divided. Deduce the character of the phase paths by qualitative arguments.

31. Sketch the phase diagram for the competing species x and y for which

$$\dot{x} = (1 - x^2 - y^2)x, \quad \dot{y} = (1.1 - x - y)y.$$

32. A space satellite is in free flight on the line joining, and between, a planet

(mass m_1) and its moon (mass m_2), which are at a fixed distance a apart. Show that

$$-\frac{\gamma m_1}{x^2} + \frac{\gamma m_2}{(a-x)^2} = \ddot{x}$$

where x is the distance of the satellite from the planet and γ is the gravitational constant. Show that the equilibrium point is unstable according to the linear approximation.

33. The system

$$\dot{V}_1 = -\sigma V_1 + f(E - V_2), \quad \dot{V}_2 = -\sigma V_2 + f(E - V_1), \quad \sigma > 0, \quad E > 0$$

represents (Andronov and Chaikin 1949) a model of a triggered sweeping circuit for an oscilloscope. The conditions on $f(u)$ are: $f(u)$ continuous, $-\infty < u < \infty$, $f(-u) = -f(u)$, $f(u)$ tends to a limit as $u \to \infty$, and $f'(u)$ is monotonic decreasing (see Figure 3.19).

Show by a geometrical argument that there is always at least one equilibrium point, (V_0, V_0) say, and that when $f'(E - V_0) < \sigma$ it is the only one; and deduce by taking the linear approximation that it is a stable node. (Note that $f'(E - v) = -df(E - v)/dv$.)

Show that when $f'(E - V_0) > \sigma$ there are two others, at $(V', (1/\sigma)f(E - V'))$ and $((1/\sigma)f(E - V'), V')$ respectively for some V'. Show that these are stable nodes, and that the one at (V_0, V_0) is a saddle point.

34. The response of a certain biological oscillator, (x, y), $x \geqslant 0$, $y \geqslant 0$, to a stimulus measured by a constant b satisfies the system

$$\dot{x} = x - ay + b, \quad \dot{y} = x - cy \text{ for } x \geqslant 0, \quad y \geqslant 0;$$

$$\dot{y} = -cy \text{ for } x = 0.$$

Show that when $c < 1$ and $4a > (1 + c)^2$ then there exists a limit cycle, part of which lies on the y axis, whose period is independent of b. Sketch the corresponding solution function. (Varying the amount of stimulus translates the response in time without affecting the period.)

35. Figure 2.11 represents a circuit for activating an electric arc A which has the voltage-current characteristic shown. Show that $L\dot{I} = V - V_a(I)$, $RC\dot{V} = -RI - V + E$. By forming the linear approximating equations near the equilibrium points find the conditions on E, L, C, R, and V'_a for stable working.

36. The equation for the current in the circuit of Fig. 2.12(a) is

$$LC\ddot{x} + RC\dot{x} + x = I.$$

Neglect the grid current, and assume that I depends only on the relative grid potential e_g: $I = I_s$ (saturation current) for $e_g > 0$ and $I = 0$ for $e_g < 0$. Assume also that $M > 0$, so that $e_g \gtrless 0$ according as $\dot{x} \gtrless 0$. Find the nature of the phase

FIG. 2.11.

paths. By considering their successive intersections with the x axis show that a limit cycle is approached from all initial conditions.

37. For the circuit in Figure 2.12(a) assume that the relation between I and e_g is as in Fig. 2.13: $I = f(e_g + ke_p)$, where e_g and e_p are the relative grid and plate potentials, $k > 0$ is a constant, and in the neighbourhood of the point of inflection, $f(u) = I_0 + au - bu^3$, where $a > 0$, $b > 0$. Deduce the equation for x when the operating point is the point of inflection. Find when the origin is a stable or an unstable point of equilibrium. (A form of Rayleigh's equation (1.40) is obtained, implying an unstable or a stable limit cycle respectively.)

38. Figure 2.14(a) represents two identical D.C. generators connected in parallel, with inductance and resistance L, r. R is the resistance of the load. Show that the equations for the currents are

$$L\frac{di_1}{dt} = -(r+R)i_1 - Ri_2 + E(i_1), \quad L\frac{di_2}{dt} = -Ri_1 - (r+R)i_2 + E(i_2).$$

(a)

(b)

FIG. 2.12.

FIG. 2.13.

Assuming that $E(i)$ has the characteristics indicated by Fig. 2.14(b) show that
(i) when $E'(0) < r$ the state $i_1 = i_2 = 0$ is stable and is otherwise unstable;
(ii) when $E'(0) > r$ there is a stable state $i_1 = -i_2$ (no current flows to R);
(iii) when $E'(0) > r + 2R$ there is a state with $i_1 = i_2$, which may be unstable.

39. Show that the Emden–Fowler equation of astrophysics

$$(\xi^2 \eta')' + \xi^\lambda \eta^n = 0$$

is equivalent to the predator–prey model

$$\dot{x} = -x(1+x+y), \quad \dot{y} = y(\lambda+1+nx+y)$$

after the change of variable

$$x = \xi\eta'/\eta, \quad y = \xi^{\lambda-1}\eta^n/\eta', \quad t = \log|\xi|.$$

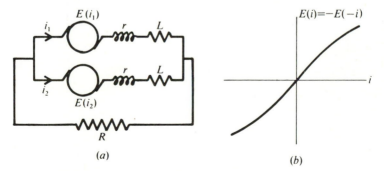

(a)

(b)

FIG. 2.14.

40. Show that Blasius' equation

$$\eta''' + \eta\eta'' = 0$$

is transformed by

$$x = \eta\eta'/\eta'', \quad y = \eta'^2/\eta\eta'', \quad t = \log|\eta'|$$

into

$$\dot{x} = x(1 + x + y), \quad \dot{y} = y(2 + x - y).$$

41. Consider the family of linear systems

$$\dot{x} = X\cos\alpha - Y\sin\alpha, \quad \dot{y} = X\sin\alpha + Y\cos\alpha$$

where

$$X = ax + by, \quad Y = cx + dy,$$

and a, b, c, d are constants and α is a parameter. Show that the equilibrium point at the origin passes through the sequence stable node, stable spiral, centre, unstable spiral, unstable node, as α varies over range π.

42. Show that, given $X(x, y)$, the system equivalent to the equation $\ddot{x} + h(x, \dot{x}) = 0$ is

$$\dot{x} = X(x, y), \quad \dot{y} = -\left\{h(x, X) + X\frac{\partial X}{\partial x}\right\} \Big/ \frac{\partial X}{\partial y}.$$

43. The following system models two species with populations N_1 and N_2 competing for a common food supply:

$$\dot{N}_1 = \{a_1 - d_1(bN_1 + cN_2)\}N_1,$$
$$\dot{N}_2 = \{a_2 - d_2(bN_1 + cN_2)\}N_2.$$

Classify the equilibrium points of the system. Show that if $a_1d_2 > a_2d_1$ then the species N_2 dies out and the species N_1 approaches a limiting size (Volterra's Exclusion Principle).

3 Geometrical and computational aspects of the phase diagram

IN THIS CHAPTER we discuss several topics which are useful in sketching phase diagrams. The 'index' of an equilibrium point provides supporting information on its nature and complexity which is particularly useful in strongly nonlinear cases where the linear approximation is zero. Secondly, the phase diagram does not give a complete picture of the solutions; it is not sufficiently specific about the behaviour of paths 'at infinity' beyond the boundaries of any diagram; but we show various projections which include 'points at infinity' and give the required overall view. Thirdly, a difficult question is to determine whether there are any limit cycles and roughly where they are; this question is treated again in Chapter 11, but here we give some elementary conditions for their existence or non-existence. Finally, having obtained all the information our methods allow about the geometrical layout of the phase diagram we may want to compute a number of typical paths, and some suggestions for carrying this out are made in Section 3.4.

3.1. The index of a point

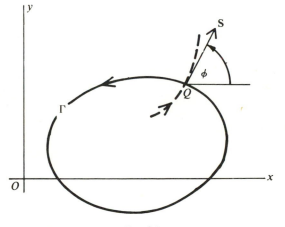

Fig. 3.1.

Given the system

$$\dot{x} = X(x, y), \quad \dot{y} = Y(x, y), \tag{3.1}$$

let Γ be a smooth closed curve consisting of ordinary points of (3.1). Let Q be a point on Γ (Fig. 3.1); then there is one and only one phase path passing through Q. The paths (without implication as to direction) belong to the family described by the equation

$$\frac{dy}{dx} = \frac{Y(x, y)}{X(x, y)}. \tag{3.2}$$

In time $\delta t > 0$ the coordinates of the point Q, (x_Q, y_Q), will increase by $\delta x, \delta y$ respectively, where

$$\delta x \approx X(x_Q, y_Q)\delta t, \quad \delta y \approx Y(x_Q, y_Q)\delta t.$$

Therefore the vector $S = (X, Y)$ is tangential to the path through the point, and points in the direction of increasing t. Its inclination can be measured by the angle ϕ measured anti-clockwise from the positive direction of the x axis to the direction of S, so that

$$\tan \phi = Y/X. \tag{3.3}$$

When the value of ϕ at one point has been decided, the value for other points is settled by requiring ϕ to be a continuous function of position, except that ϕ will not in general have its original value on returning to its starting point after a full cycle: the value may differ by $2\pi n$, where n is an integer.

Example 3.1   *Trace the variation of the vector* S *and the angle* ϕ *when* $X(x, y)$ *= $(2x^2 - 1)$, $Y(x, y) = 2xy$, and Γ is the unit circle centred at the origin.*

Let θ be the polar angle of the representative point, and put $x = \cos \theta$, $y = \sin \theta$ on Γ. Then $(X, Y) = (\cos 2\theta, \sin 2\theta), 0 \leqslant \theta \leqslant 2\pi$. This vector is displayed in Fig. 3.2. The angle ϕ takes, for example, the values $0, \frac{1}{2}\pi, 2\pi, 4\pi$ at A, B, C, and A' as we track anticlockwise round the circle.

In any complete revolution, ϕ increases by 4π.

In every case the change in ϕ must be a multiple of 2π;

$$[\phi]_\Gamma = 2\pi I_\Gamma \tag{3.4}$$

say, where I_Γ is an integer, positive, negative, or zero. I_Γ is called the *index of Γ with respect to the vector field (X, Y), Γ being described anticlockwise.*

FIG. 3.2.

An algebraical representation of I_Γ is obtained as follows. Suppose that the curve Γ is described anti-clockwise once by the position vector **r**, where

$$\mathbf{r}(s) = (x(s), y(s)), \quad s_0 \leqslant s \leqslant s_1, \tag{3.5}$$

and s is a parameter. From (3.3),

$$\frac{\mathrm{d}}{\mathrm{d}s}(\tan \phi) = \frac{\mathrm{d}}{\mathrm{d}s}\left(\frac{Y}{X}\right),$$

or, after some reduction,

$$\frac{\mathrm{d}\phi}{\mathrm{d}s} = \frac{X\dot{Y} - Y\dot{X}}{X^2 + Y^2}. \tag{3.6}$$

(The dot denotes differentiation with respect to s.) Then from (3.4)

$$I_\Gamma = \frac{1}{2\pi} \int_{s_0}^{s_1} \frac{\mathrm{d}\phi}{\mathrm{d}s}\,\mathrm{d}s = \frac{1}{2\pi} \int_{s_0}^{s_1} \frac{X\dot{Y} - Y\dot{X}}{X^2 + Y^2}\,\mathrm{d}s. \tag{3.7}$$

Alternatively, we can write (see Fig. 3.3)

$$I_\Gamma = \frac{1}{2\pi} \oint_{\Gamma_R} \frac{X\mathrm{d}Y - Y\mathrm{d}X}{X^2 + Y^2}. \tag{3.8}$$

As $\mathbf{r}(s) = (x(s), y(s))$ describes Γ, $\mathbf{R}(s) = (X, Y)$, regarded as a position vector on a plane with axes X, Y, describes another curve, Γ_R say. Γ_R is closed, since **R** returns to its original value after a complete cycle. From eqn (3.4) Γ_R encircles the origin I_Γ times, anti-clockwise when I_Γ is positive and clockwise when it is negative. This is illustrated in Fig. 3.3 for a particular case.

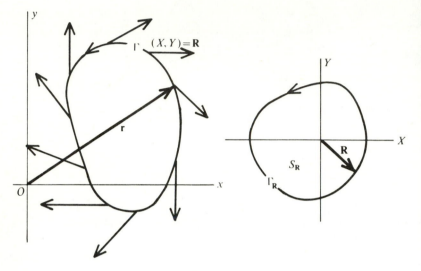

FIG. 3.3.

THEOREM 3.1   *Suppose that, on and inside Γ, X, Y and their first derivatives are continuous and X and Y are not simultaneously zero. (In other words there is no equilibrium point there.) Then I_Γ is zero.*

Proof.   Green's theorem for plane curves applied to (3.8) gives

$$I_\Gamma = \frac{1}{2\pi} \int \int_{S_R} \left[\frac{\partial}{\partial X} \frac{X}{X^2 + Y^2} + \frac{\partial}{\partial Y} \frac{Y}{X^2 + Y^2} \right] dX \, dY,$$

where S_R is the region enclosed by Γ_R. The application of the theorem is justified by the given conditions. It is easily verified that the integrand is identically zero.

COROLLARY TO THEOREM 3.1   *Let Γ be a closed curve, and Γ' a closed curve inside Γ, then if on Γ, Γ' and the region between them there is no equilibrium point, and if X, Y and their first derivatives are continuous there, then $I_\Gamma = I_{\Gamma'}$.*

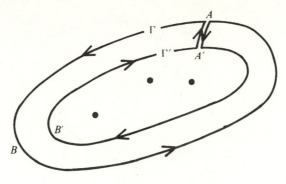

FIG. 3.4.

Proof Let AA' be a line connecting Γ and Γ', and consider the composite contour, $\mathscr{C}$, described by $ABAA'B'A'A$ in Fig. 3.4. Since $\mathscr{C}$ contains no equilibrium points, $I_{\mathscr{C}} = 0$ by the Theorem. But by (3.4)

$$0 = I_{\mathscr{C}} = \frac{1}{2\pi} \oint_{\mathscr{C}} \mathrm{d}\phi = \frac{1}{2\pi}\left(\oint_{\Gamma} \mathrm{d}\phi + \oint_{AA'} \mathrm{d}\phi - \oint_{\Gamma'} \mathrm{d}\phi + \oint_{A'A} \mathrm{d}\phi \right)$$

where

$$\oint_{\Gamma'} \mathrm{d}\phi \quad \text{and} \quad \oint_{\Gamma} \mathrm{d}\phi$$

represent integrals taken in the anti-clockwise direction. Since

$$\oint_{A'A} \mathrm{d}\phi = -\oint_{AA'} \mathrm{d}\phi,$$

we obtain

$$0 = \frac{1}{2\pi}\left(\oint_{\Gamma} \mathrm{d}\phi - \oint_{\Gamma'} \mathrm{d}\phi \right)$$

so that

$$I_{\Gamma} = I_{\Gamma'}.$$

This theorem shows that the index I_{Γ} of the vector field (X, Y) with respect to Γ is to a large extent independent of Γ, and enables the index to be associated with special points rather than with contours. *If the smoothness conditions on the field (X, Y) are satisfied in a region containing a single equilibrium point, then any simple closed curve Γ surrounding the point generates the same number I_{Γ}. We therefore drop the suffix Γ, and say that I is the index of the equilibrium point.*

For an alternative form of proof of these results see Exercise 43.

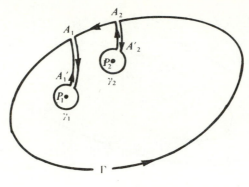

FIG. 3.5.

THEOREM 3.2 *If Γ surrounds n equilibrium points $P_1, P_2, \ldots, P_n$ then*

$$I_\Gamma = \sum_{i=1}^{n} I_i$$

where I_i is the index of the point P_i, $i = 1, 2, \ldots, n$.

Proof We illustrate the proof for the case of two equilibrium points, at P_1, P_2. Construct a contour $\mathscr{C}$ consisting of Γ, two 'bridges' $A_1 A_1'$ and $A_2 A_2'$, and curves γ_1 and γ_2 surrounding P_1 and P_2 respectively, as shown in Fig. 3.5. The interior of $\mathscr{C}$ does not contain any equilibrium points, so $I_\mathscr{C} = 0$. But also

$$I_\mathscr{C} = \frac{1}{2\pi} \oint_\Gamma \mathrm{d}\phi - \frac{1}{2\pi} \oint_{\gamma_1} \mathrm{d}\phi - \frac{1}{2\pi} \oint_{\gamma_2} \mathrm{d}\phi,$$

since the integrals over the 'bridges' cancel to zero. Therefore

$$I_\Gamma = I_1 + I_2.$$

The easiest way to calculate the index in a particular case is to use the following result.

THEOREM 3.3 *Let p be the number of times $Y(x, y)/X(x, y)$ changes from $+\infty$ to $-\infty$, and q the number of times it changes from $-\infty$ to $+\infty$, on Γ. Then $I_\Gamma = \frac{1}{2}(p - q)$.*

Proof Since $\tan \phi = Y/X$, we are simply counting the number of times the direction (X, Y) is vertical (up or down), and associating a direction of rotation. We could instead examine the points where $\tan \phi$ is zero, that is, where Y is zero. Then if P and Q are numbers of changes in $\tan \phi$

from (negative/positive) to (positive/negative) respectively across the zero of Y, $I_\Gamma = \frac{1}{2}(P - Q)$.

Example 3.2 *Find the index of the equilibrium point $(0,0)$ of $\dot{x} = y^3, \dot{y} = x^3$.*

By stretching the theory a little we can let Γ be the square with sides $x = \pm 1$, $y = \pm 1$. Start from $(1, 1)$ and go round anti-clockwise. On $y = 1$, $\tan \phi = x^3$, with no infinities; similarly on $y = -1$. On $x = -1$, $\tan \phi = -y^{-3}$ with a change from $-\infty$ to $+\infty$, and on $x = 1$, $\tan \phi = y^{-3}$, with a change from $-\infty$ to $+\infty$. Thus $p = 0$, $q = 2$ and the index is -1.

Example 3.3 *Find the index of the equilibrium point at $(0,0)$ of the system $\dot{x} = y - \frac{1}{2}xy - 3x^2, \dot{y} = -xy - \frac{3}{2}y^2$.*

There is another equilibrium point at $(-\frac{1}{4}, \frac{1}{6})$. The contour must not enclose this, so choose a square bounded by $x = \pm\frac{1}{8}$, $y = \pm\frac{1}{8}$. We have

$$\tan \phi = -y(x + \tfrac{3}{2}y)/(y - \tfrac{1}{2}xy - 3x^2)$$

and the numerator is simple, so it is easy to consider the zeros of $\tan \phi$.

On AB and CD in Fig. 3.6, $\tan \phi$ is never zero. On BC, $\tan \phi$ is zero at $y = 1/12$, passing from negative to positive, and at $y = 0$, passing from negative to positive. On DA, $\tan \phi$ is zero at $y = -1/12$, passing from positive to negative, and at $y = 0$, passing from negative to positive. Therefore $P = 3$ and $Q = 1$, so $I = 1$. The equilibrium point is a type for which linearization gives no information.

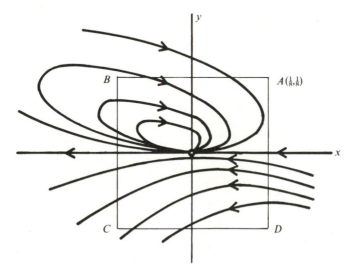

FIG. 3.6.

If we already know the nature of the equilibrium point, the index is readily found by simply drawing a figure and following the angle round. The following shows the indices of the elementary types met in Chapter 2.

(i) *A saddle point*   The change in ϕ in a single circuit of the curve Γ surrounding the saddle point is -2π, and the index is therefore -1.

(ii) *A centre*   The index is $+1$. (This is irrespective of the direction of the paths.)

FIG. 3.7. Saddle point, index (-1)

FIG. 3.8. Centre, index 1

(iii) *A spiral* (*stable or unstable*)   The index is $+1$.

(iv) *A node* (*stable or unstable*)   The index is $+1$.

A simple practical way of finding the index of a curve Γ uses a watch. Set the watch to 12.00 (say) and put the watch at a point on Γ with the minute hand pointing in the direction of the phase path there. Move the

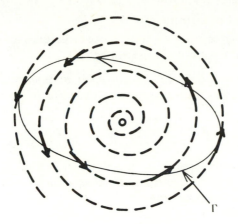

Fig. 3.9. Spiral, index 1

Fig. 3.10. Node, index 1

watch without rotation around Γ always turning the direction of the
minute hand along the phase path. After one counterclockwise circuit
of Γ, the index is given by the gain (-1 for each hour gained) or loss, of
the hour hand. For example the 'time' after one circuit of the centre is
11.00, an index of $+1$; for the saddle point the 'time' is 1.00, an index of
-1.

3.2. The index at infinity

In the next two sections we show some techniques for getting a more
global view of the pattern of trajectories.

DEFINITION   *The index, I_∞, of the point at infinity. Introduce new
coordinates x_1, y_1 by the transformation (inversion in the origin)*

$$x_1 = \frac{x}{x^2 + y^2}, \quad y_1 = -\frac{y}{x^2 + y^2}$$

*(or, in polar coordinates, let $\theta_1 = -\theta$ and $r_1 = 1/r$). The index of the
origin, $x_1 = y_1 = 0$, for the transformed equation is called the index at
infinity for the original equation.*

In order to prove a result concerning the index of the point at infinity,
it is convenient to write

$$z = x + iy, \quad Z = X + iY,$$

and the differential equation system becomes

$$\frac{dz}{dt} = Z. \tag{3.9}$$

(Z is not, of course, a regular function of z in general.) Note that ϕ (eqn
(3.3)) is given by

$$\phi = \arg Z. \tag{3.10}$$

The transformation in the definition is equivalent to the transformation

$$z_1 = z^{-1}$$

and (3.9) then becomes

$$-\frac{1}{z_1^2} \frac{dz_1}{dt} = Z,$$

or

$$\frac{dz_1}{dt} = -z_1^2 Z = Z_1 \qquad (3.11)$$

say. We require the index of (3.11) at $z_1 = 0$.

THEOREM 3.4 *The index* I_∞ *for the system* $x = X(x, y)$, $y = Y(x, y)$, *having a finite number n of equilibrium points with indices* $I_i, i = 1, 2, \ldots, n$, *is given by*

$$I_\infty = 2 - \sum_{i=1}^{n} I_i. \qquad (3.12)$$

Proof To obtain the index at the origin, surround $z_1 = 0$ by a simple closed contour $\mathscr{C}_1$ containing no other equilibrium points of the system. Under the transformation $z = z_1^{-1}$, $\mathscr{C}_1$ described anti-clockwise maps into $\mathscr{C}$ described clockwise, and the exterior of $\mathscr{C}_1$ maps into the interior of $\mathscr{C}$ (Fig. 3.11). Thus $\mathscr{C}$ embraces all the equilibrium points of the original system.

Now let

$$z_1 = r_1 e^{i\theta_1} \qquad (3.13)$$

on $\mathscr{C}_1$; and on $\mathscr{C}_1$ and $\mathscr{C}$ respectively let

$$Z_1 = \rho_1 e^{i\phi_1}, \quad Z = \rho e^{i\phi},$$

where $\rho_1 = |Z_1|, \rho = |Z|$. Then

$$I_\infty = \frac{1}{2\pi} [\phi_1]_{\mathscr{C}_1}.$$

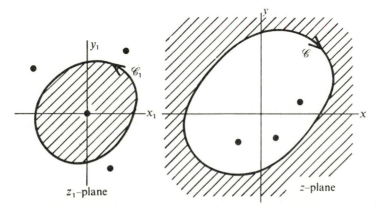

FIG. 3.11.

But from (3.11)

$$Z_1 = \rho_1 e^{i\phi_1} = -r_1^2 e^{2i\theta_1} \rho e^{i\phi} = r_1^2 \rho e^{i(2\theta_1 + \phi + \pi)}.$$

Therefore

$$I_\infty = \frac{1}{2\pi}[2\theta_1 + \phi + \pi]_{\mathscr{C}_1}$$

$$= \frac{1}{2\pi}\{2[\theta_1]_{\mathscr{C}_1} + [\phi]_{\mathscr{C}_1}\} = \frac{1}{2\pi}\{2[\theta_1]_{\mathscr{C}_1} - [\phi]_{\mathscr{C}}\}$$

$$= \frac{1}{2\pi}\left\{4\pi - \sum_{i=1}^{n} I_i\right\}$$

by Theorem 3.2, since also $\mathscr{C}$ is described clockwise. The result follows.

COROLLARY TO THEOREM 3.4   *Under the conditions stated in the theorem, the sum of all the indices, including that of the point at infinity, is 2.*

Example 3.4   *Find the index at infinity of the system*

$$\dot{x} = x - y, \quad \dot{y} = x - y^2.$$

We show a different approach from that of Examples 3.2 and 3.3. Putting $x = r\cos\theta, y = r\sin\theta$ we have

$$Z = (r\cos\theta - r\sin\theta) + i(r\cos\theta - r^2\sin^2\theta),$$

and the transformation $z = z_1^{-1}$ gives $r = r_1^{-1}, \theta = -\theta_1$, so from (3.11)

$$Z_1 = -r_1^2(\cos 2\theta_1 + i\sin 2\theta_1)\{r_1^{-1}(\cos\theta_1 + \sin\theta_1)$$
$$+ ir_1^{-1}(\cos\theta_1 - r_1^{-1}\sin^2\theta_1)\}. \quad (3.14)$$

We require the index of eqn (3.11) at the point $r_1 = 0$. To evaluate it choose Γ_1 to be a circle centred at the origin, having arbitrarily small radius equal to r_1. Then we need to consider, in (3.14), only the terms of $O(1)$, the terms of $O(r_1)$ making no contribution in the limit $r_1 \to 0$. Then on Γ

$$Z_1 = (\cos 2\theta_1 + i\sin 2\theta_1)\sin^2\theta_1 + O(r_1) = \sin^2\theta_1 e^{2i\theta_1} + O(r_1).$$

As shown in Fig. 3.12 the direction, ϕ_1, of Z_1 is equal to $2\theta_1$, and the index at infinity is therefore equal to 2.

It can be confirmed that the other equilibrium points are a spiral or centre at $x = 0, y = 0$ with index 1, and a saddle at $x = 1, y = 1$ with index -1. The sum of the indices is therefore 2, as required by Theorem 3.4.

3.3.   The phase diagram at infinity

Phase diagrams such as we have shown are incomplete since there is always an area outside the picture which we cannot see. When the phase

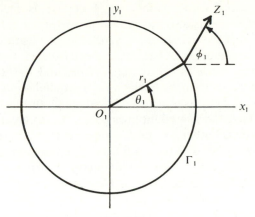

FIG. 3.12.

diagram has a simple structure this may not be a serious loss, but in complicated cases the overall structure may be obscure. We therefore show how the whole phase diagram, including the infinitely remote parts, can be displayed.

Projection on to a sphere

In Fig. 3.13, $\mathscr{S}$ is a sphere of unit radius touching the phase plane $\mathscr{P}$ at O. Its centre is O^*. A representative point P on $\mathscr{P}$ projects from O^* on to P' on the sphere. Paths on $\mathscr{P}$ become paths on $\mathscr{S}$; such features as

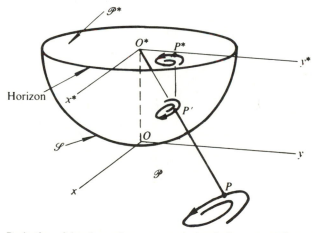

FIG. 3.13. Projection of the phase plane $\mathscr{P}$ on to the hemisphere $\mathscr{S}$, and from $\mathscr{S}$ on to the diametrical plane $\mathscr{P}^*$

spirals, nodes, and saddle points on $\mathscr{P}$ appear as such on $\mathscr{S}$. The orientation of curves is also unchanged provided the sphere is viewed from inside. Local distortion occurs increasingly for features far from the origin, radial distance on $\mathscr{P}$ becoming greatly scaled down on $\mathscr{S}$. The horizontal great circle is called the *horizon*, and the happenings 'at infinity' on $\mathscr{P}$ occur near the horizon of $\mathscr{S}$. Radial straight lines on $\mathscr{P}$ project on to great circles of $\mathscr{S}$ in a vertical plane, cutting the horizon at right angles. Non-radial straight lines project on to great circles which cut the horizon at other angles. To represent on another plane the resulting pattern of lines on the lower hemisphere, we may simply take its orthogonal projection on to the diametrical plane $\mathscr{P}^*$ containing the horizon. A topological analysis (Andronov *et al.* 1973) requires points on $\mathscr{P}$ to be mapped *twice* onto $\mathscr{S}$: to a point P on $\mathscr{P}$ there are two intersections of O^*P with $\mathscr{S}$, and the upper hemisphere is the reflection of the lower in the centre. Our interest here, however, is chiefly with obtaining pictures.

Example 3.5 $\dot{x} = y, \dot{y} = -4x - 5y$

   This represents a heavily damped linear oscillator. On the phase plane $\mathscr{P}$, $(0,0)$ is a stable node. $y = -x$, $y = -4x$ are two phase paths, and the results of Chapter 2 show that all other paths are tangent to the first at the origin and parallel to the second at infinity. The appearance of the plane $\mathscr{P}^*$ with axes x^*, y^* is shown in Fig. 3.14.

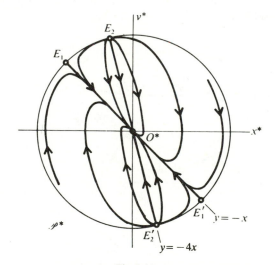

Fig. 3.14.

The paths $y = -x, y = -4x$ on the phase plane $\mathscr{P}$ are radial straight lines and project into lines of longitude through O on the sphere. Their projections on to the diametrical plane are therefore again straight lines $E_1 O, E_1' O, E_2 O, E_2' O$ of Fig. 3.14.

Since all other paths become parallel to $y = -4x$ at large distance from the origin, their projections enter the points E_2, E_2' on the sphere as shown. Other mappings of the hemisphere on to a plane can be devised.

Detail on the horizon

In order to study the paths on the hemisphere near the horizon we will project the paths found on $\mathscr{S}$ from O^* on to the plane $\mathscr{U}: x = 1$. We thus have, simultaneously, a projection of the paths in the phase plane $\mathscr{P}$ on to $\mathscr{U}$. u, z are axes with origin at N, where $\mathscr{S}$ touches $\mathscr{U}: u$ is parallel to y and z is positive downwards, so u, z is right-handed viewed from $\dot{O}^*$. Points in $\mathscr{P}$ for $x < 0$ go into the left half of the lower hemisphere, then through O^* on to $\mathscr{U}$ for $z < 0$; and $\mathscr{P}$ for $x > 0$ goes into the right half of $\mathscr{S}$, then on to $\mathscr{U}$ for $z > 0$. Points at infinity on $\mathscr{P}$ go to the horizon of $\mathscr{S}$, then to $z = 0$ on $\mathscr{U}$. The points at infinity on $\mathscr{P}$ in direction $\pm y$ go to points at infinity on $\mathscr{U}$, as does the point O. The topological features of $\mathscr{P}$ are preserved on $\mathscr{U}$, except that the orientation of closed paths on the part of $\mathscr{S}$ corresponding to $x < 0$ is reversed (see Fig. 3.15).

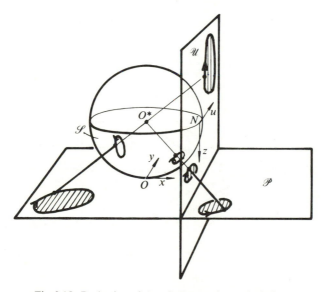

Fig. 3.15. Projection of $\mathscr{P}$ and $\mathscr{S}$ on to the vertical plane $\mathscr{U}$

It can be shown easily that the transformation from $\mathscr{P}$ to $\mathscr{U}$ is given by

$$u = y/x, \quad z = 1/x, \tag{3.15}$$

with inverse

$$x = 1/z, \quad y = u/z.$$

Example 3.6 *Examine the system $\dot{x} = y$, $\dot{y} = -4x - 5y$ at infinity (excluding the directions $\pm y$).*

(This is the system of Example 3.5.) The transformation (3.15) leads to

$$\dot{u} = (x\dot{y} - y\dot{x})/x^2 = -(4x^2 + 5xy + y^2)/x^2 = -4 - 5u - u^2$$

and, similarly,

$$\dot{z} = -\dot{x}/x^2 = -uz.$$

There are equilibrium points at $E_1 : u = -1, z = 0$, and at $E_2 : u = -4, z = 0$ and it can be confirmed in the usual way that E_1 is a saddle and E_2 a node. The pattern of the paths is shown in Fig. 3.16, and should be compared with Fig. 3.14.

To examine the phase plane at infinity in the directions $\pm y$, $\mathscr{S}$ is projected on another plane $\mathscr{V} : y = 1$. The corresponding axes v, z, right-handed viewed from O^*, have their origin at the point of tangency; v points in the direction of $-x$ and z downward. The required transformation is

$$v = x/y, \quad z = 1/y,$$

or   (3.16)

$$x = v/z, \quad y = 1/z.$$

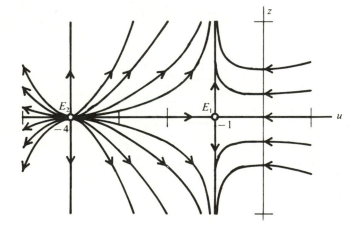

Fig. 3.16. Phase paths near the horizon

3.4. Limit cycles and other closed paths

As we have seen (Section 1.5), a *limit cycle* is an isolated periodic solution of an autonomous system, represented in the phase plane by an isolated closed path. The neighbouring paths are, by definition, not closed, but spiral into or away from the limit cycle $\mathscr{C}$ as shown in Fig. 3.17. In the case illustrated, which is a *stable limit cycle*, the device represented by the system (which might be, for example, an electrical circuit), will spontaneously drift into the corresponding periodic oscillation from a wide range of initial states. The existence of limit cycles is therefore a feature of great practical importance.

Linear systems with constant coefficients cannot exhibit limit cycles. Since normally we cannot solve the nonlinear equations which might have them it is important to be able to establish by indirect means whether a limit cycle is present in the phase diagram. If one exists, then various approximate methods can be used to locate it. In this section we give some simple indications and counter-indications, the question of existence being dealt with in more detail in Chapter 11.

The index of a limit cycle $\mathscr{C}$ is 1 since the vector (X, Y) is tangential to $\mathscr{C}$ at every point on it, and the change in ϕ around $\mathscr{C}$ is 2π. By Theorem 3.2, therefore, *a necessary condition for $\mathscr{C}$ to be a limit cycle is that the sum of the indices at the equilibrium points enclosed by $\mathscr{C}$ is 1.*

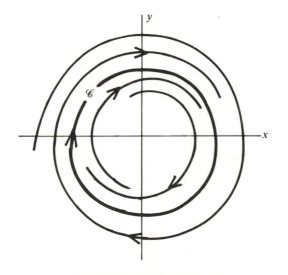

FIG. 3.17. An idealized limit cycle

This result applies equally to any type of periodic solution and by using it we can at least determine some cases when a limit cycle *cannot* exist. For example, a limit cycle cannot surround a region containing no equilibrium points, nor one containing only a saddle. However, consider the system

$$\dot{x} = y, \quad \dot{y} = (1 - x^2)x/(1 + x^2).$$

This has a saddle point at the origin, index -1, and centres at $(0, 1)$, $(0, -1)$, with indices 1. The sum of the indices is 1, and it is therefore possible that there might be a limit cycle surrounding all three points. (In fact, however, the paths surrounding them are *all* closed.)

The following result is due to Bendixson (Cesari 1971) and is called Bendixson's Negative Criterion:

THEOREM 3.5   *There are no closed paths in a simply-connected domain of the phase plane on which $\partial X/\partial x + \partial Y/\partial y$ is of one sign.*

Proof   We make the usual assumptions about the smoothness of the vector field (X, Y) necessary to justify the application of the divergence theorem. Suppose that there is a closed path $\mathscr{C}$ in the region $\mathscr{D}$ where $\partial X/\partial x + \partial Y/\partial y$ is of one sign (see Fig. 3.18); we shall show that this assumption leads to a contradiction. By the divergence theorem,

$$\int\int_{\mathscr{S}} \left(\frac{\partial X}{\partial x} + \frac{\partial Y}{\partial y} \right) dx\, dy = \int_{\mathscr{C}} (X, Y).\mathbf{n}\, ds,$$

where $\mathscr{S}$ is the interior of $\mathscr{C}$, $\mathbf{n}$ is the unit outward normal, and ds is an undirected line element of $\mathscr{C}$. Since $\mathscr{C}$ is a path, (X, Y) is normal to $\mathbf{n}$, so the integral on the right is zero. But since the integrand on the left is of one sign its integral cannot be zero. Therefore $\mathscr{C}$ cannot be a path.

An extension of this result, called Dulac's test, is given in Exercise 23.

Example 3.7   *Show that the equation $\ddot{x} + f(x)\dot{x} + g(x) = 0$ can have no periodic solution whose phase path lies in a region where f is of one sign. (Such regions have only 'positive damping' or 'negative damping'.)*

The equivalent system is $\dot{x} = y$, $\dot{y} = -f(x)y - g(x)$, so that $(X, Y) = (y, -f(x)y - g(x))$, and

$$\frac{\partial X}{\partial x} + \frac{\partial Y}{\partial y} = -f(x),$$

which is of one sign whenever f is.

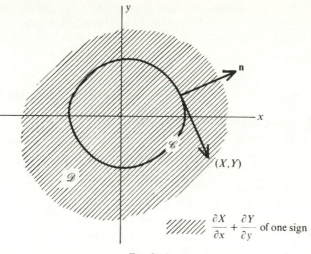

$$\frac{\partial X}{\partial x} + \frac{\partial Y}{\partial y} \text{ of one sign}$$

FIG. 3.18.

Example 3.8 The equations

$$\dot{V}_1 = -\sigma V_1 + f(E - V_2) = X(V_1, V_2), \text{ say;}$$

$$\dot{V}_2 = -\sigma V_2 + f(E - V_1) = Y(V_1, V_2), \text{ say;}$$

describe the voltages generated over the deflection plates in a simplified triggered sweeping circuit for an oscilloscope (Andronov and Chaikin 1949). E and σ are positive constants, and f has the shape shown in Fig. 3.19. f is a continuous, odd function, tending to a limit as $x \to \infty$, and f' is positive and decreasing on $x > 0$. It can be shown (Chapter 2, Exercise 33) that there is an equilibrium point, at $V_1 = V_2, = V_0$, say. If, moreover, $f'(E - V_0) > \sigma$, then there

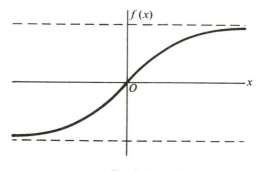

FIG. 3.19.

are two others, and in this case (V_0, V_0) is a saddle point and the other two are stable nodes. The sum of the indices is 1, so periodic solutions with paths surrounding the three points are not ruled out by this test. However,

$$\frac{\partial X}{\partial V_1} + \frac{\partial Y}{\partial V_2} = -2\sigma,$$

which is of one sign everywhere, so in fact there are no periodic solutions.

3.5. Computation of the phase diagram

Detailed plotting of phase paths can give a much enhanced understanding of the character of the solutions, even when we have obtained a general idea of the paths from analytical and geometrical arguments. We shall give no discussion of technical points in numerical analysis here (see, for example, Cohen 1973), but simply indicate that very unsophisticated methods can be made to work well so long as obvious precautions are taken and analytical reasoning is not lost sight of. We envisage calculations being done at a computer terminal, possibly with limited program and storage space, in an elementary language such as BASIC.

We consider autonomous equations of the form

$$\dot{x} = P(x, y), \quad \dot{y} = Q(x, y). \tag{3.17}$$

(P and Q are used since x and X are indistinguishable in computer programming.) In plotting a phase path we proceed step-by-step in some way from the initial point x_0, y_0:

$$x(t_0) = x_0, \quad y(t_0) = y_0. \tag{3.18}$$

The simplest step-by-step method is Euler's, using t as a supplementary variable, which we now describe and use as the basis of discussion.

Let the length of the time step be h, assumed small; and let $(x_0, y_0), (x_1, y_1), (x_2, y_2), \ldots$, be the points on the phase path at times $t_0, t_0 + h, t_0 + 2h, \ldots$. Assume that (x_n, y_n) is known. Then, writing $\dot{x}(t_0 + nh) \simeq (x_{n+1} - x_n)/h$, $\dot{y}(t_0 + nh) = (y_{n+1} - y_n)/h$, (3.17) becomes approximately

$$(x_{n+1} - x_n)/h = P(x_n, y_n), \quad (y_{n+1} - y_n)/h = Q(x_n, y_n),$$

or, after rearrangement,

$$x_{n+1} = x_n + hP(x_n, y_n), \quad y_{n+1} = y_n + hQ(x_n, y_n). \tag{3.19}$$

(x_{n+1}, y_{n+1}) is now recorded and we proceed to the next step. The danger of progressively increasing error is present, even when h is very small so

FIG. 3.20. Equal time steps near equilibrium points

that good accuracy is ostensibly being obtained: the reader should see, for example, Cohen (1973) for a discussion of error.

The point which concerns us here is the general unsuitability of t (time) as the parameter for the step-by-step construction of phase paths. The difficulty is shown schematically in Fig. 3.20, where a path is shown approaching a node, and passing near a saddle. Notionally equal time steps are marked. Since $\dot{x}$, $\dot{y}$, or P, Q, are small near an equilibrium point, progress is very uneven.

It is preferable to use arc length, s, as the parameter, which gives equally spaced points along the arc. Since $\delta s^2 = \delta x^2 + \delta y^2$, (3.17) with parameter s becomes

$$\frac{\mathrm{d}x}{\mathrm{d}s} = \frac{P}{\sqrt{(P^2+Q^2)}} = U, \quad \frac{\mathrm{d}y}{\mathrm{d}s} = \frac{Q}{\sqrt{(P^2+Q^2)}} = V \qquad (3.20)$$

(s is measured increasing in the direction of t increasing). Letting h now represent the basic step length, (3.20) leads to the iterative scheme

$$x_{n+1} = x_n + hU(x_n, y_n), \quad y_{n+1} = y_n + hV(x_n, y_n) \qquad (3.21)$$

with the initial values (x_0, y_0) given.

Improved processes, such as the Runge–Kutta method (Cohen 1973) are generally to be preferred, and can easily be programmed. It is desirable to have a program which prints out only after a predetermined number of calculation steps. In this way it can be arranged, for example, to produce a point every centimetre of arc along a phase path, the interval being reduced if the path is turning rapidly.

In a practical plot it is helpful, if possible, to locate and classify the equilibrium points and to know at least whether limit cycles are likely to appear. In particular, saddle points are difficult to locate closely by

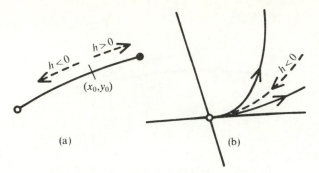

FIG. 3.21. Illustrating reversal of step sign

plotting paths. The location of equilibrium points may be obtained by algebraic or numerical solution of the equations $P(x, y) = Q(x, y) = 0$. Classification on the basis of linear approximation may not be possible, for example when P or Q have a zero linear expansion, or when the linear approximation predicts a centre (which may become a spiral in the presence of nonlinear terms; end of Section 2.7 and Exercise 7, Chapter 2), or in certain other cases. In such cases many exploratory plots may have to be made near the equilibrium point before an adequate characterization is obtained.

By reversing the sign of h in the input to suitable programs, a plot 'backwards' from the starting point is obtained. This is useful to complete a path (Fig. 3.21(a)) or to get close to an unstable node (Fig. 3.21(b)).

It may be difficult to distinguish with confidence between certain cases, for example between a node whose paths are seen to wind around

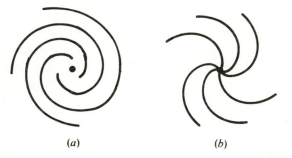

FIG. 3.22. (a) Small-scale plot (spiral?); (b) large-scale plot (node?)

it and a genuine spiral (Fig. 3.22). A display which appears to represent a spiral may, on scaling up the neighbourhood of the origin, begin to resemble a node, and conversely. This consideration applies, in principle, to all numerical and plotted data: it is impossible to deduce from the data alone that some qualitative deviation does not appear between the plotted points. However, in the case of equilibrium points and limit cycles there may be genuine uncertainty, and it is preferable to establish their nature analytically and to determine the directions of separatrices when this is possible.

The connections between equilibrium points are often unexpected and interesting: see, for example, Figs. 7.4, 7.12(*b*), 7.23(*a*), and 3.26.

A limit cycle is a feature requiring some care. The 'model' of a limit cycle which it is useful to have in mind is a pattern of spirals inside and outside the cycle as in Fig. 3.17. The figure shows a stable limit cycle, for which all nearby solutions drift towards the limit cycle. If they all recede

FIG. 3.23. Limit cycle for van der Pol's equation $\ddot{x} + (x^2 - 1)\dot{x} + x = 0$

(reverse the arrows) the limit cycle is unstable. A semi-stable limit cycle can also occur, in which the paths approach on one side and recede on the other.

Plotting a limit cycle may be difficult. Both the limit cycle and the neighbouring paths can diverge very considerably from the idealized pattern of Fig. 3.17; for example Fig. 3.23 shows the limit cycle for van der Pol's equation with a moderate value for the parameter. In general there is no way of finding a single point, exactly lying on the cycle, from which to start; the cycle has to be located by 'squeezing' it between inner and outer spirals. Clearly it is helpful to reverse the sign of t (or of h in the program input) if necessary, so as to approach rather than recede from the cycle during the process of locating it.

Exercises

1. By considering the variation of path direction on closed curves round the equilibrium points, find the index in each case of Fig. 3.24.

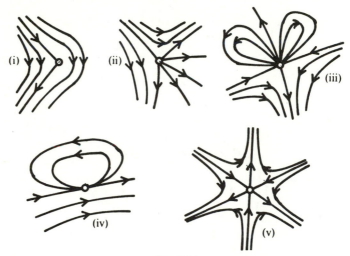

Fig. 3.24.

2. The motion of a damped pendulum is described by the equations

$$\dot{\theta} = \omega, \quad \dot{\omega} = -k\omega - v^2 \sin \theta,$$

where k (> 0) and v are constants. Find the indices of all the equilibrium states.

3. Find the index of the equilibrium points of the following systems: (i) $\dot{x} = 2xy$, $\dot{y} = 3x^2 - y^2$; (ii) $\dot{x} = y^2 - x^4$, $\dot{y} = x^3 y$; (iii) $\dot{x} = x - y$, $\dot{y} = x - y^2$.

4. For the linear system $\dot{x} = ax + by$, $\dot{y} = cx + dy$, where $ad - bc \neq 0$, obtain the index at the origin by evaluating eqn (3.7), showing that it is equal to $\text{sgn}(ad - bc)$. (Hint: choose Γ to be the ellipse $(ax + by)^2 + (cx + dy)^2 = 1$.)

5. The equation of motion of a bar restrained by springs (see Fig. 3.25) and attracted by a parallel current-carrying conductor is

$$\ddot{x} + c\{x - \lambda/(a - x)\} = 0,$$

where c, a, and λ are positive constants. Sketch the phase paths for $-x_0 < x < a$, and find the indices of the equilibrium points for all $\lambda > 0$.

FIG. 3.25.

6. Show that the equation

$$\ddot{x} - \varepsilon(1 - x^2 - \dot{x}^2)\dot{x} + x = 0$$

has an equilibrium point of index 1 at the origin of the phase plane x, y with $\dot{x} = y$. (It also has a limit cycle, $x = \cos t$.) Use eqn (3.7), with Γ a circle of radius a to show that, for all a,

$$\int_0^{2\pi} \frac{d\theta}{1 - 2\varepsilon(1 - a^2)\sin\theta\cos\theta + \varepsilon^2(1 - a^2)^2\sin^2\theta} = 2\pi.$$

7. A limit cycle encloses N nodes, F spirals, C centres, and S saddle points only, all of the type of Chapter 2. Show that $N + F + C - S = 1$.

8. Given the system

$$\dot{x} = X(x, y)\cos\alpha - Y(x, y)\sin\alpha, \quad \dot{y} = X(x, y)\sin\alpha + Y(x, y)\cos\alpha,$$

where α is a parameter, prove that the index of a simple closed curve which does not meet an equilibrium point is independent of α. (See also Exercise 41, Chapter 2.)

9. Suppose that the system $\dot{\mathbf{x}} = \mathbf{X}(\mathbf{x})$ has a finite number of equilibrium points, each of which is either a node, a centre, a spiral, or a saddle point, of the elementary types discussed in Chapter 2. Show that the total number of nodes, centres, and spirals is equal to the total number of saddle points plus two.

10. Obtain differential equations describing the behaviour of the linear system, $\dot{x} = ax + by$, $\dot{y} = cx + dy$, at infinity. Analyse the system $\dot{x} = 2x - y$, $\dot{y} = 3x - 2y$ near the horizon.

11. A certain system is known to have exactly two equilibrium points, both saddle points. Sketch phase diagrams in which
(i) a separatrix connects the saddle points;
(ii) no separatrix connects them.

12. Deduce the index at infinity for the system $\dot{x} = x - y$, $\dot{y} = x - y^2$ by calculating the indices of the equilibrium points.

13. Use the geometrical picture of the field (X, Y) in the neighbourhood of an ordinary point to confirm Theorem 3.1.

14. Suppose that, for the two plane systems $\dot{\mathbf{x}}_1 = \mathbf{X}_1(\mathbf{x}_1)$, $\dot{\mathbf{x}}_2 = \mathbf{X}_2(\mathbf{x}_2)$, and for a given closed curve Γ, there is no point on Γ at which $\mathbf{X}_1$ and $\mathbf{X}_2$ are opposite in direction. Show that the index of Γ is the same for both systems.

The system $\dot{x} = y^2$, $\dot{y} = xy$ has a saddle point at the origin. Show that the index of the origin for the system $\dot{x} = y^2 + cx^2$, $\dot{y} = xy - cy^2$ is likewise -1 for $|c|$ small enough.

15. Use Exercise 14 to show that the index of the equilibrium point $x = 0$, $\dot{x} = 0$ for the equation $\ddot{x} + \sin x = 0$ on the usual phase plane has index 1, by comparing the equation $\ddot{x} + x = 0$.

16. The system
$$\dot{x} = ax + by + P(x, y), \quad \dot{y} = cx + dy + Q(x, y)$$
has an isolated equilibrium point at $(0, 0)$, and $P(x, y) = O(r^2)$, $Q(x, y) = O(r^2)$ as $r \to 0$, where $r^2 = x^2 + y^2$. Assuming that $ad - bc \neq 0$, show that the origin has the same index as its linear approximation.

17. Show that, on the phase plane with $\dot{x} = y$, $\dot{y} = Y(x, y)$, Y continuous, the index I_Γ of any simple closed curve Γ can only be 1, -1, or zero.

Let $\dot{x} = y$, $\dot{y} = f(x, \lambda)$, f, f' continuous, represent a parameter-dependent system with parameter λ. Show that, at a bifurcation point (Section 1.7), where an equilibrium point divides as λ varies, the sum of the indices of the equilibrium points resulting from the splitting is unchanged. (Hint: the integrand in eqn (3.7) is continuous.)

Deduce that the equilibrium points for the system $\dot{x} = y$, $\dot{y} = -\lambda x + x^3$ consist of a saddle point for $\lambda < 0$, and a centre and two saddles for $\lambda > 0$.

18. Prove a similar result to that of Exercise 17 for the system $\dot{x} = y$,

$\dot{y} = f(x, y, \lambda)$. Deduce that the system $\dot{x} = y$, $\dot{y} = -\lambda x - ky - x^3$, $(k > 0)$, has a saddle point at $(0,0)$ when $\lambda < 0$, which bifurcates into a stable spiral or node and two saddle points as λ becomes positive.

19. A system is known to have three closed paths, $\mathscr{C}_1$, $\mathscr{C}_2$, and $\mathscr{C}_3$, such that $\mathscr{C}_2$ and $\mathscr{C}_3$ are interior to $\mathscr{C}_1$ and such that $\mathscr{C}_2, \mathscr{C}_3$ have no interior points in common. Show that there must be at least one equilibrium point in the domain bounded by $\mathscr{C}_1, \mathscr{C}_2$, and $\mathscr{C}_3$.

20. For each of the following systems you are given some information about phase paths and equilibrium points. Sketch phase diagrams consistent with these requirements.
   (i) $x^2 + y^2 = 1$ is a phase path, $(0,0)$ a saddle point, $(\pm\frac{1}{2}, 0)$ centres.
   (ii) $x^2 + y^2 = 1$ is a phase path, $(-\frac{1}{2}, 0)$ a saddle point, $(0,0)$ and $(\frac{1}{2}, 0)$ centres.
   (iii) $x^2 + y^2 = 1$, $x^2 + y^2 = 2$ are phase paths, $(0, \pm\frac{3}{2})$ stable spirals, $(\pm\frac{3}{2}, 0)$ saddle points, $(0,0)$ a stable spiral.

21. Consider the system

$$\dot{x} = y(z-2), \quad \dot{y} = x(2-z)+1, \quad x^2+y^2+z^2 = 1$$

which has exactly two equilibrium points, both of which lie on the unit sphere. Project the phase diagram on to the plane $z = -1$ through the point $(0,0,1)$. Deduce that $I_\infty = 0$ on this plane (consider the projection of a small circle on the sphere with its centre on the z axis). Explain, in general terms, why the sum of the indices of the equilibrium points on a sphere is two.

For a certain problem, the phase diagram on the sphere has centres and saddle points only, and it has exactly two saddle points. How many centres has the phase diagram?

22. Show that the following systems have no periodic solutions:
   (i) $\dot{x} = y + x^3$, $\quad \dot{y} = x + y + y^3$;
   (ii) $\dot{x} = y$, $\quad \dot{y} = -(1 + x^2 + x^4)y - x$.

23. (Dulac's Test) For the system $\dot{x} = X(x,y)$, $\dot{y} = Y(x,y)$, show that there are no closed paths in a simply-connected region in which $(\partial(\rho X)/\partial x) + (\partial(\rho Y)/\partial y)$ is of one sign, where $\rho(x,y)$ is any function having continuous first partial derivatives.

24. Explain in general terms how Dulac's test (Exercise 23) and Bendixson's test may be extended to cover the cases when $(\partial(\rho X)/\partial x) + (\partial(\rho Y)/\partial y)$ is of one sign except on isolated points or curves within a simply-connected region.

25. For the second-order system $\dot{x} = X(x)$, $\operatorname{div}(X) = 0$ and $\operatorname{curl}(X) = 0$ in a simply-connected region $\mathscr{D}$. Show that the system has no closed paths in $\mathscr{D}$. Deduce that

$$\dot{x} = y + 2xy, \quad \dot{y} = x + x^2 - y^2$$

has no periodic solutions.

26. In Exercise 25 show that $\mathrm{div}(\mathbf{X}) = 0$ may be replaced by $\mathrm{div}(\psi\mathbf{X}) = 0$, where $\psi(x, y)$ is of one sign in $\mathcal{D}$.

27. By using Dulac's test (Exercise 23) with $\rho = \mathrm{e}^{-2x}$, show that

$$\dot{x} = y, \quad \dot{y} = -x - y + x^2 + y^2$$

has no periodic solutions.

28. Use Dulac's test (Exercise 23) to show that $\dot{x} = x(y-1), \dot{y} = x + y - 2y^2$, has no periodic solutions.

29. Show that the following systems have no periodic solutions:
    (i) $\dot{x} = y, \dot{y} = 1 + x^2 - (1-x)y$;
    (ii) $\dot{x} = -(1-x)^3 + xy^2, \dot{y} = y + y^3$;
    (iii) $\dot{x} = 2xy + x^3, \dot{y} = -x^2 + y - y^2 + y^3$;
    (iv) $\dot{x} = x, \dot{y} = 1 + x + y^2$;
    (v) $\dot{x} = y, \dot{y} = -1 - x^2$;
    (vi) $\dot{x} = 1 - x^3 + y^2, \dot{y} = 2xy$;
    (vii) $\dot{x} = y, \dot{y} = (1 + x^2)y + x^3$.

30. Let $\mathcal{D}$ be a doubly-connected domain in the x, y plane. Show that, if $\rho(x, y)$ has continuous first partial derivatives and $\mathrm{div}(\rho\mathbf{X})$ is of constant sign in $\mathcal{D}$, then the system has not more than one closed path in $\mathcal{D}$. (An extension of Dulac's theorem, Exercise 23.)

31. A system has exactly two limit cycles with one lying interior to the other and with no equilibrium points between them. Can the limit cycles be described in opposite senses?

Obtain the equations of the phase paths of the system

$$\dot{r} = \sin \pi r, \quad \dot{\theta} = \cos \pi r$$

as described in polar coordinates (r, θ). Sketch the phase diagram.

32. Using Bendixson's theorem (Section 3.4) show that the response amplitudes a, b for the van der Pol equation in the 'van der Pol plane' (Chapter 7), described by the equations

$$\dot{a} = \tfrac{1}{2}\varepsilon(1 - \tfrac{1}{4}r^2)a - \frac{\omega^2 - 1}{2\omega}b, \quad \dot{b} = \tfrac{1}{2}\varepsilon(1 - \tfrac{1}{4}r^2)b + \frac{\omega^2 - 1}{2\omega}a + \frac{\Gamma}{2\omega},$$
$$r = \sqrt{(a^2 + b^2)},$$

have no closed paths in the circle $r < \sqrt{2}$.

33. Let $\mathscr{C}$ be a closed path for the system $\dot{\mathbf{x}} = \mathbf{X}(\mathbf{x})$, having $\mathcal{D}$ as its interior. Show that

$$\int_{\mathcal{D}} \mathrm{div}(\mathbf{X})\,\mathrm{d}x\,\mathrm{d}y = 0.$$

34. Assume that van der Pol's equation in the phase plane

$$\dot{x} = y, \quad \dot{y} = -\varepsilon(x^2 - 1)y - x$$

has a single closed path, which, for ε small, is approximately a circle of radius a. Use the result of Exercise 33 to show that approximately

$$\int_{-a}^{a} \int_{-\sqrt{(a^2 - x^2)}}^{\sqrt{(a^2 - x^2)}} (x^2 - 1)\, dy\, dx = 0,$$

and so deduce a.

35. Following Exercises 33 and 34, deduce a condition on the amplitudes of periodic solutions of

$$\ddot{x} + \varepsilon h(x, \dot{x})\dot{x} + x = 0, \quad |\varepsilon| \ll 1.$$

36. For the system

$$\ddot{x} + \varepsilon h(x, \dot{x})\dot{x} + g(x) = 0,$$

suppose that $g(0) = 0$ and $g'(x) > 0$. Let $\mathscr{C}$ be a closed path in the phase plane (as all paths must be) for the equation

$$\ddot{x} + g(x) = 0$$

having interior $\mathscr{D}$. Use the result of Exercise 33 to deduce that for small ε, $\mathscr{C}$ approximately satisfies

$$\int_{\mathscr{D}} \{h(x, y) + h_y(x, y)y\}\, dx\, dy = 0.$$

Adapt this result to the equation

$$\ddot{x} + \varepsilon(x^2 - \alpha)\dot{x} + \sin x = 0,$$

with ε small, $0 < \alpha \ll 1$, and $|x| < \frac{1}{2}\pi$. Show that a closed path (a limit cycle) is given by

$$y^2 = 2A + 2\cos x$$

where A satisfies

$$\int_{-\cos^{-1}(-A)}^{\cos^{-1}(-A)} (x^2 - \alpha)\sqrt{\{2A + 2\cos x\}}\, dx = 0.$$

37. For the system

$$\dot{\mathbf{x}} = \mathbf{X}_0 + \varepsilon\mathbf{X}_1$$

$\mathrm{div}(\mathbf{X}_0) = 0$, and ε is small. Suppose that the system has a centre at the origin when $\varepsilon = 0$. Adapt the method of Exercise 33 to find approximately the periodic solutions, if any, of the system. For the case when there are only limit-cycles, suggest a condition for stability of these. (The system $\dot{\mathbf{x}} = \mathbf{X}_0$ is a Hamiltonian

system: for the condition for a centre see Exercise 27, Chapter 2.)
Examine the system

$$\dot{x} = -2y, \quad \dot{y} = x + \varepsilon(x^2 - 1)y$$

from this point of view.

38. The equation $\ddot{x} + F_0 \tanh k(\dot{x} - 1) + x = 0, F_0 > 0, k \gg 1$, can be thought of as a plausible continuous representation of the type of Coulomb friction problem of Section 1.6. Show, however, that the only equilibrium point is a stable spiral at the origin, and that there are no periodic solutions.

39. Show that the third-order system

$$\dot{x}_1 = x_2, \quad \dot{x}_2 = -x_1, \quad \dot{x}_3 = 1 - (x_1^2 + x_2^2)$$

has no equilibrium points but nevertheless has closed paths (periodic solutions).

40. Sketch the phase diagram for the quadratic system

$$\dot{x} = 2xy, \quad \dot{y} = y^2 - x^2.$$

41. Locate the equilibrium points of the system

$$\dot{x} = x(x^2 + y^2 - 1), \quad \dot{y} = y(x^2 + y^2 - 1),$$

and sketch the phase diagram.

42. In the system $\dot{x} = X(x, y), \dot{y} = Y(x, y)$, X and Y are relatively prime polynomials of degree two in x, y. Show that three equilibrium points can never be collinear.

43. Produce an argument to show that if a closed curve Γ is continuously deformed without passing over an equilibrium point, then the index $[\phi]_\Gamma$ (Section 3.1) varies continuously. Deduce the theorems of Section 3.1 from this consideration.

The following exercises are to be carried out on a computer.

44. Plot a phase diagram, showing the main features of the phase plane, for the equation $\ddot{x} + \varepsilon(1 - x^2 - \dot{x}^2)\dot{x} + x = 0$, $\varepsilon = 0.1$ and $\varepsilon = 5$.

45. Plot a phase diagram for the damped pendulum equation $\ddot{x} + 0.15\dot{x} + \sin x = 0$. See Fig. 3.26.

46. The system

$$\dot{x} = -\frac{1}{2\omega}y\{(\omega^2 - 1) - \tfrac{3}{4}\beta(x^2 + y^2)\}, \quad \dot{y} = \frac{1}{2\omega}x\{(\omega^2 - 1) - \tfrac{3}{4}\beta(x^2 + y^2)\} + \frac{\Gamma}{2\omega}$$

occurs in the theory of the forced oscillations of a pendulum (eqns (7.16), (7.17)). Obtain the phase diagram when $\omega = 0.975, \Gamma = 0.005$.

FIG. 3.26. Phase diagram for the pendulum equation with small damping

47. A population of rabbits $R(t)$ and foxes $F(t)$ live together in a certain territory. The combined birth and death rate of the rabbits due to 'natural' causes is $\alpha_1 > 0$, and the additional deaths due to their being eaten by foxes is introduced through an 'encounter factor' β_1, so that

$$\frac{dR}{dt} = \alpha_1 R - \beta_1 RF.$$

The foxes die of old age with death rate $\beta_2 > 0$, and the live birth rate is sustained through an encounter factor α_2, so that (compare Section 2.2)

$$\frac{dF}{dt} = \alpha_2 RF - \beta_2 F.$$

Plot the phase diagram, when $\alpha_1 = 10$, $\beta_1 = 0.2$, $\alpha_2 = 4 \times 10^{-5}$, $\beta_2 = 0.2$. Also plot the solution curves $R(t)$ and $F(t)$ (these are oscillatory, having the same period but different phase).

48. The system

$$\dot{x} = \tfrac{1}{2}e(1 - \tfrac{1}{4}r^2)x - \frac{\omega^2 - 1}{2\omega} y, \quad (r^2 = x^2 + y^2),$$

$$\dot{y} = \frac{\omega^2 - 1}{2\omega} x + \tfrac{1}{2}e(1 - \tfrac{1}{4}r^2)y + \frac{\Gamma}{2\omega},$$

occurs in the theory of forced oscillations of the van der Pol equation (eqns (7.34a, b)). Plot phase diagrams for the cases
(i) $e = 1$, $\Gamma = 0.75$, $\omega = 1.2$;
(ii) $e = 1$, $\Gamma = 2.0$, $\omega = 1.6$.

49. The equation for a tidal bore on a shallow stream is

$$\varepsilon \frac{d^2\eta}{d\xi^2} - \frac{d\eta}{d\xi} + \eta^2 - \eta = 0$$

where (in appropriate dimensions), η is the height of the free surface, and $\xi = x - ct$ where c is the wave speed. For $0 < \varepsilon \ll 1$, find the equilibrium points of the equation and classify them according to their linear approximations.

Plot the phase paths in the plane of η, w, where

$$\frac{d\eta}{d\xi} = w, \quad \varepsilon\frac{dw}{d\xi} = \eta + w - \eta^2$$

and show that a separatrix from the saddle point at the origin reaches the other equilibrium point. Interpret this observation in terms of the shape of the wave.

Fig. 3.27. Phase diagram for a Coulomb friction problem

50. Determine the nature of the equilibrium point, and compute the phase diagram for the Coulomb friction type problem

$$\ddot{x} + x = F(\dot{x})$$

where

$$F(y) = \begin{cases} -6.0(y-1), & |y-1| \leq 0.4 \\ \cdot\cdot[1+1.4\exp\{-0.5|y-1|+0.2\}]\,\text{sgn}\,(y-1), & |y-1| \geq 0.4. \end{cases}$$

(See Fig. 3.27, and compare the simpler case shown in Figs. 1.15(b) and 1.16.)

51. Compute the phase diagram for the system whose polar representation is

$$\dot{r} = r(1-r), \quad \dot{\theta} = \sin^2\left(\tfrac{1}{2}\theta\right).$$

52. Compute the phase diagrams for the following systems: (i) $\dot{x} = 2xy$, $\dot{y} = y^2 - x^2$; (ii) $\dot{x} = 2xy$, $\dot{y} = x^2 - y^2$; (iii) $\dot{x} = x^3 - 2x^2y$, $\dot{y} = 2xy^2 - y^3$.

4 Averaging methods

CONSIDER an equation of the form $\ddot{x} + \varepsilon h(x, \dot{x}) + x = 0$ where ε is small, and assume that it has at least one periodic solution, so that its phase diagram contains either a limit cycle or a centre. Such an equation is in a sense 'close' to the linear equation $\ddot{x} + x = 0$, and the paths will be close to circles. It should be possible to take advantage of this fact in constructing approximate solutions, in a way similar to that in which a linear approximation gives a guide to the nature of the phase paths near an equilibrium point. However, a solution may differ very little from some solution of the linear equation but it may still be sufficient to prevent the corresponding path's being closed (similarly, an equilibrium point which is a centre in the linear approximation may in fact be a spiral: see Chapter 2, Exercise 7). In order to determine periodic solutions we use a process which essentially assumes the existence of a closed path which is nearly circular but has unknown radius, then utilize the consistency conditions on the radius implied by the differential equation. Thus we make direct use of the term of order ε in the equation in order to establish the consistency. In doing this we simplify the function h by in effect considering only the first harmonic in the variation of $\dot{x}h(x, \dot{x})$ over the cycle; this can also be interpreted as approximation in the sense of taking smoothed-out values of $\dot{x}h(x, \dot{x})$ over a cycle.

Elaborations of the method permit the determination of the approximate period and stability of limit cycles, the shape of the spiral paths around limit cycles, and amplitude–frequency relations in the case of a centre. Similar estimates are successful even in very unpromising cases where ε is not small and the supporting arguments are no longer plausible (see Example 4.12 and Section 4.5). In such cases we can say that we are guided to the appropriate sense of how best to fit a circle to a non-circular path by arguments which are valid when ε is small.

4.1. An energy-balance method for limit cycles

The nonlinear character of isolated periodic oscillations makes their detection and construction difficult. Here we discuss limit cycles and other periodic solutions in the phase plane $\dot{x} = y$, $\dot{y} = Y(x, y)$, which allows the mechanical interpretation of Section 1.6 and elsewhere. Consider the family of equations of the form

$$\ddot{x} + \varepsilon h(x, \dot{x}) + x = 0 \tag{4.1}$$

(note that the equation $\ddot{x} + \varepsilon h(x, \dot{x}) + \omega^2 x = 0$ can be put into this form by the change of variable $\tau = \omega t$). Then on the phase plane we have

$$\dot{x} = y, \quad \dot{y} = -\varepsilon h(x, y) - x. \tag{4.2}$$

Assume that $|\varepsilon| \ll 1$, so that the nonlinearity is small, and that $h(0, 0) = 0$, so that the origin is an equilibrium point. Suppose we have reason to think that there is at least one periodic solution with phase path surrounding the origin: an appraisal of the regions of the phase plane in which energy loss and energy gain take place might give grounds for expecting a limit cycle.

When ε is zero, eqn (4.1) becomes $\ddot{x} + x = 0$, with solutions $x(t) = a \cos (t + \alpha)$ where a, α are constants. These solutions appear as circles of radius a on the phase plane, and the motion has period 2π. We do not lose any generality so far as the phase-plane representation is concerned if we limit consideration to the case

$$a > 0, \quad \alpha = 0,$$

since all other solutions simply correspond to a change of time origin.

Suppose now that ε is small but not zero. Then we should expect that *the limit cycle assumed above will be close to one of the circles*

$$x^2 + y^2 = a^2 \tag{4.3}$$

*generated by the **linearized equation** (with $\varepsilon = 0$), and that its period will be very nearly 2π.* In order to decide which circle most nearly satisfies the conditions for a limit cycle, we tighten up the argument of Section 1.5. Assume that the period of the postulated limit cycle is very nearly 2π, and that it is represented approximately (ignoring the unimportant phase angle) by

$$x(t) = a \cos t, \quad a > 0. \tag{4.4}$$

Then $\dot{x}(t) = y(t) = -a \sin t$. From (1.35) (with εh for h) the change in energy over a cycle is given approximately by

$$\mathscr{E}(2\pi) - \mathscr{E}(0) = -\varepsilon \int_0^{2\pi} (-a \sin t) h(a \cos t, -a \sin t) \, dt. \tag{4.5}$$

(Obviously integration over any interval of length 2π gives the same result since the integrand is periodic.) But the change in energy must be zero over the cycle, since x and $\dot{x}$ return to their original values in the periodic motion. Therefore (4.5) implies that on a limit cycle, and, indeed, for any periodic solution,

$$\int_0^{2\pi} h(a\cos t, -a\sin t)\sin t \, dt = 0, \qquad (4.6)$$

an equation which, in principle, determines a in the case of a limit cycle, and is satisfied identically around a centre.

Example 4.1 *Find the approximate amplitude for the limit cycle of van der Pol's equation*

$$\ddot{x} + \varepsilon(x^2 - 1)\dot{x} + x = 0. \qquad (4.7)$$

Here,

$$h(x, \dot{x}) = (x^2 - 1)\dot{x}.$$

Assuming that $x \approx a\cos t$, the energy balance equation (4.6) gives

$$\int_0^{2\pi} (a^2\cos^2 t - 1)\sin t \sin t \, dt = 0, \qquad (4.8)$$

leading to $\frac{1}{4}a^2 - 1 = 0$, with positive solution $a = 2$. Figure 4.1 shows the calculated limit cycle for $\varepsilon = 0.1$.

By an extension of this argument the stability of the limit cycle can also be determined. Taking the model of a limit cycle as conforming to Fig. 3.17 we should expect that paths close enough to the limit cycle, though spiralling gradually, will also be given approximately by $x = a\cos t$, $y = -a\sin t$ where a is nearly constant. On a path segment corresponding to $0 \leqslant t \leqslant 2\pi$ the change in energy is approximately

$$\mathscr{E}(2\pi) - \mathscr{E}(0) = \varepsilon a \int_0^{2\pi} h(a\cos t, -a\sin t)\sin t \, dt = g(a), \qquad (4.9)$$

say. Let $a = a_0 > 0$ on the limit cycle, so that $g(a_0) = 0$. If the cycle is stable, then on interior segments $(a < a_0)$ energy must be gained, and on exterior segments $(a > a_0)$ energy must be lost: that is, there exists $\delta > 0$ such that

$$g(a) > 0 \quad \text{when} \quad a_0 - \delta < a < a_0$$

and

$$g(a) < 0 \quad \text{when} \quad a_0 < a < a_0 + \delta.$$

Positive damping Negative damping Positive damping

FIG. 4.1. Phase diagram, with limit cycle, for $\ddot{x} + 0.1(x^2 - 1)\dot{x} + x = 0$

Therefore

$$g'(a_0) < 0 \tag{4.10}$$

is a sufficient condition for stability.

These arguments can be carried through without using the concept of energy, in the manner of Example 1.7. In place of $\mathscr{E}(t)$ the function $\frac{1}{2}y^2 + \frac{1}{2}x^2$ is made explicit, and the change in $\mathscr{E}$ over a cycle appears clearly as a change in distance from the origin.

Example 4.2 *Check the stability of the limit cycle in Example* 4.1.

From eqn (4.14)

$$g(a) = -\varepsilon a^2 \int_0^{2\pi} (a^2 \cos^2 t - 1) \sin^2 t \, dt = -\varepsilon a^2 \pi (\tfrac{1}{4} a^2 - 1).$$

Therefore

$$g'(a) = -\varepsilon \pi a (a^2 - 2)$$

and

$$g'(2) = -4\pi\varepsilon.$$

The cycle is therefore stable when $\varepsilon > 0$, unstable when $\varepsilon < 0$.

Direct differentiation of (4.9) gives an alternative criterion. From (4.9),

$$g'(a) = \varepsilon \int_0^{2\pi} h \sin t \, dt + \varepsilon a \int_0^{2\pi} \frac{dh}{da} \sin t \, dt$$

$$= \varepsilon [-h \cos t]_0^{2\pi} + \varepsilon \int_0^{2\pi} \frac{dh}{dt} \cos t \, dt + \varepsilon a \int_0^{2\pi} \frac{dh}{da} \sin t \, dt$$

(integrating by parts). But by the chain rule

$$\frac{dh}{dt} = -ah_1 \sin t - ah_2 \cos t$$

and

$$\frac{dh}{da} = h_1 \cos t - h_2 \sin t$$

(where $h_1(u,v) = \partial h(u,v)/\partial u$ and $h_2(u,v) = \partial h(u,v)/\partial v$). Therefore

$$g'(a) = -\varepsilon a \int_0^{2\pi} h_2(a\cos t, -a\sin t)\,dt.$$

For *stability* we require $g'(a_0) < 0$ which is equivalent to

$$\varepsilon \int_0^{2\pi} h_2(a_0\cos t, -a_0\sin t)\,dt > 0, \qquad (4.11)$$

where a_0 is a positive solution of (4.6).

Example 4.3   *Check the stability of the limit cycle in Examples* 4.1.
  $h(x,y) = (x^2-1)y$ and $h_2(x,y) = x^2-1$. Equation (4.11), with $a_0 = 2$, gives

$$\varepsilon \int_0^{2\pi} (4\cos^2 t - 1)\,dt = 2\pi\varepsilon$$

and the cycle is stable if $\varepsilon > 0$.

4.2.   Amplitude and frequency estimates

As described in the last section, a periodic solution of the equation

$$\ddot{x} + \varepsilon h(x,\dot{x}) + x = 0 \qquad (4.12)$$

or of the system

$$\dot{x} = y, \quad \dot{y} = -\varepsilon h(x,y) - x, \qquad (4.13)$$

with $h(0,0) = 0$ and $|\varepsilon| \ll 1$ is expected to appear in the phase plane as a small distortion, or perturbation, of one of the circular paths of the linearized equation $\ddot{x} + x = 0$.

In Fig. 4.2, P is a representative point for a phase path of the system and $a(t)$ (> 0), $\theta(t)$ are its polar coordinates. Note that the positive direction of θ is opposite to the direction of the path. Also $x = a\cos\theta$ and $y = a\sin\theta$ (not $(-a\sin\theta)$). Since $a^2 = x^2 + y^2$,

$$a\dot{a} = x\dot{x} + y\dot{y} = -\varepsilon yh(x,y)$$

so that

$$\dot{a} = -\varepsilon h(a\cos\theta, a\sin\theta)\sin\theta. \qquad (4.14)$$

Fig. 4.2.

Similarly, since $\tan \theta = y/x$ we find that

$$\dot\theta = -1 - \frac{\varepsilon}{a} h(a \cos\theta, a \sin\theta) \cos\theta. \tag{4.15}$$

These equations are exact. We also deduce that

$$\frac{\mathrm{d}a}{\mathrm{d}\theta} = \frac{\dot a}{\dot\theta} = \frac{-\varepsilon h(a \cos\theta, a \sin\theta) \sin\theta}{-1 - \dfrac{\varepsilon}{a} h(a \cos\theta, a \sin\theta) \cos\theta},$$

$$\approx \varepsilon h(a \cos\theta, a \sin\theta) \sin\theta \tag{4.16}$$

accurate to order ε.

Since the motion is periodic, the interval $2\pi \geqslant \theta \geqslant 0$ corresponds to a single period (measured forwards in time: see Fig. 4.2). a returns to its original value in the period, so

$$\int_0^{2\pi} \frac{\mathrm{d}a}{\mathrm{d}\theta} \mathrm{d}\theta = 0 = \varepsilon \int_0^{2\pi} h(a \cos\theta, a \sin\theta) \sin\theta \, \mathrm{d}\theta. \tag{4.17}$$

a in the integral is a function of θ through (4.16). Since $\mathrm{d}a/\mathrm{d}\theta = O(\varepsilon)$, then over one period

$$a(\theta) = a_0 + O(\varepsilon), \tag{4.18}$$

where a_0 is the value, say, at the beginning of the period. In (4.17), therefore, by retaining only the dominant term we have

$$\int_0^{2\pi} h(a_0 \cos\theta, a_0 \sin\theta) \sin\theta \, \mathrm{d}\theta = 0. \tag{4.19}$$

This is eqn (4.6) again and provides, in principle, a means for estimating a_0. The first approximation on the phase plane is then

$$x^2 + y^2 = a_0^2. \tag{4.20}$$

To find the perturbed *period*, T, we have

$$T = \int_0^T dt = \int_{2\pi}^0 \left(\frac{d\theta}{dt}\right)^{-1} d\theta$$

(see Fig. 4.2 for the correspondence between the limits of integration). Therefore, from (4.15),

$$T = \int_0^{2\pi} \frac{d\theta}{1 + \dfrac{\varepsilon}{a} h(a\cos\theta, a\sin\theta)\cos\theta}$$

$$\approx 2\pi - (\varepsilon/a) \int_0^{2\pi} h(a\cos\theta, a\sin\theta)\cos\theta \, d\theta$$

(by expanding the integrand and retaining terms of order ε). Finally from (4.18)

$$T \approx 2\pi - \frac{\varepsilon}{a_0} \int_0^{2\pi} h(a_0\cos\theta, a_0\sin\theta)\cos\theta \, d\theta \tag{4.21}$$

to the same order, where a_0 is given by (4.19). The *frequency* ω is given to order ε by

$$\omega = 2\pi/T \approx 1 + \frac{\varepsilon}{2\pi a_0} \int_0^{2\pi} h(a_0\cos\theta, a_0\sin\theta)\cos\theta \, d\theta. \tag{4.22}$$

Example 4.4 *Obtain the frequency, to order ε, of the limit cycle of* $\ddot{x} + \varepsilon(x^2 - 1)\dot{x} + x = 0.$

Since $h(x, \dot{x}) = (x^2 - 1)\dot{x}$ and $a_0 = 2$ to order ε (Example 4.3), eqn (4.22) gives

$$\omega = 1 + \frac{\varepsilon}{4\pi} \int_0^{2\pi} (4\cos^2\theta - 1)(+2\sin\theta)\cos\theta \, d\theta = 1 + \text{zero}.$$

Therefore the frequency is 1 with error $O(\varepsilon^2)$.

Example 4.5 *Obtain the relation between frequency and amplitude for the pendulum equation* $\ddot{x} + \sin x = 0.$

Write the equation in the form $\ddot{x} + (\sin x - x) + x = 0$. Then for amplitudes of motion not too large we may write

$$\varepsilon h(x, \dot{x}) = \sin x - x, \quad 0 < \varepsilon \ll 1,$$

since $\sin x - x = -\frac{1}{6}x^3 + \dots$, and $|\sin x - x| \ll 1$ so long as, say, $|x| < 1$. Equation (4.19) becomes

$$0 = \int_0^{2\pi} [\sin (a_0 \cos \theta) - a_0 \cos \theta] \sin \theta \, d\theta.$$

The integrand is anti-symmetric (odd) about $\theta = \pi$, so this equation is satisfied by all a_0, as expected.

For the frequency, (4.22) becomes

$$\omega = 1 + \frac{1}{2\pi a_0} \int_0^{2\pi} [\sin (a_0 \cos \theta) - a_0 \cos \theta] \cos \theta \, d\theta.$$

For the small amplitudes concerned, we may write $\sin x = x - \frac{1}{6}x^3$. Then

$$\omega = 1 - \frac{a_0^2}{12\pi} \int_0^{2\pi} \cos^4 \theta \, d\theta = 1 - \tfrac{1}{16}a_0^2,$$

which is the required relationship for a_0 not too large. The frequency is seen to decrease with amplitude.

Example 4.6 *Find approximately the amplitude and frequency of the limit cycle for Rayleigh's equation $\ddot{x} + \varepsilon(\frac{1}{3}\dot{x}^3 - \dot{x}) + x = 0$.*

Here, $h(x, \dot{x}) = (\frac{1}{3}\dot{x}^2 - 1)\dot{x}$. Equation (4.19) becomes

$$\int_0^{2\pi} (\tfrac{1}{3}a_0^2 \sin^2 \theta - 1) \sin^2 \theta \, d\theta = 0,$$

which can be verified to have solution $a_0 = 2$. The frequency equation (4.15) becomes

$$\omega = 1 + \frac{\varepsilon}{2\pi} \int_0^{2\pi} (\tfrac{1}{3}a_0^2 \sin^2 \theta - 1) \sin \theta \cos \theta \, d\theta.$$

The integral is zero, since the integrand is odd. The frequency is thus unchanged from unity to this order.

4.3. Slowly-varying amplitude: nearly periodic solutions

Equations (4.14), (4.15), and (4.16) apply to any solutions of $\ddot{x} + \varepsilon h(x, \dot{x}) + x = 0$ which are nearly circles in the phase plane, and not just to the periodic solutions considered in Section 4.1. However, they cannot in general be solved exactly, and we shall show how simpler, approximating, equations can be constructed.

We shall assume that a is nearly constant over a 'cycle', whether it is closed or not; specifically, that its rate of change with t or θ is of order ε. Then a is said to be a *slowly-varying amplitude*. The character of such solutions on the phase-plane is of slowly-winding, nearly circular spirals.

The right-hand side of (4.14) contains the function

$h(a \cos \theta, a \sin \theta) \sin \theta$, where a is a function of θ in any motion. We represent $h \sin \theta$ as a Fourier-type series over a 'cycle' $0 \leqslant \theta \leqslant 2\pi$ in the following way. Write

$$h[a(\theta) \cos \theta, a(\theta) \sin \theta] \sin \theta$$

$$= p_0[a(\theta)] + \sum_{n=1}^{\infty} \{p_n[a(\theta)] \cos n\theta + q_n[a(\theta)] \sin n\theta\}, \quad (4.23)$$

where $p_n(a)$ and $q_n(a)$ are obtained by treating a as constant in the usual definitions for Fourier series coefficients:

$$p_0[a(\theta)] = \frac{1}{2\pi} \int_0^{2\pi} h[a(\theta) \cos u, a(\theta) \sin u] \sin u \, du; \quad (4.24)$$

and for $n \geqslant 1$

$$p_n[a(\theta)] = \frac{1}{\pi} \int_0^{2\pi} h[a(\theta) \cos u, a(\theta) \sin u] \sin u \cos nu \, du,$$

$$q_n[a(\theta)] = \frac{1}{\pi} \int_0^{2\pi} h[a(\theta) \cos u, a(\theta) \sin u] \sin u \sin nu \, du$$

Note that (4.23) is true exactly; the series represents the function on the left although the coefficients are not constant; also the equality is not restricted to the interval $0 \leqslant \theta \leqslant 2\pi$.

Equation (4.16) becomes, for all θ,

$$\frac{da}{d\theta} = \varepsilon p_0(a) + \varepsilon \sum_{n=1}^{\infty} [p_n(a) \cos n\theta + q_n(a) \sin n\theta]. \quad (4.25)$$

The terms under the summation sign can be looked on as nearly harmonic 'inputs' or forcing terms applied to a differential equation $(da/d\theta) - \varepsilon p_0(a) = 0$. We shall show that the *increment* in a over every 'cycle' $\theta_0 \leqslant \theta \leqslant \theta_0 + 2\pi$ is nearly independent of whether these terms are present or not.

By integrating (4.25) over this interval we have, for any θ,

$$a(\theta_0 + 2\pi) - a(\theta_0) = \varepsilon \int_{\theta_0}^{\theta_0 + 2\pi} p_0(a) \, d\theta + \varepsilon \sum_{n=1}^{\infty} \int_{\theta_0}^{\theta_0 + 2\pi} [p_n(a) \cos n\theta + q_n(a) \sin n\theta] \, d\theta.$$

Over such a cycle, a is constant to order ε (eqn (4.18)), so to this order the terms under the summation sign vanish, and

$$a(\theta_0 + 2\pi) - a(\theta_0) = 2\pi \varepsilon p_0(a)$$

where a is a representative value of $a(\theta)$ on the cycle. Therefore the simplified equation

$$\frac{da}{d\theta} = \varepsilon p_0(a) \tag{4.26}$$

at least gives the increments in a per cycle correctly to order ε, though the actual shape of the spiral curves it predicts will lack contributions from the higher harmonics in (4.25). Note that $p_0(a)$ is approximately the mean value of $h[a(\theta)\cos\theta, a(\theta)\sin\theta]\sin\theta$ over a cycle, so we have replaced the right side of (4.16) by its mean value. This process characterizes a group of procedures called *averaging methods* (Krylov and Bogoliubov 1949).

The simplified equation corresponding to (4.14) is obtained by writing

$$\frac{da}{dt} = \frac{da}{d\theta}\frac{d\theta}{dt} = -\frac{da}{d\theta} + O(\varepsilon^2)$$

from (4.15), so that, approximately,

$$\frac{da}{dt} = -\varepsilon p_0(a) \tag{4.27}$$

where p_0 is given by (4.24). Similarly (4.15) becomes

$$\frac{d\theta}{dt} = -1 - \frac{\varepsilon}{a} r_0(a), \tag{4.28}$$

where

$$r_0(a) = \frac{1}{2\pi} \int_0^{2\pi} h[a\cos u, a\sin u]\cos u\, du, \tag{4.29}$$

and a is treated as constant for the integration.

Example 4.7 *Find approximate solutions to van der Pol's equation*

$$\ddot{x} + \varepsilon(x^2 - 1)\dot{x} + x = 0,$$

for small positive ε.

From (4.24),

$$p_0(a) = \frac{a}{2\pi} \int_0^{2\pi} (a^2\cos^2 u - 1)\sin^2 u\, du = \tfrac{1}{2}a(\tfrac{1}{4}a^2 - 1).$$

We have the approximate equation, (4.27), for the radial coordinate:

$$\frac{da}{dt} = -\tfrac{1}{2}\varepsilon a(\tfrac{1}{4}a^2 - 1).$$

The constant solution $a = 2$ corresponds to the limit cycle. The equation separates, giving

$$\int \frac{da}{a(a^2-4)} = -\tfrac{1}{8}\varepsilon(t+C)$$

where C is a constant. Thus

$$-\tfrac{1}{4}\log a + \tfrac{1}{8}\log|a^2-4| = -\tfrac{1}{8}\varepsilon(t+C).$$

When $a(0) = a_0$, the solution is

$$a(t) = 2/\{1-(1-4/a_0^2)e^{-\varepsilon t}\}^{1/2},$$

which tends to 2 as $t \to \infty$. It is easy to verify from eqn (4.29) that $r_0(a) = 0$. Equation (4.28) therefore gives $\theta(t) = -t + \theta_0$ where θ_0 is the initial polar angle. The frequency of the spiral motion is therefore the same as that of the limit cycle to our degree of approximation. Finally, the required approximate solutions are given by

$$x(t) = a(t)\cos\theta(t) = \frac{2\cos(t-\theta_0)}{\{1-(1-4/a_0^2)e^{-\varepsilon t}\}^{1/2}}.$$

Example 4.8   *Find the approximate phase paths for the equation*

$$\ddot{x} + \varepsilon(|\dot{x}|-1)\dot{x} + x = 0.$$

We have $h(x,y) = (|y|-1)y$. Therefore, from (4.24),

$$p_0(a) = \frac{a}{2\pi}\int_0^{2\pi} (|a\sin\theta|-1)\sin^2\theta\, d\theta$$

$$= \frac{a}{2\pi}\left(\int_0^{\pi}(a\sin\theta-1)\sin^2\theta\, d\theta + \int_{\pi}^{2\pi}(-a\sin\theta-1)\sin^2\theta\, d\theta\right)$$

$$= \frac{a}{2\pi}2\int_0^{\pi}(a\sin\theta-1)\sin^2\theta\, d\theta = \frac{1}{\pi}a(\tfrac{4}{3}a - \tfrac{1}{2}\pi - \tfrac{1}{6}).$$

There is a limit cycle when $p_0(a) = 0$; that is, at $a = \frac{3}{8}(\pi + \tfrac{1}{3}) = a^*$, say. Equation (4.26) becomes

$$\frac{da}{d\theta} = \frac{4\varepsilon}{3\pi}a(a-a^*).$$

The path satisfying $a = a_0$ at $\theta = \theta_0$ is given in polar coordinates by

$$a(\theta) = a^*\Big/\left(1-(1-a^*/a_0)\exp\left[\varepsilon\frac{4a^*}{3\pi}(\theta-\theta_0)\right]\right).$$

Alternatively, eqn (4.27) gives a as a function of t.

4.4. Periodic solutions: harmonic balance

One of the easiest methods for estimating periodic solutions in simple cases is illustrated by the following examples.

Example 4.9 *Find an approximation to the amplitude and frequency of the limit cycle of van der Pol's equation* $\ddot{x} + \varepsilon(x^2 - 1)\dot{x} + x = 0$.

Assume an approximate solution $x = a \cos \omega t$. (The prescription of zero value for the phase is not a limitation; in the case of an autonomous equation we are, in effect, simply choosing the time origin so that $\dot{x}(0) = 0$.) Write the equation in the form

$$\ddot{x} + x = -\varepsilon(x^2 - 1)\dot{x}.$$

Upon substituting the assumed form of solution we obtain

$$(-\omega^2 + 1)a \cos \omega t = -\varepsilon(a^2 \cos^2 \omega t - 1)(-a\omega \sin \omega t)$$
$$= \varepsilon a\omega(\tfrac{1}{4}a^2 - 1)\sin \omega t + \tfrac{1}{4}\varepsilon a^4 \omega \sin 3\omega t,$$

after some reduction. The right-hand side is just the Fourier series for $\varepsilon h(x, \dot{x})$: it is easier to get it this way than to work with the equations of the last section. Now, ignore the presence of the term in $\sin 3\omega t$, and match terms in $\cos \omega t$, $\sin \omega t$. We find

$$1 - \omega^2 = 0$$

from the $\cos \omega t$ term, and

$$\tfrac{1}{4}a^2 - 1 = 0$$

from the $\sin \omega t$ term. The second equation gives $a = \pm 2$, and the first $\omega = \pm 1$ (which are equivalent), as in several earlier examples.

Example 4.10 *Obtain the amplitude-frequency relation for the approximate pendulum equation* $\ddot{x} + x - \tfrac{1}{6}x^3 = 0$.

Assume an approximate solution of the form $x = a \cos \omega t$. We obtain, after substituting in the equation:

$$-\omega^2 a \cos \omega t + a \cos \omega t - \tfrac{1}{6}a^3(\tfrac{3}{4}\cos \omega t - \tfrac{1}{4}\cos 3\omega t) = 0.$$

Ignore the higher harmonic and collect the coefficients of $\cos \omega t$. We find that

$$a[(1 - \omega^2) - \tfrac{1}{8}a^2] = 0.$$

Therefore the frequency-amplitude relation becomes

$$\omega = \sqrt{(1 - \tfrac{1}{8}a^2)} \approx 1 - \tfrac{1}{16}a^2$$

for amplitude not too large (compare Example 4.5).

For the general equation

$$\ddot{x} + \varepsilon h(x, \dot{x}) + x = 0, \tag{4.30}$$

suppose that there is a periodic solution close to $a \cos \omega t$ and that h has a

Fourier series

$$h(x, \dot{x}) \approx h(a \cos \omega t, -a\omega \sin \omega t)$$
$$= A_1(a)\cos \omega t + B_1(a)\sin \omega t + \text{higher harmonics}, \quad (4.31)$$

with the constant term absent. Then (4.30) becomes

$$(1 - \omega^2)a \cos \omega t + \varepsilon A_1(a) \cos \omega t + \varepsilon B_1(a) \sin \omega t + \text{higher harmonics} = 0.$$
$$(4.32)$$

This equation can hold for all t only if

$$(1 - \omega^2)a + \varepsilon A_1(a) = 0, \quad B_1(a) = 0, \quad (4.33)$$

which determines a and ω.

It will not be possible in general to ensure that full matching is obtained in (4.32): since the solution is not exactly of the form $a \cos \omega t$ we should ideally have represented it, too, by a Fourier series and matching could then in principle be completed. A way of justifying the non-matching of the higher harmonics is to regard them as additional, neglected input (or forcing terms) to the linear equation $\ddot{x} + x = 0$. The periodic response of this equation to a force $K \cos n\omega t$ is equal to $-K \cos n\omega t/(n^2\omega^2 - 1)$, and is therefore of rapidly diminishing magnitude as n increases.

The method can also be applied to nonlinear equations with a periodic forcing term, and is used extensively in Chapter 7.

4.5. The equivalent linear equation by harmonic balance

The method of harmonic balance can be adapted to construct a linear substitute for the original nonlinear equation; the process can be described as psuedo-linearization. We illustrate the process by an example.

Example 4.11   *Obtain approximate solutions to the equation*

$$\ddot{x} + \varepsilon(x^2 - 1)\dot{x} + x = 0.$$

The phase diagram for this by now familiar equation (see Examples 4.1, 4.4, 4.7, 4.9) consists of a limit cycle and spirals slowly converging on it when ε is small and positive. We shall approximate to $\varepsilon(x^2 - 1)\dot{x} = \varepsilon h(x, \dot{x})$ on one 'cycle' of spiral: this nonlinear term is already small and a suitable approximation retaining the right characteristics should be sufficient. Suppose, then, that we assume (for the purpose of approximating h only) that $x = a \cos \omega t$ and $\dot{x} = -a\omega \sin \omega t$, where a, ω are considered constant on the 'cycle'. Then

$$\varepsilon(x^2 - 1)\dot{x} = -\varepsilon(a^2 \cos^2 \omega t - 1)a\omega \sin \omega t$$
$$= -\varepsilon a\omega(\tfrac{1}{4}a^2 - 1)\sin \omega t - \tfrac{1}{4}\varepsilon a^3 \sin 3\omega t. \tag{4.34}$$

(see Example 4.9). As in the harmonic balance method, we neglect the effect of the higher harmonic, and see that since $-a\omega \sin \omega t = \dot{x}$ on the cycle, eqn (4.34) can be written

$$\varepsilon(x^2 - 1)\dot{x} \approx \varepsilon(\tfrac{1}{4}a^2 - 1)\dot{x}. \tag{4.35}$$

We now replace this in the differential equation to give the linear equation

$$\ddot{x} + \varepsilon(\tfrac{1}{4}a^2 - 1)\dot{x} + x = 0. \tag{4.36}$$

This equation is peculiar in that it contains a parameter of its own solutions, namely the amplitude a. If $a = 2$ the damping term vanishes and the solutions are periodic of the form $2\cos t$; hence $\omega = 1$. This is an approximation to the limit cycle.

The non-periodic solutions are spirals in the phase plane. Consider the motion for which $x(0) = a_0$, $\dot{x}(0) = 0$: for the next few 'cycles' a_0 will serve as the amplitude used in the above approximation, so (4.40) may be written

$$\ddot{x} + \varepsilon(\tfrac{1}{4}a_0^2 - 1)\dot{x} + x = 0.$$

With the initial conditions given, the solution is

$$x(t) = a_0 \exp\left[-\tfrac{1}{2}\varepsilon(\tfrac{1}{4}a_0^2 - 1)t \right] \cos\left[1 - \tfrac{1}{4}\varepsilon^2(\tfrac{1}{4}a_0^2 - 1) \right]^{1/2}t.$$

It can be confirmed, by expanding both in powers of εt, that so long as $\varepsilon t \ll 1$ this 'solution' agrees with the approximate solution found in Example 4.7. Damped or negatively damped behaviour of the solutions occurs according to whether $a_0 > 2$ or $a_0 < 2$; that is, according to whether we start inside or outside the limit cycle.

Example 4.12 *Find the frequency–amplitude relation for the equation* $\ddot{x} + \mathrm{sgn}\,(x) = 0$ *and compare it with the exact result.*

It can be shown that the system $\dot{x} = y$, $\dot{y} = -\mathrm{sgn}(x)$ has a centre at the origin: the behaviour is that of a particle on a spring, with a symmetrical restoring force of constant magnitude.

Suppose the oscillations have the form $a\cos \omega t$ approximately. Then $\mathrm{sgn}(x) = \mathrm{sgn}\,(a\cos \omega t)$. This has the shape shown in Fig. 4.3(b). The period is $2\pi/\omega$ and we shall approximate to $\mathrm{sgn}\,(a\cos \omega t)$ by the first term in its Fourier series on the interval $(0, 2\pi/\omega)$:

$$\mathrm{sgn}\,(a\cos \omega t) = A_1(a)\cos \omega t + \text{higher harmonics}$$

(there is no sine term, since the function is even), where

FIG. 4.3. (a) Graph of sgn (x); (b) graph of sgn $(a \cos \omega t)$

$$A_1(a) = \frac{\omega}{\pi} \int_0^{2\pi/\omega} \mathrm{sgn}\,(a\cos\omega t)\cos\omega t\, \mathrm{d}t$$

$$= \frac{1}{\pi} \int_0^{2\pi} \mathrm{sgn}\,(a\cos u)\cos u\, \mathrm{d}u$$

$$= \frac{1}{\pi} \left(\int_0^{\pi/2} \cos u\, \mathrm{d}u - \int_{\pi/2}^{3\pi/2} \cos u\, \mathrm{d}u + \int_{3\pi/2}^{2\pi} \cos u\, \mathrm{d}u \right) = \frac{4}{\pi}.$$

Therefore sgn (x) is replaced by $(4/\pi)\cos\omega t$, which in turn is replaced by $(4/\pi)x/a$. The approximate equation is then

$$\ddot{x} + \frac{4}{\pi a} x = 0.$$

The solution, having amplitude a, of the form $a\cos\omega t$ is

$$x(t) = a\cos\left(\frac{4}{\pi a}\right)^{1/2} t.$$

Therefore

$$\omega = 2/\sqrt{(\pi a)}$$

and the period T is given by

$$T = 2\pi/\omega = \pi^{3/2}a^{1/2} \approx 5.568\sqrt{a}.$$

The exact period can be found as follows. When $x > 0$, $\ddot{x} = -1$. The solution for $x > 0$ is therefore

$$x(t) = -\tfrac{1}{2}t^2 + \alpha t + \beta.$$

The solution for which $x(0) = 0$ and $\dot{x}(\tfrac{1}{4}T_e) = 0$, where T_e is the exact period, is

$-\frac{1}{2}t^2 + \frac{1}{4}T_e t$. Further, $x = a$ when $t = \frac{1}{4}T_e$, so that

$$T_e = 4\sqrt{(2a)} \approx 5.657\sqrt{a},$$

which compares very favourably with the above estimate.

Example 4.13 *Obtain the period–amplitude relation for the Poisson–Boltzmann-type equation* $\ddot{x} + (e^x - e^{-x}) = 0$.

When $x = a \cos \omega t$,

$$e^x - e^{-x} = e^{a \cos \omega t} - e^{-a \cos \omega t} = A_1(a) \cos \omega t + \text{higher harmonics}$$

(there is no $\sin \omega t$ term since the function is even), where

$$A_1(a) = \frac{\omega}{\pi} \int_0^{2\pi/\omega} (e^{a \cos \omega t} - e^{-a \cos \omega t}) \cos \omega t \, dt$$

$$= \frac{1}{\pi} \int_0^{2\pi} (e^{a \cos u} - e^{-a \cos u}) \cos u \, du.$$

These integrals are expressible in terms of the modified Bessel function I_1 which has the integral representation (Abramowitz and Stegun 1965)

$$I_1(z) = \frac{1}{\pi} \int_0^\pi e^{z \cos u} \cos u \, du,$$

and we find that

$$A_1(a) = 4I_1(a).$$

Neglecting the higher harmonics, we write the equation as

$$\ddot{x} + (4I_1(a)/a)x = 0.$$

The frequency ω is therefore given by

$$\omega = 2\sqrt{(I_1(a)/a)}$$

and the period T by

$$T = \pi\sqrt{(a/I_1(a))}.$$

Note that a is unrestricted in these equations—it can be verified that $x = 0$ is a centre for the original equation (see Exercise 21, Chapter 1).

Exercises

1. By transforming to polar coordinates, find the limit cycles of the systems
(i) $\dot{x} = y + x(1 - x^2 - y^2)$, $\dot{y} = -x + y(1 - x^2 - y^2)$;
(ii) $\dot{x} = (x^2 + y^2 - 1)x - y\sqrt{(x^2 + y^2)}$, $\dot{y} = (x^2 + y^2 - 1)y + x\sqrt{(x^2 + y^2)}$;
and investigate their stability.

2. Consider the system

$$\dot{x} = y + xf(r^2), \quad \dot{y} = -x + yf(r^2),$$

where $r^2 = x^2 + y^2$ and $f(u)$ is continuous on $u \geqslant 0$. Show that r satisfies

$$\frac{d(r^2)}{dt} = 2r^2 f(r^2).$$

If $f(r^2)$ has n zeros, at $r = r_k, k = 1, 2, \ldots, n$, how many periodic solutions has the system? Discuss their stability in terms of the sign of $f'(r_k^2)$.

3. Apply the energy balance method of Section 4.1 to each of the following equations, and find the amplitude and stability of any limit cycles:

   (i) $\ddot{x} + \varepsilon(x^2 + \dot{x}^2 - 1)\dot{x} + x = 0$;

  (ii) $\ddot{x} + \varepsilon(\frac{1}{3}\dot{x}^3 - \dot{x}) + x = 0$;

 (iii) $\ddot{x} + \varepsilon(x^4 - 1)\dot{x} + x = 0$;

 (iv) $\ddot{x} + \varepsilon \sin(x^2 + \dot{x}^2) \operatorname{sgn}(\dot{x}) + x = 0$;

  (v) $\ddot{x} + \varepsilon(|x| - 1)\dot{x} + x = 0$;

 (vi) $\ddot{x} + \varepsilon(\dot{x} - 3)(\dot{x} + 1)\dot{x} + x = 0$;

(vii) $\ddot{x} + \varepsilon(x - 3)(x + 1)\dot{x} + x = 0$.

4. For the equation $\ddot{x} + \varepsilon(x^2 + \dot{x}^2 - 4)\dot{x} + x = 0$, the solution $x = 2\cos t$ is a limit cycle. Test its stability, using the method of Section 4.1, and obtain an approximation to the paths close to the limit cycle by the method of slowly-varying amplitude. Show that the period of the limit cycle is equal to 2π, with error of order ε^2.

5. For the equation $\ddot{x} + \varepsilon(|x| - 1)\dot{x} + x = 0$, find approximately the amplitude of the limit cycle and its period, and the polar equations for the phase paths near the limit cycle.

6. Repeat Exercise 5 with Rayleigh's equation, $\ddot{x} + \varepsilon(\frac{1}{3}\dot{x}^3 - \dot{x}) + x = 0$.

7. Find approximately the radius of the limit cycle, and its period, for the equation

$$\ddot{x} + \varepsilon(x^2 - 1)\dot{x} + x - \varepsilon x^3 = 0.$$

8. Show that the frequency–amplitude relation for the pendulum equation, $\ddot{x} + \sin x = 0$, is $\omega^2 = 2J_1(a)/a$, using the methods of Section 4.4 or 4.5. (J_1 is the Bessel function of order 1, with representations

$$J_1(a) = \frac{2}{\pi} \int_0^{\frac{1}{2}\pi} \sin(a\cos u)\cos u \, du = \sum_{n=0}^{\infty} \frac{(-1)^n (\frac{1}{2}a)^{2n+1}}{n!(n+1)!}.\bigg)$$

Show that, for small amplitudes,

$$\omega = 1 - \tfrac{1}{16}a^2.$$

9. In the equation

$$\ddot{x} + \varepsilon h(x, \dot{x}) + g(x) = 0$$

suppose that $g(0) = 0$, and that in some interval $|x| < \delta$, g is continuous and strictly increasing. Show that the origin for the equation $\ddot{x} + g(x) = 0$ is a centre. Let $\zeta(t, a)$ represent its periodic solutions near the origin, where a is a parameter which distinguishes the solutions, say the amplitude. Also, let $T(a)$ be the corresponding period.

By using an energy balance argument show that the periodic solutions of the original equation satisfy

$$\int_0^{T(a)} h(\zeta, \dot{\zeta})\dot{\zeta}\, dt = 0.$$

Apply this equation to obtain the amplitude of the limit cycle of the equation

$$\ddot{x} + \varepsilon(x^2 - 1)\dot{x} + \operatorname{sgn} x = 0.$$

(Confirm that the function ζ defined by

$$\zeta(t, a) = \begin{cases} \sqrt{(2a)}t + \tfrac{1}{2}t^2, & -2\sqrt{(2a)} \leqslant t \leqslant 0; \\ \sqrt{(2a)}t - \tfrac{1}{2}t^2, & 0 \leqslant t \leqslant 2\sqrt{(2a)}; \end{cases}$$

$$\zeta(t + 4\sqrt{(2a)}, a) = \zeta(t, a), \quad -\infty < t < \infty$$

is a periodic solution of $\ddot{x} + \operatorname{sgn} x = 0$ with amplitude a.)

10. For the following equations, show that, for small ε, the amplitude $a(t)$ satisfies approximately the equation given.
   (i) $\ddot{x} + \varepsilon(x^4 - 2)\dot{x} + x = 0$, $\quad 16\dot{a} = -\varepsilon a(a^4 - 16)$;
   (ii) $\ddot{x} + \varepsilon \sin(x^2 + \dot{x}^2)\operatorname{sgn}(\dot{x}) + x = 0$, $\quad \pi\dot{a} = -\varepsilon 2a \sin(a^2)$;
   (iii) $\ddot{x} + \varepsilon(x^2 - 1)\dot{x}^3 + x = 0$, $\quad 16\dot{a} = -\varepsilon a^3(a^2 - 6)$.

11. Verify that the equation

$$\ddot{x} + \varepsilon h(x^2 + \dot{x}^2 - 1)\dot{x} + x = 0$$

where f is continuous, $h(u) < 0$ for $u < 0$, $h(0) = 0$, and $h(u) > 0$ for $u > 0$, has the periodic solutions $x = \cos(t + \alpha)$ for any α. Using the method of slowly varying amplitude show that this solution is a stable limit cycle when $\varepsilon > 0$.

12. Find, by the method of Section 4.5 the equivalent linear equation for

$$\ddot{x} + \varepsilon(x^2 + \dot{x}^2 - 1)\dot{x} + x = 0.$$

Show that it gives the limit cycle exactly. Obtain from the linear equation the equations of the nearby spiral paths, and compare these (to order ε) with those predicted by the method of slowly-varying amplitude.

13. Use the method of equivalent linearization to find the amplitude and frequency of the limit cycle of the equation

$$\ddot{x} + \varepsilon(|x| - 1)|\dot{x}|\dot{x} + x + \varepsilon x^3 = 0.$$

14. The equation $\ddot{x} + x^3 = 0$ has a centre at the origin of the phase plane.
(i) Substitute $x = a \cos \omega t$ to find by the harmonic balance method the frequency–amplitude relation $\omega = \sqrt{3}a/2$.
(ii) Construct, by the method of equivalent linearization, the associated linear equation, and show how the processes (i) and (ii) are equivalent.
(iii) Compare the linearization obtained in (ii) with that obtained by fitting x^3 on $[-a, a]$ by a linear function, $x^3 \approx \beta x$, by the method of least squares, and compare the amplitude–frequency relation. Explain why such a linearization is not likely to be so reliable as that obtained in (ii).

15. The displacement x of a relativistic oscillator satisfies

$$m_0 \ddot{x} + k(1 - (\dot{x}/c)^2)^{3/2} x = 0.$$

Show that the equation becomes $\ddot{x} + (\alpha/a)x = 0$ when linearized with respect to the approximate solution $x = a \cos \omega t$ by the method of equivalent linearization, where

$$\alpha = \frac{1}{\pi} \int_0^{2\pi} \frac{ka}{m_0} \cos^2\theta \left(1 - \frac{a^2 \omega^2}{c^2} \sin^2\theta\right)^{3/2} d\theta.$$

Confirm that, when $a^2 \omega^2/c^2$ is small, the period of the oscillations is given approximately by

$$2\pi \sqrt{\left(\frac{m_0}{k}\right)\left(1 + \frac{3a^2 k}{16 m_0 c^2}\right)}.$$

16. Show that the phase paths of the equation

$$\ddot{x} + (x^2 + \dot{x}^2)x = 0$$

are given by

$$e^{-x^2}(y^2 + x^2 - 1) = \text{constant}.$$

Show that the surface $e^{-x^2}(y^2 + x^2 - 1) = z$ has a maximum at the origin, and deduce that the origin is a centre.

Use the method of harmonic balance to obtain the frequency–amplitude relation $\omega^2 = 3a^2/(4 - a^2)$ for $a < 2$, assuming solutions of the approximate form $a \cos \omega t$. Verify that $\cos t$ is an exact solution, and that $\omega = 1$, $a = 1$ is predicted by harmonic balance.

Plot some exact phase paths to indicate where the harmonic balance method is likely to be unreliable.

17. Show, by the method of harmonic balance, that the frequency–amplitude relation for the periodic solutions of the approximate form $a \cos \omega t$, for

$$\ddot{x} - x + \alpha x^3 = 0, \quad \alpha > 0,$$

is $\omega^2 = 3\alpha a^2 - 1$.

By analysing the phase diagram, explain the lower bound $2/\sqrt{(3\alpha)}$ for the amplitude of periodic motion. Find the equation of the phase paths, and compare where the separatrix cuts the x-axis, with the amplitude $2/\sqrt{(3\alpha)}$.

18. Apply the method of harmonic balance to the equation

$$\ddot{x} + x - \alpha x^2 = 0, \quad \alpha > 0,$$

using the approximate form of solution $x = c + a \cos \omega t$ to show that, for α small,

$$\omega^2 = 1 - 2\alpha c, \quad c = \frac{1}{2\alpha} - \frac{1}{2\alpha}\sqrt{(1 - 2\alpha^2 a^2)}.$$

Deduce the frequency–amplitude relation

$$\omega = (1 - 2\alpha^2 a^2)^{1/4}, \quad a < 1/\sqrt{2\alpha}.$$

Explain, in general terms, why an upper bound on the amplitude is to be expected.

19. Apply the method of harmonic balance to the equation

$$\ddot{x} - x + x^3 = 0$$

in the neighbourhood of the centre at $x = 1$, using the approximate form of solution $x = 1 + c + a \cos \omega t$. Deduce that the mean displacement, frequency, and amplitude are related by

$$\omega^2 = 3c^2 + 6c + 2 + \tfrac{3}{4}a^2, \quad 2c^3 + 6c^2 + c(4 + 3a^2) + 3a^2 = 0.$$

20. Consider the van der Pol equation with nonlinear restoring force

$$\ddot{x} + \varepsilon(x^2 - 1)\dot{x} + x - \alpha x^2 = 0,$$

where ε and α are small. By assuming solutions approximately of the form $x = c + a \cos \omega t$, show that the mean displacement, frequency, and amplitude are related by

$$c \approx 4\alpha, \quad \omega \approx 1 - 4\alpha^2, \quad a \approx 2(1 - 8\alpha^2).$$

21. Suppose that the nonlinear system

$$\dot{\mathbf{x}} = \mathbf{p}(\mathbf{x}), \quad \mathbf{x} = \begin{pmatrix} x \\ y \end{pmatrix},$$

has an isolated equilibrium point at $\mathbf{x} = \mathbf{0}$, and that solutions exist which are approximately of the form

$$\tilde{\mathbf{x}} = B\begin{pmatrix} \cos \omega t \\ \sin \omega t \end{pmatrix}, \quad B = \begin{pmatrix} a & b \\ c & d \end{pmatrix}.$$

Adapt the method of equivalent linearization to this problem by approximating $\mathbf{p}(\tilde{\mathbf{x}})$ by its first harmonic terms:

$$\mathbf{p}\{\tilde{\mathbf{x}}(t)\} = C\begin{pmatrix} \cos \omega t \\ \sin \omega t \end{pmatrix},$$

where C is a matrix of the Fourier coefficients. It is assumed that

$$\int_0^{2\pi/\omega} \mathbf{p}\{\tilde{\mathbf{x}}(t)\}\,\mathrm{d}t = \mathbf{0}.$$

Substitute in the system to show that

$$BU = C, \quad U = \begin{pmatrix} 0 & -\omega \\ \omega & 0 \end{pmatrix}.$$

Deduce that the equivalent linear system is

$$\dot{\mathbf{x}} = BUB^{-1}\mathbf{x} = CUC^{-1}\mathbf{x},$$

when B and C are nonsingular.

22. Use the method of Exercise 21 to construct a linear system equivalent to the van der Pol equation

$$\dot{x} = y, \quad \dot{y} = -x - \varepsilon(x^2 - 1)y.$$

23. Apply the method of Exercise 21 to construct a linear system equivalent to

$$\begin{pmatrix} \dot{x} \\ \dot{y} \end{pmatrix} = \begin{pmatrix} \varepsilon & 1 \\ -1 & \varepsilon \end{pmatrix}\begin{pmatrix} x \\ y \end{pmatrix} + \begin{pmatrix} 0 \\ -\varepsilon x^2 y \end{pmatrix},$$

and show that there exists a limit cycle on which the amplitudes α and β of x and y are given by $\alpha \approx \sqrt{8}$, $\beta \approx \alpha$ for small ε.

24. Apply the method of Exercise 21 to the predator–prey equation (see Section 2.2)

$$\dot{x} = x - xy, \quad \dot{y} = -y + xy,$$

in the neighbourhood of the equilibrium point $(1,1)$, by using the displaced solutions

$$x = 1 + \lambda + a\cos \omega t + b\sin \omega t, \quad y = 1 + \mu + c\cos \omega t + d\sin \omega t.$$

Show that, for small amplitudes,
  (i) the amplitudes of x and y are equal;
  (ii) $\lambda = \mu = \frac{1}{2}(-1 + \sqrt{(1 + 2\alpha^2)})$ where α is the amplitude;
(iii) $\omega = (1 + 2\alpha^2)^{1/4}$.

25. Show that the amplitude–frequency relation for oscillations of the equation

$$\ddot{x} = x^2 - x^3$$

in the neighbourhood of $x = 1$ is $\omega^2 = 1 - \frac{13}{4}a^2$. (Take as the approximate solutions the form $x = c + a\cos \omega t$.)

26. Use the method of Section 4.2 to obtain approximate solutions of the equation

$$\ddot{x} + \varepsilon\dot{x}^3 + x = 0, \quad |\varepsilon| \ll 1.$$

27. Suppose that the equation $\ddot{x} + f(x)\dot{x} + g(x) = 0$ has a periodic solution. Represent the equation in the phase plane given by

$$\dot{x} = y - F(x), \quad \dot{y} = -g(x)$$

where

$$F(x) = \int_0^x f(u)\,du$$

(the Liénard plane). Let

$$v(x, y) = \tfrac{1}{2}y^2 + \int_0^x g(u)\,du,$$

and by considering du/dt on the closed path $\mathscr{C}$ show that

$$\int_{\mathscr{C}} F(x)\,dy = 0$$

(Liénard's criterion).

Deduce that on the assumption that van der Pol's equation $\ddot{x} + \varepsilon(x^2 - 1)\dot{x} + x = 0$ has a periodic solution approximately of the form $x = A\cos t$, then $A \approx 2$.

5 Perturbation methods

IN THE PREVIOUS chapter approximations to periodic solutions of differential equations were achieved by in effect fitting a harmonic oscillation to the required solution. The methods of this chapter are for use when the nonlinearity is small, and generate—in principle—complete representations of periodic solutions of certain types of both autonomous and forced equations as power series in a small parameter. All the essentials are present in treating the pendulum equation, upon which much of this chapter is based.

5.1. Outline of the direct method

The equation for forced oscillations of a pendulum is

$$\ddot{x} + \omega_0^2 \sin x = F \cos \omega t, \tag{5.1}$$

where x is the inclination, $\omega_0 > 0$ is the natural circular frequency for small-amplitude swings, and $F \cos \omega t$ is the forcing term. We can suppose without loss of generality that $\omega > 0$, also that $F > 0$, since $F < 0$ implies a phase difference which can be eliminated by change of time origin and an appropriate modification of initial conditions.

We put $\sin x \approx x - \frac{1}{6}x^3$ to allow for moderately large swings, which is good to 1% for $|x| < 1$ radian. Equation (5.1) then becomes

$$\ddot{x} + \omega_0^2 x - \frac{1}{6}\omega_0^2 x^3 = F \cos \omega t, \tag{5.2}$$

which is nonlinear. We shall see that the effect of the nonlinearity is not confined to occasions when large swings actually occur.

Standardize the form of (5.2) by writing

$$\tau = \omega t, \quad \Omega^2 = \omega_0^2/\omega^2 \quad (\Omega > 0), \quad \Gamma = F/\omega^2. \tag{5.3}$$

Then

$$x'' + \Omega^2 x - \frac{1}{6}\Omega^2 x^3 = \Gamma \cos \tau, \tag{5.4}$$

where the dashes represent differentiation with respect to τ. The argument depends on the nonlinear terms being in some sense small (which does not mean that the resulting phenomena are not spectacular): here we assume that $\frac{1}{6}\Omega^2$ is 'small' and write

$$\tfrac{1}{6}\Omega^2 = \varepsilon_0. \tag{5.5}$$

The equation becomes

$$x'' + \Omega^2 x - \varepsilon_0 x^3 = \Gamma \cos \tau. \tag{5.6}$$

Instead of taking (5.6) as it stands, with ω, Γ, ε_0 as constants, we consider the *family* of differential equations

$$x'' + \Omega^2 x - \varepsilon x^3 = \Gamma \cos \tau, \tag{5.7}$$

where ε is a *parameter occupying an interval* I_ε which includes $\varepsilon = 0$. When $\varepsilon = \varepsilon_0$ we recover (5.6), and when $\varepsilon = 0$ we obtain the *linearized equation corresponding to the family* (5.7):

$$x'' + \Omega^2 x = \Gamma \cos \tau. \tag{5.8}$$

The solutions of (5.7) are now to be thought of as functions of both ε and τ and we will write $x(\varepsilon, \tau)$.

The most elementary version of the *perturbation method* is to attempt a representation of the solutions of (5.7) in the form of a power series in ε:

$$x(\varepsilon, \tau) = x_0(\tau) + \varepsilon x_1(\tau) + \varepsilon^2 x_2(\tau) + \dots, \tag{5.9}$$

whose coefficients $x_i(\tau)$ are functions only of τ. To form equations for $x_i(\tau)$, $i = 0, 1, 2, \dots$, substitute the series (5.9) into eqn (5.7):

$$(x_0'' + \varepsilon x_1'' + \dots) + \Omega^2 (x_0 + \varepsilon x_1 + \dots) - \varepsilon (x_0 + \varepsilon x_1 + \dots)^3 = \Gamma \cos \tau.$$

Since this is assumed to hold for *every member* of the family (5.7), that is for *every* ε on I_ε the coefficients of powers of ε must balance and we obtain

$$x_0'' + \Omega^2 x_0 = \Gamma \cos \tau, \tag{5.10a}$$

$$x_1'' + \Omega^2 x_1 = x_0^3, \tag{5.10b}$$

$$x_2'' + \Omega^2 x_2 = 3x_0^2 x_1, \tag{5.10c}$$

and so on.

We shall be concerned only with *periodic solutions having the period, 2π, of the forcing term.* (There are periodic solutions having other periods: see Chapter 7; also, by a modification of the present procedure as in Section 5.3, we shall find yet more solutions having period 2π.) Then, for all ε on I_ε and for all τ,

$$x(\varepsilon, \tau + 2\pi) = x(\varepsilon, \tau) \tag{5.11}$$

which, by (5.9), implies that for all τ

$$x_i(\tau + 2\pi) = x_i(\tau), \quad i = 0, 1, 2, \dots. \tag{5.12}$$

Equations (5.10) and (5.12) provide the solutions required. The details will be worked out in Section 5.2; for the present, note that (5.10a) is the same as the 'linearized equation' (5.8): necessarily so, since putting $\varepsilon_0 = 0$ in (5.6) implies putting $\varepsilon = 0$ in (5.9). The major term in (5.9) is therefore a periodic solution of the linearized equation (5.8). It is therefore clear that this process restricts us to finding the solutions of the nonlinear equation which are close to (or branch from, or bifurcate from) the solution of the linearized equation. The method will not expose any other periodic solutions. The zero-order solution $x_0(\tau)$ is known as the *generating solution* for the family of equations (5.7).

5.2. Forced oscillations far from resonance

Suppose that

$$\Omega \neq \text{an odd integer.} \tag{5.13}$$

We now solve (5.10) subject to the periodicity condition (5.12). The solutions of (5.10a) are

$$a_0 \cos \Omega\tau + b_0 \sin \Omega\tau + \frac{\Gamma}{\Omega^2 - 1} \cos \tau, \tag{5.14}$$

where a_0, b_0 are constants. If Ω is not an integer *the only solution having period* 2π *is obtained by putting* $a_0 = b_0 = 0$ *in* (5.14); that is

$$x_0(\tau) = \frac{\Gamma}{\Omega^2 - 1} \cos \tau. \tag{5.15}$$

(For Ω even the following argument must be modified.) Equation (5.10b) then becomes

$$x_1'' + \Omega^2 x_1 = \left(\frac{\Gamma}{\Omega^2 - 1}\right)^3 \cos^3 \tau = \frac{\Gamma^3}{(\Omega^2 - 1)^3} (\tfrac{3}{4}\cos \tau + \tfrac{1}{4}\cos 3\tau).$$

The only solution with period 2π is given by

$$x_1(\tau) = \frac{3}{4} \frac{\Gamma^3}{(\Omega^2 - 1)^4} \cos \tau + \frac{1}{4} \frac{\Gamma^3}{(\Omega^2 - 1)^3(\Omega^2 - 9)} \cos 3\tau \tag{5.16}$$

for reasons similar to those above.

The first two terms of the expansion (5.9) give

$$x(\varepsilon, \tau) = \frac{\Gamma}{\Omega^2 - 1} \cos \tau + \varepsilon \left(\frac{3}{4} \frac{\Gamma^3}{(\Omega^2 - 1)^4} \cos \tau \right.$$

$$\left. + \frac{1}{4} \frac{\Gamma^3}{(\Omega^2 - 1)^3(\Omega^2 - 9)} \cos 3\tau \right) + O(\varepsilon^2). \tag{5.17}$$

The series continues with terms of order ε^2, ε^3 and involves the harmonics $\cos 5\tau$, $\cos 7\tau$, and so on. For the pendulum, $\varepsilon = \varepsilon_0 = \frac{1}{6}\Omega^2$, by (5.5).

The method obviously fails if Ω^2 takes one of the values $1, 9, 25, \ldots$, since certain terms would then be infinite, and this possibility is averted by condition (5.13). However, the series would not converge well if Ω^2 were even close to one of these values, and so a few terms of the series would describe $x(\varepsilon, \tau)$ only poorly. Such values of Ω correspond to conditions of *near-resonance*. $\Omega \approx 1$ is such a case; not surprisingly, since it is a value close to resonance for the linearized equation. The other critical values correspond to resonances peculiar to the nonlinear equation caused by the higher harmonics present in $x^3(\varepsilon, \tau)$, which can be regarded as 'feeding-back' into the linear equation like forcing terms. This idea is explored in Chapter 7.

Example 5.1 *Obtain an approximation to the forced response, of period 2π, for the equation $x'' + \frac{1}{4}x + 0.1x^3 = \cos \tau$.*

Consider the family $x'' + \frac{1}{4}x + \varepsilon x^3 = \cos \tau$. Assume that $x(\varepsilon, \tau) = x_0(\tau) + \varepsilon x_1(\tau)$ $+ \ldots$. The periodicity requires $x_0(\tau), x_1(\tau), \ldots$, to have period 2π. The equations for x_0, x_1 are

$$x_0'' + \frac{1}{4}x_0 = \cos \tau,$$

$$x_1'' + \frac{1}{4}x_1 = -x_0^3.$$

The only periodic solution of the first equation is

$$x_0(\tau) = -\frac{4}{3}\cos \tau,$$

and the second becomes

$$x_1'' + \frac{1}{4}x_1 = \frac{16}{9}\cos \tau + \frac{16}{27}\cos 3\tau.$$

The periodic solution is

$$x_1(\tau) = -\frac{64}{27}\cos \tau - \frac{64}{945}\cos 3\tau.$$

Therefore

$$x(\varepsilon, \tau) = -\frac{4}{3}\cos \tau - \varepsilon(\frac{64}{27}\cos \tau + \frac{64}{945}\cos 3\tau) + O(\varepsilon^2).$$

With $\varepsilon = 0.1$,

$$x(\varepsilon, \tau) \approx -1.570\cos \tau - 0.006\cos 3\tau.$$

5.3. Forced oscillations near resonance with weak excitation

Not only does the method of the previous sections fail to give a good representation near critical values of Ω, but it does not reveal all the solutions having the period 2π of the forcing term. The only solution it can detect is the one which bifurcates from the one-and-only solution having period 2π of the linear eqn (5.8). More solutions can be found by, so to speak, offering the nonlinear equation a greater range of choice of possible generating solutions by modifying the structure so that a different linearized equation is obtained.

We will generalize the pendulum equation to include small damping:

$$\ddot{x} + k\dot{x} + \omega_0^2 x - \tfrac{1}{6}\omega_0^2 x^3 = F \cos \omega t. \tag{5.18}$$

Corresponding to eqn (5.4) we have

$$x'' + Kx' + \Omega^2 x - \varepsilon_0 x^3 = \Gamma \cos \tau$$

where

$$\tau = \omega t, \quad \Omega^2 = \omega_0^2/\omega^2, \quad \varepsilon_0 = \tfrac{1}{6}\Omega^2, \quad K = k/\omega, \quad \Gamma = F/\omega^2. \tag{5.19}$$

Assume that Γ is small (weak excitation), and K is small (small damping) and therefore put

$$\Gamma = \varepsilon_0 \gamma, \quad K = \varepsilon_0 \kappa, \quad (\gamma, \kappa > 0). \tag{5.20}$$

Suppose also that Ω is close to one of the critical values $1, 3, 5, \ldots$ say that $\Omega \approx 1$ (near to resonance), so we write

$$\Omega^2 = 1 + \varepsilon_0 \beta. \tag{5.21}$$

Equation (5.18) becomes

$$x'' + x = \varepsilon_0(\gamma \cos \tau - \kappa x' - \beta x + x^3). \tag{5.22}$$

Now consider the family of equations

$$x'' + x = \varepsilon(\gamma \cos \tau - \kappa x' - \beta x + x^3), \tag{5.23}$$

with parameter ε, in which γ, κ, β retain the constant values given by (5.20) and (5.21). When $\varepsilon = \varepsilon_0$ we return to (5.22). When $\varepsilon = 0$ we obtain the new linearized equation $x'' + x = 0$. *This has infinitely many solutions with period 2π, offering a wide choice of generating solutions* (compare (5.8), in which only one presents itself).

Now assume that $x(\varepsilon, \tau)$ may be expanded in the form

$$x(\varepsilon, \tau) = x_0(\tau) + \varepsilon x_1(\tau) + \varepsilon^2 x_2(\tau) + \ldots, \tag{5.24}$$

where (by the same argument as led up to (5.12)), for all τ

$$x_i(\tau + 2\pi) = x_i(\tau), \quad i = 0, 1, 2, \ldots. \tag{5.25}$$

Substitute (5.24) into (5.23). By the argument leading to (5.10) we have

$$x_0'' + x_0 = 0, \tag{5.26a}$$

$$x_1'' + x_1 = \gamma \cos \tau - \kappa x_0' - \beta x_0 + x_0^3, \tag{5.26b}$$

$$x_2'' + x_2 = -x_0 x_1' - \beta x_1 + 3x_0^2 x_1, \tag{5.26c}$$

and so on.

The solution of (5.26a) is

$$x_0(\tau) = a_0 \cos \tau + b_0 \sin \tau \tag{5.27}$$

for every a_0, b_0. Now put (5.27) into (5.26b). After writing

$$\cos^3 \tau = \tfrac{3}{4} \cos \tau + \tfrac{1}{4} \cos 3\tau, \quad \sin^3 \tau = \tfrac{3}{4} \sin \tau - \tfrac{1}{4} \sin 3\tau$$

and making other similar reductions† we have

$$\begin{aligned}
x_1'' + x_1 &= \{\gamma - \kappa b_0 + a_0[-\beta + \tfrac{3}{4}(a_0^2 + b_0^2)]\} \cos \tau \\
&\quad + \{\kappa a_0 + b_0[-\beta + \tfrac{3}{4}(a_0^2 + b_0^2)]\} \sin \tau \\
&\quad + \tfrac{1}{4} a_0(a_0^2 - 3b_0^2) \cos 3\tau + \tfrac{1}{4} b_0(3a_0^2 - b_0^2) \sin 3\tau.
\end{aligned} \tag{5.28}$$

This is the crucial stage: the solution $x_1(\tau)$ is required to have period 2π, but *unless the coefficients of* cos τ *and* sin τ *in* (5.28) *are zero, there are no periodic solutions, since any solution would contain terms of the form* $\tau \cos \tau, \tau \sin \tau$. Such non-periodic or otherwise undesirable constituents of a solution are usually called secular terms. We eliminate the secular terms by *requiring* the coefficients of cos τ and sin τ to be zero, which decides the acceptable values of a_0, b_0:

$$\kappa a_0 - b_0\{\beta - \tfrac{3}{4}(a_0^2 + b_0^2)\} = 0, \tag{5.29a}$$

$$\kappa b_0 + a_0\{\beta - \tfrac{3}{4}(a_0^2 + b_0^2)\} = \gamma. \tag{5.29b}$$

These equations settle the values of a_0 and b_0 as follows. Let r_0 be the amplitude of the generating solution:

$$r_0 = \sqrt{(a_0^2 + b_0^2)} > 0. \tag{5.30}$$

Square and add eqns (5.29a) and (5.29b), obtaining

$$r_0^2\{\kappa^2 + (\beta - \tfrac{3}{4}r_0^2)^2\} = \gamma^2. \tag{5.31}$$

---

† Often, the easiest way to carry out such calculations is to write the solution of (5.26a) as $A_0 e^{i\tau} + \bar{A}_0 e^{-i\tau}$. Then

$$x_0^3 = (A_0 e^{i\tau} + \bar{A}_0 e^{-i\tau})^3 = A_0^3 e^{3i\tau} + 3A_0^2 \bar{A}_0 e^{i\tau} + \text{complex conjugate}.$$

The substitution $A_0 = \tfrac{1}{2}(a_0 - ib_0)$, a_0 real, b_0 real, is left to the very end.

When (5.31) is solved for r_0, a_0 and b_0 can be obtained from (5.29) and (5.30).

Equation (5.31) will be analysed in the next section. For the present, note that there may be as many as three positive values of r_0^2 (hence of $r_0 > 0$) satisfying (5.31). This indicates that for certain values of κ, β, γ there may be three distinct solutions of (5.18), (5.22), or (5.23), each bifurcating from one of three distinct generating solutions.

Having selected a pair of values a_0 and b_0 satisfying (5.29), we solve (5.28) to give

$$x_1(\tau) = a_1 \cos \tau + b_1 \sin \tau - \tfrac{1}{32} a_0 (a_0^2 - 3b_0^2) \cos 3\tau - \tfrac{1}{32} b_0 (3a_0^2 - b_0^2) \sin 3\tau,$$

where a_1 and b_1 are any constants. This expression is substituted into eqn (5.26c); the requirement that $x_2(\tau)$ should have period 2π provides equations determining a_1, b_1, as before. In this and subsequent stages the determining equations for a_i, b_i are linear, so no further multiplicity of solutions is introduced.

5.4. The amplitude equation for the undamped pendulum

Suppose that the damping coefficient is zero: in (5.18), (5.19), (5.23)

$$k = K = \kappa = 0. \tag{5.32}$$

Instead of seeking r_0 through (5.31), the coefficients a_0, b_0 can be found directly from (5.29): the only solutions are given by

$$b_0 = 0 \tag{5.33a}$$

$$a_0(\beta - \tfrac{3}{4}a_0^2) = \gamma. \tag{5.33b}$$

We shall consider in detail only the pendulum case. The original parameters ω, ω_0, and F of eqn (5.18) can be restored through (5.3), (5.20), and (5.21). Equation (5.33b) becomes

$$a_0(\omega^2 - \omega_0^2 + \tfrac{1}{8}\omega_0^2 a_0^2) = -F. \tag{5.34}$$

The solutions a_0 can be obtained by drawing a cubic curve

$$z = a_0(\omega^2 - \omega_0^2 + \tfrac{1}{8}\omega_0^2 a_0^2) = f(a_0) \tag{5.35}$$

on a graph with axes a_0, z for each fixed value of ω and ω_0, then finding the intersections with the lines $z = -F$ ($F > 0$) for various F, as in Fig. 5.1.

The main features revealed are as follows.

FIG. 5.1. Illustrating eqn (5.34)

(i) When $\omega^2 > \omega_0^2$ (Fig. 5.1(c)), there is exactly one periodic oscillation possible. When F is small, this approximates to the linear response (5.15), and to the 'corrected linear response' (5.17), unless $\omega^2 \approx \omega_0^2$ (very near to resonance) in which case it is considerably different (Fig. 5.1(b)). These responses are 180° out of phase with the forcing term.

(ii) When $\omega^2 < \omega_0^2$ (Fig. 5.1(a)), there is a single response, 180° out of phase, when F is relatively large. When F is smaller there are three distinct periodic responses, two in phase and one out of phase with the forcing term. The response marked 'A' in Fig. 5.1(a) corresponds to the response (5.15) of the linearized equation, and to the corrected linear response (5.17).

(iii) All three of the responses described in (ii) will be of small amplitude if the intersections of the curve $z = f(a_0)$ with the a_0 axis (Fig. 5.1(a)) occur close to the origin. These intersections are at $a_0 = 0$, $a_0 = \pm 2\sqrt{2}(1 - \omega^2/\omega_0^2)^{1/2}$. Three intersections are actually achieved provided $F < \frac{4}{3}\sqrt{\frac{2}{3}}\omega_0^2(1 - \omega^2/\omega_0^2)^{3/2}$ (recalling that F is positive). Therefore, by choosing ω^2/ω_0^2 sufficiently near to 1 (near to resonance) and F correspondingly small, the amplitudes of all the three responses can be made as small as we wish. In particular, they can all be confined in what we should normally regard as the linear range of amplitude for the pendulum.

(iv) Despite there being no damping, there are steady, bounded oscillations even when $\omega = \omega_0$ (unlike the linearized case). The non-linearity controls the amplitude in the following way. Amplitude increases indefinitely if the forcing term remains in step with the natural oscillation and reinforces it cycle by cycle. However (e.g. Example 4.10), the frequency of the natural oscillation varies with amplitude due to the nonlinearity and does not remain in step with the forcing term.

(v) Whether a steady oscillation is set up or approached at all, and if it is, which of the possible modes is adopted, depends on the initial conditions of the problem, which are not considered at all here (see Chapter 7).

(vi) Whether a particular mode can be sustained in practice depends on its stability, of which an indication can be got as follows. If, in the neighbourhood of amplitude a_0, the forcing amplitude required to sustain a_0 increases/decreases as a_0 increases/decreases, we expect a stable solution, in which an accidental small disturbance of amplitude cannot be sustained and amplified. If, however, F increases/decreases as a_0 decreases/increases, the conditions are right for growth of the disturbance and instability results. Further justification is given for this argument in Section 9.5. In anticipation of the analysis of Chapters 7 and 9, the stable and unstable branches are indicated in Figs. 5.1 and 5.2.

The nature of the solutions of (5.34) as a function of the parameters ω, ω_0, with F given, can be exhibited on a single 'response diagram', Fig. 5.2. The figure can be plotted directly by writing (5.34) in the form

$$\omega = \sqrt{\{\omega_0^2(1 - \tfrac{1}{8}a_0^2) - F/a_0\}}.$$

For each value of $F > 0$ the amplitudes lie on two smooth branches, a typical pair being shown as heavy lines on the figure. F increases on 'contours' increasingly distant from the curve $F = 0$ (which gives the amplitude–frequency curve for free oscillations) on both sides of it.

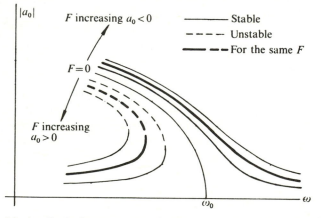

Fig. 5.2. Amplitude–frequency curves for the undamped pendulum, eqn (5.34)

Example 5.2 *Investigate the forced periodic solutions of the equation*

$$x'' + (9 + \varepsilon\beta)x - \varepsilon x^3 = \Gamma \cos \tau$$

where ε is small and β, Γ are not too large.

This is a case where $\Omega^2 = 9 + \varepsilon\beta$ (eqn (5.7)) has a value which causes the direct expansion (5.17) to fail. The given equation may be rewritten

$$x'' + 9x = \Gamma \cos \tau + \varepsilon(x^3 - \beta x).$$

Write $x(\varepsilon, \tau) = x_0(\tau) + \varepsilon x_1(\tau) + \ldots$, where $x_0(\tau)$, $x_1(\tau), \ldots$ have period 2π. Then

$$x_0'' + 9x_0 = \Gamma \cos \tau,$$

$$x_1'' + 9x_1 = x_0^3 - \beta x_0,$$

and so on. These have the same form as (5.10), but since 9 is the square of an integer, the first equation has solutions of period 2π of the form

$$x_0(\tau) = a_0 \cos 3\tau + b_0 \sin 3\tau + \tfrac{1}{8}\Gamma \cos \tau.$$

When this is substituted into the equation for x_1, terms in $\cos 3\tau$, $\sin 3\tau$ emerge on the right-hand side, preventing periodic solutions unless their coefficients are zero. The simplest way to make the substitution is to write

$$x_0(\tau) = A_0 e^{3i\tau} + \bar{A}_0 e^{-3i\tau} + \tfrac{1}{16}\Gamma e^{i\tau} + \tfrac{1}{16}\Gamma e^{-i\tau},$$

where $A_0 = \tfrac{1}{2}a_0 - \tfrac{1}{2}ib_0$. We find

$$x_0^3 - \beta x_0 = \left[\left(\frac{\Gamma^3}{16^3} + \frac{6\Gamma^2}{16^2} A_0 + 3A_0^2\bar{A}_0 - \beta A_0 \right) e^{3i\tau} + \text{complex conjugate} \right]$$
$$+ \text{other harmonics}.$$

Therefore we require

$$A_0 \left(3A_0\bar{A}_0 - \beta + \frac{6\Gamma^2}{16^2} \right) = -\frac{\Gamma^3}{16^3}.$$

This implies that A_0 is real: $b_0 = 0$, $A = \tfrac{1}{2}a_0$, and the equation for a_0 is

$$a_0 \left(\tfrac{3}{4}a_0^2 - \beta + \frac{6\Gamma^2}{16^2} \right) + \frac{\Gamma^3}{16^3} = 0.$$

5.5. The amplitude equation for a damped pendulum

The amplitude equation (5.31) translated into the parameters of eqn (5.18) by (5.19), (5.20), and (5.21), becomes

$$r_0^2\{k^2\omega^2 + (\omega^2 - \omega_0^2 + \tfrac{1}{8}\omega_0^2 r_0^2)^2\} = F^2. \tag{5.36}$$

Only solutions with $r_0 > 0$ are valid (see eqn (5.30)). Solving the quadratic equation for ω^2 given by (5.36) we find that

$$\omega^2 = \tfrac{1}{2}\{2\omega_0^2(1 - \tfrac{1}{8}r_0^2) - k^2 \pm \sqrt{\{k^4 - 4\omega_0^2 k^2(1 - \tfrac{1}{8}r_0^2) + 4F^2/r_0^2\}}\}.$$

FIG. 5.3. Amplitude–frequency curves for the damped pendulum, eqn (5.36).
On each curve F is constant

Typical response curves are shown in Fig. 5.3 for fixed values of k and ω_0, and selected values of $F > 0$. k was chosen fairly 'small' so that comparison might be made with the undamped case shown in Fig. 5.2. The figure is similar to Fig. 5.2, with some important differences. There are no longer two distinct branches corresponding to each value of F: the branches join up to make continuous curves. In the near neighbourhood of the point $\omega = \omega_0$, $r_0 = 0$ the curves do not turn over: that is to say, a certain minimum value of F is required before it is possible to have three alternative forced responses (see Exercise 10). The response curves for small F represent the approximately linear response, as can be seen from (5.31).

The 'jump effect' associated with the response curves is presented in detail in Section 7.3.

5.6. Soft and hard springs

We have so far been concerned with equations of the type

$$\ddot{x} + k\dot{x} + cx + \varepsilon x^3 = F\cos\omega t,$$

which arise from an approximation to the pendulum equation. In the general case this is called a *Duffing's equation* with a forcing term. Now consider

$$\ddot{x} + k\dot{x} + cx + \varepsilon g(x) = F \cos \omega t, \tag{5.37}$$

where $k > 0$, $c > 0$, and $|\varepsilon| \ll 1$. This can be interpreted as the equation of forced motion with damping of a particle on a spring which provides a restoring force which is almost but not quite linear. This is a useful physical model, since it leads us to expect that the features of linear motion will be to some extent preserved; for example, if $k = 0$ and $F = 0$ we guess that solutions for small enough amplitude will be oscillatory, and that the effect of small damping is to reduce the oscillations steadily to zero.

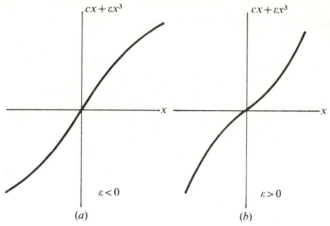

FIG. 5.4. Restoring force functions for (*a*) a soft spring; (*b*) a hard spring

Important classes of cases are represented by the example $g(x) = x^3$, with $\varepsilon > 0$ and $\varepsilon < 0$. The corresponding (symmetrical) restoring forces are illustrated in Figs. 5.4(*a*) and (*b*). When ε is negative the restoring force becomes progressively weaker in extension than for the linear spring: such a spring is called *soft*. A pendulum is modelled by a soft spring system. When ε is positive the spring becomes stiffer and is called a *hard spring*.

The nature of the response diagrams for the forced oscillations, period $2\pi/\omega$, of (5.37) in the case of a soft spring ($\varepsilon < 0$), a linear spring ($\varepsilon = 0$), and a hard spring ($\varepsilon > 0$) are shown in Fig. 5.5.

There are various ways in which the restoring force can be *unsymmetrical*: an important case is shown in Fig. 5.6, where $g(x) = -x^2$ and $\varepsilon > 0$.

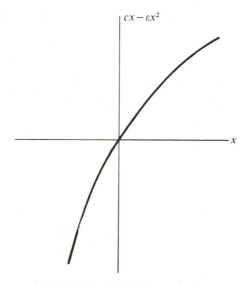

Fig. 5.5. Amplitude–frequency curves for Duffing's equation $\ddot{x} + k\dot{x} + cx + \varepsilon x^3 = F \cos \omega t$, $k > 0, c > 0$

Fig. 5.6. Unsymmetrical restoring force

The spring is 'soft' for $x > 0$ and 'hard' for $x < 0$. Exercise 22 shows that for free oscillations ($F = 0$ in (5.37)) the effect is to shift the centre of oscillation to small positive x values, as the diagram suggests; the same occurs when the oscillation is forced as is shown in the following Example. The phenomenon is called *rectification*.

Example 5.3 *Obtain approximately the solutions of period 2π of the equation* $\ddot{x} + \Omega^2 x - \varepsilon x^2 = \Gamma \cos t, \varepsilon > 0$.

Firstly, suppose that Ω is not close to an integer (far from resonance). Writing

$x(\varepsilon, t) = x_0(t) + \varepsilon x_1(t) + \ldots$, the required sequence of equations is

$$\ddot{x}_0 + \Omega^2 x_0 = \Gamma \cos t,$$

$$\ddot{x}_1 + \Omega^2 x_1 = x_0^2,$$

and so on. The periodic solution of the first is

$$x_0(t) = \frac{\Gamma}{\Omega^2 - 1} \cos t,$$

and the second becomes

$$\ddot{x}_1 + \Omega^2 x_1 = \frac{\Gamma^2}{2(\Omega^2 - 1)^2}(1 + \cos 2t)$$

whose solutions are

$$x_1(t) = \frac{\Gamma^2}{2\Omega^2(\Omega^2 - 1)^2} + \frac{\Gamma^2}{2(\Omega^2 - 1)^2 (\Omega^2 - 4)} \cos 2t + a_1 \cos \Omega t + b_1 \sin \Omega t.$$

For $x_1(t)$ to have period 2π, $a_1 = b_1 = 0$: therefore

$$x(\varepsilon, t) = \frac{\Gamma}{\Omega^2 - 1} \cos t + \frac{\varepsilon \Gamma^2}{2\Omega^2(\Omega^2 - 1)^2} + \frac{\varepsilon \Gamma^2}{2(\Omega^2 - 1)^2 (\Omega^2 - 4)} \cos 2t.$$

Now suppose $\Omega \approx 1$, so put

$$\Omega^2 = 1 + \varepsilon\beta, \quad \Gamma = \varepsilon\gamma.$$

The equation becomes

$$\ddot{x} + x = \varepsilon(\gamma \cos t + x^2 - \beta x).$$

Assume that $x(\varepsilon, \tau) = x_0(\tau) + \varepsilon x_1(\tau) + \ldots$. The first two equations are

$$\ddot{x}_0 + x_0 = 0$$

$$\ddot{x}_1 + x_1 = \gamma \cos t + x_0^2 - \beta x_0.$$

The solutions to the first, of period 2π, are $a_0 \cos t + b_0 \sin t$, and the second becomes

$$\ddot{x}_1 + x_1 = \tfrac{1}{2}(a_0^2 + b_0^2) + (\gamma - \beta a_0)\cos t - \beta b_0 \sin t + \tfrac{1}{2}(a_0^2 - b_0^2)\cos 2t + a_0 b_0 \sin 2t.$$

Secular terms vanish only if

$$b_0 = 0, \quad a_0 = \gamma/\beta.$$

Then

$$x(\varepsilon, t) \approx \frac{\gamma}{\beta} \cos t + \varepsilon\left(\frac{1}{2}\frac{\gamma^2}{\beta^2} - \frac{1}{6}\frac{\gamma^2}{\beta^2}\cos 2t + a_1 \cos t + b_1 \sin t\right).$$

This is exactly the same as the earlier result: only one solution is revealed.

5.7. Amplitude-phase perturbation for the pendulum equation

In Section 5.3 we stopped at the first approximation to a solution because of the rapidly developing complexity of the algebra. The following method allows a higher approximation to be obtained rather more efficiently.

Consider again the family (5.23):

$$x'' + x = \varepsilon(\gamma \cos \tau - \kappa x' - \beta x + x^3). \tag{5.38}$$

Instead of seeking, as in Section 5.3, solutions in effect of the form

$$(a_0 + \varepsilon a_1 + \ldots)\cos \tau + (b_0 + \varepsilon b_1 + \ldots)\sin \tau + \text{higher harmonics}$$

we shall arrange for them to appear in the form

$$x(\varepsilon, \tau) = (r_0 + \varepsilon r_1 + \ldots)\cos(\tau + \alpha_0 + \varepsilon \alpha_1 + \ldots) + \text{higher harmonics}, \tag{5.39}$$

where

$$\alpha = \alpha_0 + \varepsilon \alpha_1 + \ldots$$

is the phase difference between the response and forcing term. α_0 is expected to be the phase of the generating solution and r_0 its amplitude.

It is more convenient for manipulative reasons to have the unknown phase, α, appear in the equation itself. Therefore we discuss instead the equation

$$X'' + X = \varepsilon(\gamma \cos(s - \alpha) - \kappa X' - \beta X + X^3) \tag{5.40}$$

where we have put $s = \tau + \alpha$ and $X(\varepsilon, s) = x(\varepsilon, \tau)$ into (5.38).

Assume that for all small enough ε

$$\begin{aligned} X(\varepsilon, s) &= X_0(s) + \varepsilon X_1(s) + \ldots, \\ \alpha &= \alpha_0 + \varepsilon \alpha_1 + \ldots. \end{aligned} \tag{5.41}$$

We require solutions $X(\varepsilon, s)$ having the period, 2π, of the forcing term, which implies that for all s,

$$X_i(s + 2\pi) = X_i(s), \quad i = 1, 2, \ldots. \tag{5.42}$$

Finally we shall impose the extra condition $X'(\varepsilon, 0) = 0$. This is not a real restriction: we simply adjust the time origin, and hence the phase, so that it is so. Therefore

$$X_i'(0) = 0, \quad i = 0, 1, \ldots. \tag{5.43}$$

Substitute (5.41) into (5.40), writing

$$\cos(s - \alpha) = \cos(s - \alpha_0) + \varepsilon \alpha_1 \sin(s - \alpha_0) + \ldots.$$

By assembling powers of ε and equating their coefficients to zero we obtain

$$X_0'' + X_0 = 0, \qquad (5.44a)$$

$$X_1'' + X_1 = \gamma \cos(s - \alpha_0) - \kappa X_0' - \beta X_0 + X_0^3, \qquad (5.44b)$$

$$X_2'' + X_2 = \gamma \alpha_1 \sin(s - \alpha_0) - \kappa X_1' - \beta X_1 + 3X_0^2 X_1, \qquad (5.44c)$$

and so on.

The periodic solutions of (5.44a) satisfying (5.43) are

$$X_0(s) = r_0 \cos s, \quad r_0 > 0 \qquad (5.45)$$

where we choose $r_0 > 0$ by later adjusting the phase α_0. From (5.44b)

$$X_1'' + X_1 = (\gamma \cos \alpha_0 - \beta r_0 + \tfrac{3}{4} r_0^3) \cos s + (\kappa r_0 + \gamma \sin \alpha_0) \sin s + \tfrac{1}{4} r_0^3 \cos 3s. \qquad (5.46)$$

For there to be a periodic solution the secular terms (in $\cos s$ and $\sin s$) must be eliminated, so

$$\beta r_0 - \tfrac{3}{4} r_0^3 = \gamma \cos \alpha_0, \qquad (5.47a)$$

$$\kappa r_0 = -\gamma \sin \alpha_0. \qquad (5.47b)$$

By squaring and adding we obtain eqn (5.31) again:

$$r_0^2 \{\kappa^2 + (\beta - \tfrac{3}{4} r_0^2)^2\} = \gamma^2. \qquad (5.48)$$

α_0 is then obtainable from (5.47) (considering only $-\tfrac{1}{2}\pi \leqslant \alpha_0 \leqslant \tfrac{1}{2}\pi$):

$$\alpha_0 = -\sin^{-1}(\kappa r_0/\gamma). \qquad (5.49)$$

Equation (5.46) becomes

$$X_1'' + X_1 = \tfrac{1}{4} r_0^3 \cos 3s,$$

with solutions

$$X_1(s) = r_1 \cos s - \tfrac{1}{32} r_0^3 \cos 3s, \quad r_1 > 0, \qquad (5.50)$$

satisfying (5.42) and (5.43). Substitute (5.50) into (5.44c):

$$X_2'' + X_2 = (-\gamma \alpha_1 \sin \alpha_0 - \beta r_1 + \tfrac{9}{4} r_0^2 r_1 - \tfrac{3}{128} r_0^5) \cos s$$
$$+ (\gamma \alpha_1 \cos \alpha_0 + \kappa r_1) \sin s$$
$$+ (\tfrac{3}{4} r_0^2 r_1 - \tfrac{3}{64} r_0^5 + \tfrac{1}{32} \beta r_0^3) \cos 3s - \tfrac{3}{32} \kappa r_0^3 \sin 3s - \tfrac{3}{128} r_0^5 \sin 5s. \qquad (5.51)$$

Periodicity requires that the coefficients of $\cos s, \sin s$ should be zero, which leads to the result

$$r_1 = \tfrac{3}{128}r_0^5/(\tfrac{9}{4}r_0^2 - \beta + \kappa \tan \alpha_0), \tag{5.52a}$$

$$\alpha_1 = -\tfrac{3}{128}\frac{\kappa r_0^5}{\gamma \cos \alpha_0} \bigg/ \cdot (\tfrac{9}{4}r_0^2 - \beta + \kappa \tan \alpha_0). \tag{5.52b}$$

The solution to eqn (5.38) is then

$$x(\varepsilon, \tau) = (r_0 + \varepsilon r_1)\cos(\tau + \alpha_0 + \varepsilon\alpha_1) - \varepsilon r_0^2 \cos 3(\tau + \alpha_0) + O(\varepsilon^2). \tag{5.53}$$

5.8. Periodic solutions of autonomous equations (Lindstedt's method)

Consider the oscillations of the pendulum-type equation

$$\frac{d^2 x}{dt^2} + x - \varepsilon x^3 = 0. \tag{5.54}$$

For a soft spring $\varepsilon > 0$, and for a hard spring $\varepsilon < 0$. The system is conservative, and the method of Section 1.3 can be used to show that all motions of small enough amplitude are periodic.

In this case of unforced vibration, *the frequency, ω, is not known in advance*, but depends on the amplitude. It reduces to 1 when $\varepsilon = 0$. Assume that

$$\omega = 1 + \varepsilon\omega_1 + \dots, \tag{5.55}$$

$$x(\varepsilon, t) = x_0(t) + \varepsilon x_1(t) + \dots. \tag{5.56}$$

We could substitute these into (5.54) and look for solutions of period $2\pi/\omega$, but it is mechanically simpler to cause ω to appear as a multiplier by writing

$$\omega t = \tau. \tag{5.57}$$

Then (5.54) becomes

$$\omega^2 x'' + x - \varepsilon x^3 = 0. \tag{5.58}$$

By this substitution we have replaced eqn (5.54), which has unknown period, by eqn (5.58) which has known period 2π. Therefore, as before, for all τ,

$$x_i(\tau + 2\pi) = x_i(\tau), \quad i = 0, 1, \dots. \tag{5.59}$$

Equation (5.58) becomes

$$(1 + \varepsilon 2\omega_1 + \dots)(x_0'' + \varepsilon x_1'' + \dots) + (x_0 + \varepsilon x_1 + \dots) = \varepsilon(x_0 + \varepsilon x_1 + \dots)^3,$$

and by assembling powers of ε we obtain

$$x_0'' + x_0 = 0, \tag{5.60a}$$

$$x_1'' + x_1 = -2\omega_1 x_0'' + x_0^3, \tag{5.60b}$$

and so on.

To simplify the calculations we can impose the conditions

$$x(\varepsilon, 0) = a_0, \quad x'(\varepsilon, 0) = 0 \tag{5.61}$$

without loss of generality (only in the autonomous case!). This implies that

$$x_0(0) = a_0, \quad x_0'(0) = 0, \tag{5.62a}$$

and

$$x_i(0) = 0, \quad x_i'(0) = 0, \quad i = 1, 2, \ldots. \tag{5.62b}$$

The solution of (5.60a) satisfying (5.62a) is

$$x_0 = a_0 \cos \tau. \tag{5.63}$$

(5.60b) then becomes

$$x_1'' + x_1 = (2\omega_1 a_0 + \tfrac{3}{4}a_0^3) \cos \tau + \tfrac{1}{4}a_0^3 \cos 3\tau. \tag{5.64}$$

The solutions will be periodic only if

$$\omega_1 = -\tfrac{3}{8}a_0^2. \tag{5.65}$$

From (5.64), (5.65),

$$x_1(\tau) = a_1 \cos \tau + b_1 \sin \tau - \tfrac{1}{32}a_0^3 \cos 3\tau,$$

and (5.62b) implies

$$b_1 = 0, \quad a_1 = \tfrac{1}{32}a_0^3.$$

Therefore

$$x_1(\tau) = \tfrac{1}{32}a_0^3(\cos \tau - \cos 3\tau). \tag{5.66}$$

Finally, from (5.63) and (5.66),

$$x(\varepsilon, \tau) \approx a_0 \cos \tau + \tfrac{1}{32}\varepsilon a_0^3(\cos \tau - \cos 3\tau) + O(\varepsilon^2). \tag{5.67}$$

Returning to the variable t (eqn (5.57)), we have the approximation

$$x(\varepsilon, t) \approx a_0 \cos \omega t + \tfrac{1}{32}\varepsilon a_0^3(\cos \omega t - \cos 3\omega t), \tag{5.68a}$$

where

$$\omega \approx 1 - \tfrac{3}{8}\varepsilon a_0^2; \tag{5.68b}$$

this gives the dependence of frequency on amplitude.

5.9. Forced oscillation of a self-excited equation

Consider the van der Pol equation with a forcing term:

$$\ddot{x} + \varepsilon(x^2 - 1)\dot{x} + x = F\cos\omega t. \tag{5.69}$$

The unforced equation has a limit cycle with radius approximately 2 and period approximately 2π (see Sections 4.1 and 4.2). The limit cycle is generated by the balance between internal energy loss and energy generation (see Section 1.5), and the forcing term will alter this balance. If F is 'small' (weak excitation), its effect depends on whether or not ω is close to the natural frequency. If it is, it appears that an oscillation might be generated which is a perturbation of the limit cycle. If F is not small (hard excitation) or if the natural and imposed frequency are not closely similar, we should expect that the 'natural oscillation' will be extinguished, as occurs with the corresponding linear equation.

Firstly, write

$$\omega t = \tau; \tag{5.70}$$

then (5.69) becomes

$$\omega^2 x'' + \varepsilon\omega(x^2 - 1)x' + x = F\cos\tau, \tag{5.71}$$

where the dashes signify differentiation with respect to τ.

Hard excitation, far from resonance. Assume that ω is not close to an integer. In (5.71), let

$$x(\varepsilon, \tau) = x_0(\tau) + \varepsilon x_1(\tau) + \dots. \tag{5.72}$$

The sequence of equations for $x_0, x_1, \dots$ begins

$$\omega^2 x_0'' + x_0 = F\cos\tau, \tag{5.73a}$$

$$\omega^2 x_1'' + x_1 = -(x_0^2 - 1)x_0', \tag{5.73b}$$

$x_0(\tau), x_1(\tau)$ having period 2π.

The only solution of (5.73a) having period 2π is

$$x_0(\tau) = \frac{F}{1 - \omega^2}\cos\tau,$$

and therefore

$$x(\varepsilon, \tau) = \frac{F}{1 - \omega^2}\cos\tau + O(\varepsilon). \tag{5.74}$$

The solution is therefore a perturbation of the ordinary linear response and the limit cycle is suppressed as expected.

Soft excitation, far from resonance. This case is similar to the last, and is left as Exercise 20. However, this solution is normally unstable (see Section 7.4).

Soft excitation near resonance. For soft excitation write in (5.71)

$$F = \varepsilon\gamma \tag{5.75}$$

and for near-resonance

$$\omega = 1 + \varepsilon\omega_1. \tag{5.76}$$

The expansion is assumed to be

$$x(\varepsilon, \tau) = x_0(\tau) + \varepsilon x_1(\tau) + \dots. \tag{5.77}$$

Equations (5.72), (5.75), and (5.76) lead to the sequence

$$x_0'' + x_0 = 0, \tag{5.78a}$$

$$x_1'' + x_1 = -2\omega_1 x_0'' - (x_0^2 - 1)x_0' + \gamma\cos\tau, \tag{5.78b}$$

and so on. We require solutions with period 2π. Equation (5.78a) has the solutions

$$x_0(\tau) = a_0\cos\tau + b_0\sin\tau. \tag{5.79}$$

After some manipulation, (5.78b) becomes

$$x_1'' + x_1 = \{\gamma + 2\omega_1 a_0 - b_0(\tfrac{1}{4}r_0^2 - 1)\}\cos\tau$$
$$+ \{2\omega_1 b_0 + a_0(\tfrac{1}{4}r_0^2 - 1)\}\sin\tau + \text{higher harmonics}, \tag{5.80}$$

where

$$r_0 = \sqrt{(a_0^2 + b_0^2)} > 0. \tag{5.81}$$

For a periodic solution we require

$$2\omega_1 a_0 - b_0(\tfrac{1}{4}r_0^2 - 1) = -\gamma, \tag{5.82a}$$

$$2\omega_1 b_0 + a_0(\tfrac{1}{4}r_0^2 - 1) = 0. \tag{5.82b}$$

By squaring and adding these two equations we obtain

$$r_0^2\{4\omega_1^2 + (\tfrac{1}{4}r_0^2 - 1)^2\} = \gamma^2 \tag{5.83}$$

which give the possible amplitudes r_0 of the response. The structure of this equation is examined in Chapter 7 in a different connection: it is sufficient to notice here its family resemblance to (5.31) for the pendulum equation. Like (5.31), it may have as many as three real solutions for $r_0 > 0$.

5.10. The perturbation method and Fourier series

In the examples given in Sections 5.2 and 5.3 the solutions emerge as series of sines and cosines with frequencies which are integer multiples of the forcing frequency. These appeared as a result of reorganizing terms like x^3, but by making a direct attack using Fourier series we can show that this form always occurs.

Consider the more general forced equation

$$x'' + \Omega^2 x = F(\tau) - \varepsilon h(x, x') \qquad (5.84)$$

where ε is a small parameter. Suppose that F is periodic, with the time variable already scaled to give it period 2π, and that its mean value is zero so that there is zero constant term in its Fourier series representation:

$$F(\tau) = \sum_{n=1}^{\infty} (A_n \cos n\tau + B_n \sin n\tau). \qquad (5.85)$$

We shall allow the possibility that Ω is *close to some integer N* by writing

$$\Omega^2 = N^2 + \varepsilon\beta \qquad (5.86)$$

($N = 1$ in Section 5.3). The perturbation method requires that the periodic solutions emerge from periodic solutions of some appropriate linear equation. If (5.85) has a non-zero term of order N then (5.84), with $\varepsilon = 0$, is clearly not an appropriate linearization, since the forcing term has a component equal to the natural frequency N and there will be no periodic solutions. However, if we write

$$A_N = \varepsilon A, \quad B_N = \varepsilon B, \qquad (5.87)$$

the term in F giving resonance is removed from the *linearized* equation and we have a possible family of generating solutions. Now rearrange (5.84), isolating the troublesome term in F by writing

$$f(\tau) = F(\tau) - \varepsilon A \cos N\tau - \varepsilon B \sin N\tau = \sum_{n \neq N} (A_n \cos n\tau + B_n \sin n\tau). \qquad (5.88)$$

Equation (5.84) becomes

$$x'' + N^2 x = f(\tau) + \varepsilon\{-h(x, x') - \beta x + A \cos N\tau + B \sin N\tau\}. \quad (5.89)$$

The linearized equation is now $x'' + N^2 x = f(\tau)$, with no resonance.

Now write as usual

$$x(\varepsilon, \tau) = x_0(\tau) + \varepsilon x_1(\tau) + \ldots, \qquad (5.90)$$

where $x_0, x_1, \ldots$ have period 2π. By expanding h in (5.84) in powers of ε we have

$$h(x, x') = h(x_0, x_0') + \varepsilon h_1(x_0, x_0', x_1, x_1') + \ldots, \qquad (5.91)$$

where h_1 can be calculated, and by substituting (5.90) and (5.91) into (5.89) we obtain the sequence

$$x_0'' + N^2 x_0 = \sum_{n \neq N} (A_n \cos n\tau + B_n \sin n\tau), \qquad (5.92a)$$

$$x_1'' + N^2 x_1 = -h(x_0, x_0') - \beta x_0 + A \cos N\tau + B \sin N\tau, \qquad (5.92b)$$

$$x_2'' + N^2 x_2 = -h_1(x_0, x_0', x_1, x_1') - \beta x_1, \qquad (5.92c)$$

and so on. The solution of (5.92a) is

$$x_0(\tau) = a_0 \cos N\tau + b_0 \sin N\tau + \sum_{n \neq N} \frac{A_n \cos n\tau + B_n \sin n\tau}{N^2 - n^2}$$

$$= a_0 \cos N\tau + b_0 \sin N\tau + \phi(\tau), \qquad (5.93)$$

say; a_0, b_0 are constants to be determined at the next stage.

We require x_0, as determined by (5.93), to be such that (5.92b) has periodic solutions. This is equivalent to requiring that its right side has no Fourier term of order N, since such a term would lead to resonance. We require therefore that

$$\beta a_0 = -\frac{1}{\pi} \int_0^{2\pi} h(a_0 \cos N\tau + b_0 \sin N\tau + \phi(\tau),$$

$$-a_0 N \sin N\tau + b_0 N \cos N\tau + \phi'(\tau)) \cos N\tau \, d\tau + A, \qquad (5.94a)$$

$$\beta b_0 = -\frac{1}{\pi} \int_0^{2\pi} h(a_0 \cos N\tau + b_0 \sin N\tau + \phi(\tau),$$

$$-a_0 N \sin N\tau + b_0 N \cos N\tau + \phi'(\tau)) \sin N\tau \, d\tau + B, \qquad (5.94b)$$

which constitute two equations for the unknowns a_0, b_0. The reader should confirm that the resulting equations are the same as those for the first order approximation, with $N = 1$, obtained in Section 5.3.

Each equation in the sequence (5.92) has solutions containing constants a_1, b_1; a_2, b_2; whose values are similarly established at the succeeding step. However, the equations for subsequent constants are linear: the pair (5.94) are the only ones which may have several solutions.

Exercises

1. Find all the periodic solutions of

$$\ddot{x} + \Omega^2 x = \Gamma \cos t$$

for all values of Ω^2.

2. Find the first two harmonics of the solutions of period 2π of the following
   (i) $\ddot{x} - 0.5x^3 + 0.25x = \cos t$;
   (ii) $\ddot{x} - 0.1x^3 + 0.6x = \cos t$;
   (iii) $\ddot{x} - 0.1\dot{x}^2 + 0.5x = \cos t$.

3. Find a first approximation to the limit cycle for Rayleigh's equation

$$\ddot{x} + \varepsilon(\tfrac{1}{3}\dot{x}^3 - \dot{x}) + x = 0, \quad |\varepsilon| \ll 1,$$

using the method of Section 5.8.

4. Take the method of Section 5.8 to order ε to obtain solutions of period 2π, and the amplitude–frequency relation, for
   (i) $\ddot{x} - \varepsilon x\dot{x} + x = 0$;
   (ii) $(1 + \varepsilon\dot{x})\ddot{x} + x = 0$.

5. Apply the perturbation method to the equation $\ddot{x} + \Omega^2 \sin x = \cos t$ by considering $\ddot{x} + \varepsilon\Omega^2 x + \varepsilon\Omega^2(\sin x - x) = \cos t$, with $\varepsilon = 1$, and assuming that Ω is not close to an odd integer. Use the Fourier expansion

$$\sin(a\cos t) = 2 \sum_{n=0}^{\infty} (-1)^n J_{2n+1}(a)\cos\{(2n+1)t\},$$

where J_{2n+1} is the Bessel function of order $2n+1$. Confirm that the leading terms are given by

$$x = \frac{1}{\Omega^2 - 1}[1 + \Omega^2 - 2\Omega^2 J_1\{1/(\Omega^2 - 1)\}]\cos t + \frac{2}{\Omega^2 - 9}J_3\{1/(\Omega^2 - 1)\}\cos 3t.$$

Compute the coefficients for representative values of Ω^2.

6. For the equation $\ddot{x} + \Omega^2 x - 0 \cdot 1 x^3 = \cos\tau$, where $\Omega^2 \not\approx 1$, sketch the variation in the ratio of the magnitudes of the first two harmonics.

7. In the equation

$$\ddot{x} + \Omega^2 x + \varepsilon f(x) = \Gamma \cos t$$

Ω is not close to an odd integer, and $f(x)$ is an odd function of x, with expansion

$$f(a\cos t) = -a_1(a)\cos t - a_3(a)\cos 3t - \dots.$$

Derive a perturbation solution of period 2π, to order ε.

8. The Duffing equation near resonance at $\Omega = 3$, with weak excitation, is

$$\ddot{x} + 9x = \varepsilon(\gamma \cos t - \beta x + x^3).$$

Show that there are solutions of period 2π if the amplitude of the zero-order solution is 0 or $2\sqrt{(\beta/3)}$. Find the solution to order ε in the latter case.

9. From eqn (5.34), the amplitude equation for an undamped pendulum is

$$-F = a_0(\omega^2 - \omega_0^2 + \tfrac{1}{8}\omega_0^2 a_0^2).$$

When ω_0 is given, find for what values of ω there are three possible responses. (Find the stationary values of F with respect to a_0, with ω fixed. These are the points where the response curves of Fig. 5.2 turn over.)

10. From eqn (5.36), the amplitude equation for the damped pendulum is $F^2 = r_0^2\{k^2\omega^2 + (\omega^2 - \omega_0^2 + \tfrac{1}{8}\omega_0^2 r_0^2)^2\}$. By considering $d(F^2)/d(r_0^2)$, show that if $(\omega^2 - \omega_0^2)^2 \leqslant 3k^2\omega^2$, then the amplitude equation has one real root r_0 for all F.

11. Find the linear equivalent form (Section 4.5) of the expression $\ddot{x} + \Omega^2 x - \varepsilon x^3$, with respect to the periodic form $x = a \cos t$. Use the linear form to obtain the frequency–amplitude relation for the equation

$$\ddot{x} + \Omega^2 x - \varepsilon x^3 = \Gamma \cos t.$$

Solve the equation approximately by assuming that $a = a_0 + \varepsilon a_1$, and show that this agrees with the first harmonic in eqn (5.17). (Note that there may be three solutions, but that this method of solution shows only the one close to $\{\Gamma/(1 - \Omega^2)\} \cos t$.)

12. Generalize the method of Exercise 11 for the same equation by putting $x = x^{(0)} + x^{(1)} + \ldots$, where $x^{(0)}$ and $x^{(1)}$ are the first two harmonics to order ε, $a \cos t$ and $b \cos 3t$, say, in the expansion of the solution. Show that the linear form equivalent to x^3 is

$$(\tfrac{3}{4}a^2 + \tfrac{3}{4}ab + \tfrac{3}{2}b^2)x^{(0)} + (\tfrac{1}{4}a^3 + \tfrac{3}{2}a^2 b + \tfrac{3}{4}b^3)x^{(1)}/b.$$

Split the pendulum equation into the two equations

$$\ddot{x}^{(0)} + \{\Omega^2 - \varepsilon(\tfrac{3}{4}a^2 + \tfrac{3}{4}ab + \tfrac{3}{2}b^2)\}x^{(0)} = \Gamma \cos t,$$
$$\ddot{x}^{(1)} + \{\Omega^2 - \varepsilon(\tfrac{1}{4}a^3 + \tfrac{3}{2}a^2 b + \tfrac{3}{4}b^3)/b\}x^{(1)} = 0.$$

Deduce that a and b must satisfy

$$a\{\Omega^2 - 1 - \varepsilon(\tfrac{3}{4}a^2 + \tfrac{3}{4}ab + \tfrac{3}{2}b^2)\} = \Gamma,$$
$$b\{\Omega^2 - 9 - \varepsilon(\tfrac{1}{4}a^3 + \tfrac{3}{2}a^2 b + \tfrac{3}{4}b^3)\} = 0.$$

Assume that $a = a_0 + \varepsilon a_1$, $b = \varepsilon b_1$, and obtain a_0, a_1, and b_1 (giving the perturbation solution (5.17)).

13. Apply the Lindstedt method, Section 5.8, to van der Pol's equation $\ddot{x} + \varepsilon(x^2 - 1)\dot{x} + x = 0, |\varepsilon| \ll 1$. Show that the frequency of the limit cycle is given by $\omega = 1 - \frac{1}{16}\varepsilon^2 + O(\varepsilon^3)$.

14. Investigate the forced periodic solutions of period $\frac{2}{3}\pi$ for the Duffing equation in the form $\ddot{x} + (1 + \varepsilon\beta)x - \varepsilon x^3 = \Gamma \cos 3t$.

15. For the equation $\ddot{x} + x + \varepsilon x^3 = 0, |\varepsilon| \ll 1$, with $x(0) = a, \dot{x}(0) = 0$, assume an expansion of the form $x(t) = x_0(t) + \varepsilon x_1(t) + \ldots$, and carry out the perturbation process without assuming periodicity of the solution. Show that

$$x(t) = a\cos t + \varepsilon a^3\{-\tfrac{3}{8}t\sin t + \tfrac{1}{32}(\cos 3t - \cos t)\} + O(\varepsilon^2).$$

(This expansion is valid, so far as it goes. Why is it not so suitable as those already obtained for describing the solutions?)

16. Find the first few harmonics in the solution, period 2π, of $\ddot{x} + \Omega^2 x + \varepsilon x^2 = \Gamma \cos t$, by the direct method of Section 5.2. Explain the presence of a constant term in the expansion.

For what values of Ω does the expansion fail? Show how, for small values of Γ, an expansion valid near $\Omega = 1$ can be obtained.

17. Use the method of amplitude-phase perturbation (Section 5.7) to approximate to the solutions, period 2π, of $\ddot{x} + x = \varepsilon(\gamma\cos t - x\dot{x} - \beta x)$.

18. Investigate the solutions, period 2π, of $\ddot{x} + 9x + \varepsilon x^2 = \Gamma \cos t$.

19. For the damped pendulum equation with a forcing term,

$$\ddot{x} + k\dot{x} + \omega_0^2 x - \tfrac{1}{6}\omega_0^2 x^3 = F\cos\omega t,$$

show that the amplitude–frequency curves have their maxima on

$$\omega^2 = \omega_0^2(1 - \tfrac{1}{8}r_0^2) - \tfrac{1}{2}k^2.$$

20. Show that the first harmonic for the forced van der Pol equation $\ddot{x} + \varepsilon(x^2 - 1)\dot{x} + x = F\cos\omega t$ is the same for both weak and hard excitation, far from resonance.

21. The orbital equation of a planet about the sun is

$$\frac{d^2 u}{d\theta^2} + u = k(1 + \varepsilon u^2),$$

where $u = r^{-1}$ and r, θ are polar coordinates, $k = \gamma m/h^2, \gamma$ is the gravitational constant, m is the mass of the planet and h is its moment of momentum, a constant. $\varepsilon k u^2$ is the relativistic correction term.

Obtain a perturbation expansion for the solution with initial conditions $u(0) = k(e+1), \dot{u}(0) = 0$. ($e$ is the eccentricity of the unperturbed orbit, and these are initial conditions at the perihelion: the nearest point to the sun on the unperturbed orbit.) Note that the solution of the perturbed equation is not periodic, and that 'secular' terms cannot be eliminated. Show that the expansion to order ε predicts that in each orbit the perihelion advances by $2k^2\pi\varepsilon$.

22. Use the Lindstedt procedure to find the first few terms in the expansion of the periodic solutions of

$$\ddot{x} + x + \varepsilon x^2 = 0.$$

Explain the presence of a constant term in the expansion.

23. Investigate the forced periodic solutions of period 2π of the equation

$$\ddot{x} + (4 + \varepsilon\beta)x - \varepsilon x^3 = \Gamma\cos t$$

where ε is small and β and Γ are not too large. Confirm that there is always a periodic solution of the form

$$a_0\cos 2t + b_0\sin 2t + \tfrac{1}{3}\Gamma\cos t$$

where

$$a_0(\tfrac{3}{4}r_0^2 + \tfrac{1}{6}\Gamma^2 - \beta) = b_0(\tfrac{3}{4}r_0^2 + \tfrac{1}{6}\Gamma^2 - \beta) = 0.$$

6 Singular perturbation methods

THIS CHAPTER is concerned with approximating to the solutions of differential equations containing a small parameter ε in what might be called difficult cases where, for one reason or another, a straightforward expansion of the solution in powers of ε is unobtainable or unusable. Often in such cases it is possible to modify the method so that the essential features of such expansions are redeemed. Ideally, we wish to be able to take a few terms of an expansion, and to be able to say that for some small fixed numerical value of ε supplied in a practical problem the truncated series is close to the required solution *for the whole range of the independent variable* in the differential equation. Failure of this condition is the rule rather than the exception. In fact the examples of Chapter 5 do not give useful approximations if we work solely from the initial conditions (see Chapter 5, Exercise 15): for satisfactory approximation we must use the technically redundant information that the solutions are periodic.

This chapter illustrates several other methods which have been used in such cases. If one method fails, another may work, or perhaps a combination of methods may work, but generally speaking the approach is tentative, and considerable skill and intuition is necessary to get the desired result. For a treatment of the whole topic, not restricted to ordinary differential equations, see for example Nayfeh (1973), van Dyke (1964), and O'Malley (1974).

6.1. Non-uniform approximations to functions on an interval

The solutions of the differential equations we are considering,

$$\ddot{x} = f(x, \dot{x}, t, \varepsilon),$$

are functions of t and ε. Some of the problems which may arise in approximating to them can be illustrated by using functions which are not necessarily the solutions of any particular differential equation. For example, consider the function

$$x(\varepsilon, t) = e^{-\varepsilon t} \tag{6.1}$$

on $t \geqslant 0$, where ε lies in a neighbourhood of zero. The first few terms of the Taylor expansion in powers of ε are

$$1 - \varepsilon t + \tfrac{1}{2}\varepsilon^2 t^2, \tag{6.2}$$

where the error is

$$O(\varepsilon^3). \tag{6.3}$$

The error estimate (6.3) implies that for any *fixed t*, however large, we can choose ε small enough for the error to be as small as we please, and, indeed, smaller than the smallest term in the approximation, $\tfrac{1}{2}\varepsilon^2 t^2$. However, the trend of the terms shows clearly that *if ε is fixed*, at however small a value (which is the usual situation), t may be chosen large enough to destroy the approximation completely (see Fig. 6.1).

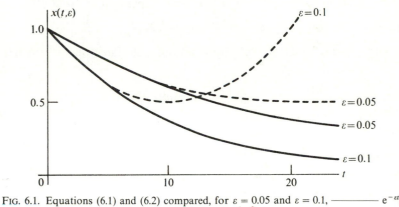

Fig. 6.1. Equations (6.1) and (6.2) compared, for $\varepsilon = 0.05$ and $\varepsilon = 0.1$, ————— $e^{-\varepsilon t}$, — — — — $1 - \varepsilon t + \tfrac{1}{2}\varepsilon t^2$

Consider now a function which appears in Section 6.4 as part of the solution to a differential equation:

$$x(\varepsilon, t) = \cos\{(1 - \varepsilon)^{1/2} t\}, \quad 0 \leqslant t < \infty, \tag{6.4}$$

where ε lies in a neighbourhood of zero. The first three terms of the Taylor series for small ε give

$$\cos t + \tfrac{1}{2}\varepsilon t \sin t + \tfrac{1}{8}\varepsilon^2 (t \sin t - t^2 \cos t) \tag{6.5}$$

with error $O(\varepsilon^3)$. Once again for t large enough the approximation fails. We can see that it fails when t is so large that εt ceases to be small. The condition that (6.5) should be an approximation *for fixed ε* is that

$$t \ll \varepsilon^{-1}. \tag{6.6}$$

If this is satisfied, then the error, which is dominated by a term like $\varepsilon^3 t^3$, is small. It is also small compared with each term in (6.5), unless t has a fortuitous value making one of these terms very small or zero.† We say that (6.5) *does not provide an approximation uniform on $t \geqslant 0$*. Failure of the ordinary perturbation method to produce a uniform approximation to the solution of a differential equation is very common, and the occasions of failure are called *singular perturbation problems*.

Formally, we require an approximation $\bar{x}(\varepsilon, t)$ to a function $x(\varepsilon, t)$, usually to be valid for an infinite range of t:

$$x(\varepsilon, t) = \bar{x}(\varepsilon, t) + E(\varepsilon, t), \quad t_0 \leqslant t < \infty,$$

with error $E(\varepsilon, t)$, where $\lim_{\varepsilon \to 0} E(\varepsilon, t) = 0$ *uniformly* on $t_0 \leqslant t < \infty$. That is to say, given any $\delta > 0$, there exists $\eta > 0$ independent of t such that

$$|\varepsilon| < \eta \Rightarrow |E(\varepsilon, t)| < \delta.$$

The general question of approximation to functions is further illustrated in Section 6.5.

6.2. Coordinate perturbation (renormalization)

Consider the equation

$$\ddot{x} + x = \varepsilon x^3. \tag{6.7}$$

The expansion

$$x(\varepsilon, t) = x_0(t) + \varepsilon x_1(t) + \dots$$

leads to

$$\ddot{x}_0 + x_0 = 0,$$
$$\ddot{x}_1 + x_1 = x_0^3,$$

and so on. The solution of the first equation is given by

$$x_0(t) = A \, e^{it} + \bar{A} \, e^{-it},$$

where A is a complex constant. Therefore

$$\ddot{x}_1 + x_1 = A^3 \, e^{3it} + 3A^2 \bar{A} \, e^{it} + 3\bar{A}^2 A \, e^{-it} + \bar{A}^3 \, e^{-3it}, \tag{6.8}$$

and secular terms begin to appear. They cannot be eliminated: only if $A = 0$ are they absent. Therefore a series similar in form to (6.5) emerges,

---

† The same conclusion is reached no matter how many terms of the Taylor series we take: if more terms are taken the approximation may be better while it lasts, but it still fails when t becomes comparable with ε^{-1}. Note that if the series is asymptotic rather than convergent as $\varepsilon \to 0$, the approximation does not necessarily improve by taking more terms (Copson 1965).

and the truncated series does not approximate $x(\varepsilon, t)$ uniformly on $t \geqslant 0$.
This problem was treated in Section 5.8. There, the difficulty was avoided by anticipating a *periodic solution* of period $2\pi/\omega$ in t, ω being unknown; we put

$$\omega = 1 + \varepsilon\omega_1 + \varepsilon^2\omega_2 + \ldots$$

where $\omega_1, \omega_2, \ldots$, are unknown constants, and then change the variable from t to τ:

$$\tau = \omega t, \tag{6.9}$$

so that the equation in τ has known period 2π (the Lindstedt procedure). Equation (6.9) introduces sufficient free constants ω_i to allow the secular terms to be eliminated. This method can be looked on as a device for obtaining uniform approximations by adapting prior knowledge of the periodicity of the solutions.

We can look at this procedure in another way. Write (6.9) in the form

$$t = \tau/\omega = \tau/(1 + \varepsilon\omega_1 + \varepsilon^2\omega_2 + \ldots) = \tau(1 + \varepsilon\tau_1 + \varepsilon^2\tau_2 + \ldots), \tag{6.10}$$

where $\tau_1, \tau_2, \ldots$, are unknown constant coefficients. Also put, as in the earlier method,

$$x(\varepsilon, t) = X(\varepsilon, \tau) = X_0(\tau) + \varepsilon X_1(\tau) + \varepsilon^2 X_2(\tau) + \ldots, \tag{6.11}$$

and substitute (6.10) and (6.11) into (6.7). We know this leads to an expansion uniform on $t \geqslant 0$ since it is equivalent to the Lindstedt procedure. However, the interpretation is now different: it appears that by (6.10) and (6.11) we have made *a simultaneous expansion in powers of ε of both the dependent and the independent variables*, generating an implicit relation between x and t through a parameter τ. The coefficients can be adjusted to eliminate terms which would give non-uniformity (in fact, the 'secular terms'), and so a uniform (and, as it turns out, a periodic) solution is obtained.

We were guided to the form (6.10) by the prior success of the Lindstedt procedure, which ensures its appropriateness. If, however, we assume no prior experience, we have to abandon the assumption of constant coefficients in (6.10). Therefore let us take, instead of the constants $\tau_1, \tau_2, \ldots$, a set of unknown functions $T_1, T_2, \ldots$, of τ, and see what happens. We have

$$x(\varepsilon, t) = X(\varepsilon, \tau) = X_0(\tau) + \varepsilon X_1(\tau) + \varepsilon^2 X_2(\tau) + \ldots \tag{6.12a}$$

and

$$t = T(\varepsilon, \tau) = \tau + \varepsilon T_1(\tau) + \varepsilon^2 T_2(\tau) + \ldots. \tag{6.12b}$$

The first term in the expansion of t remains as τ, since this is appropriate when $\varepsilon \to 0$: τ is called a *strained coordinate*, or a perturbed coordinate.

Equations (6.12*a*) and (6.12*b*) are the basis of Lighthill's method (see Section 6.3), in which the expansions are substituted directly into the differential equation. Here we show a different approach. The ordinary perturbation process is fairly easy to apply, and gives a series of the form

$$x(\varepsilon, t) = x_0(t) + \varepsilon x_1(t) + \varepsilon^2 x_2(t) + \ldots \quad (6.13)$$

A finite number of terms of this series will not generally give an approximation to $x(\varepsilon, t)$ holding uniformly for all t, but when (6.12*b*) is substituted into (6.13), it may be possible to choose $T_1(\tau)$, $T_2(\tau)$, so as to force (6.13) into a form which gives a uniform approximation. This process is called coordinate perturbation (Crocco 1972), and will be carried out in the following two examples.

Example 6.1   *Obtain an approximate solution of the equation*

$$\ddot{x} + x = \varepsilon x^3,$$

with $x(\varepsilon, 0) = 1$, $\dot{x}(\varepsilon, 0) = 0$, and error $O(\varepsilon^3)$ uniformly on $t \geq 0$, by the method of coordinate perturbation.

The expansion (6.13), together with the initial conditions, gives

$$\ddot{x}_0 + x_0 = 0, \qquad x_0(0) = 1, \quad \dot{x}_0(0) = 0;$$
$$\ddot{x}_1 + x_1 = x_0^3, \qquad x_1(0) = 0, \quad \dot{x}_1(0) = 0;$$
$$\ddot{x}_2 + x_2 = 3x_0^2 x_1, \qquad x_2(0) = 0, \quad \dot{x}_2(0) = 0;$$

and so on. Then

$$x(\varepsilon, t) = \cos t + \varepsilon(\tfrac{1}{32}\cos t + \tfrac{3}{8}t \sin t - \tfrac{1}{32}\cos 3t)$$
$$+ \varepsilon^2(\tfrac{23}{1024}\cos t + \tfrac{3}{32}t \sin t - \tfrac{9}{128}t^2 \cos t - \tfrac{3}{128}\cos 3t - \tfrac{9}{256}t \sin 3t$$
$$+ \tfrac{1}{1024}\cos 5t) + O(\varepsilon^3). \quad (6.14)$$

The expansion is clearly non-uniform on $t \geq 0$.

Now put

$$t = \tau + \varepsilon T_1(\tau) + \varepsilon^2 T_2(\tau) + \ldots \quad (6.15)$$

into (6.14), expand the terms in powers of ε, and rearrange:

$$X(\varepsilon, \tau) = \cos \tau + \varepsilon(\tfrac{1}{32}\cos \tau - T_1 \sin \tau + \tfrac{3}{8}\tau \sin \tau - \tfrac{1}{32}\cos 3\tau)$$
$$+ \varepsilon^2(\tfrac{23}{1024}\cos \tau - \tfrac{1}{2}T_1^2 \cos \tau + \tfrac{11}{32}T_1 \sin \tau - T_2 \sin \tau + \tfrac{3}{8}\tau T_1 \cos \tau$$
$$+ \tfrac{3}{32}\tau \sin \tau - \tfrac{9}{128}\tau^2 \cos \tau - \tfrac{3}{128}\cos 3\tau + \tfrac{3}{32}T_1 \sin \tau - \tfrac{9}{128}\tau^2 \cos \tau$$
$$- \tfrac{3}{128}\cos 3\tau + \tfrac{3}{32}T_1 \sin 3\tau - \tfrac{9}{256}\tau \sin 3\tau + \tfrac{1}{1024}\cos 5\tau) + O(\varepsilon^3).$$

To avoid the form $\tau \sin \tau$ in the ε term, which would give an obvious non-uniformity, define T_1 by

$$T_1(\tau) = \tfrac{3}{8}\tau. \tag{6.16}$$

The ε^2 coefficient then becomes

$$\tfrac{23}{1024} \cos \tau - T_2 \sin \tau + \tfrac{57}{256}\tau \sin \tau - \tfrac{3}{128} \cos 3\tau + \tfrac{1}{1024} \cos 5\tau,$$

and we must have

$$T_2(\tau) = \tfrac{57}{256}\tau \tag{6.17}$$

eliminating the non-uniformity $\tau \sin \tau$. On the assumption that this process could continue indefinitely we have finally

$$x = X(\varepsilon, \tau) = \cos \tau + \tfrac{1}{32}\varepsilon(\cos \tau - \cos 3\tau)$$
$$+ \varepsilon^2(\tfrac{23}{1024} \cos \tau - \tfrac{3}{128} \cos 3\tau + \tfrac{1}{1024} \cos 5\tau) + O(\varepsilon^3) \quad (6.18a)$$

where

$$t = \tau(1 + \tfrac{3}{8}\varepsilon + \tfrac{57}{256}\varepsilon^2) + O(\varepsilon^3). \tag{6.18b}$$

Example 6.2 (*Lighthill's equation.*) *Find an approximation, uniform on* $0 \leqslant t \leqslant 1$, *to the solution of*

$$(\varepsilon x + t)\dot{x} + (2 + t)x = 0, \quad \varepsilon \geqslant 0, \tag{6.19}$$

satisfying $x(\varepsilon, 1) = e^{-1}$.

Begin by noticing the general character of the solution curve for $\varepsilon > 0$.

$$\frac{dx}{dt} = -\frac{(2 + t)x}{\varepsilon x + t},$$

so $x > 0$ and $\dot{x} < 0$ at $t = 1$. x therefore increases as $t = 0$ is approached from $t = 1$. It is still finite at $t = 0$, since $\dot{x}$ is infinite only on $\varepsilon x + t = 0$, an asymptote to our solution curve shown on Fig. 6.2. Obviously $x(0)$ is large, but still finite, when $0 < \varepsilon \ll 1$.

The direct approach, writing

$$x(\varepsilon, t) = x_0(t) + \varepsilon x_1(t) + \varepsilon^2 x_2(t) + \ldots \tag{6.20}$$

leads to the sequence of linear equations

$$t\dot{x}_0 + (2 + t)x_0 = 0, \quad x_0(1) = e^{-1}; \tag{6.21a}$$

$$t\dot{x}_1 + (2 + t)x_1 = -x_0\dot{x}_0, \quad x_1(1) = 0; \tag{6.21b}$$

$$t\dot{x}_2 + (2 + t)x_2 = -\dot{x}_0 x_1 - x_0\dot{x}_1, \quad x_2(1) = 0; \tag{6.21c}$$

and so on. The solution of (6.21a) is

$$x_0(t) = e^{-t}/t^2 \tag{6.22}$$

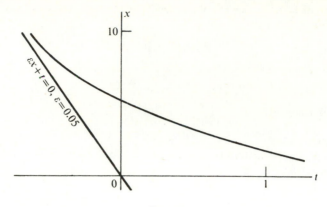

FIG. 6.2.

(predicting, incorrectly, that $x(0) = \infty$). Equation (6.21b) becomes

$$\dot{x}_1 + \left(\frac{2}{t} + 1\right)x_1 = e^{-2t}\left(\frac{2}{t^6} + \frac{1}{t^5}\right).$$

Therefore

$$x_1(t) = \frac{e^{-t}}{t^2}\int_1^t e^{-u}\left(\frac{2}{u^4} + \frac{1}{u^3}\right)du. \qquad (6.23)$$

This is of order $1/t^5$ at $t = 0$, and is even more singular than (6.22). A similar treatment of (6.21c) produces a singularity $O(1/t^8)$ as $t \to 0$. The approximation (6.20):

$$x(\varepsilon, t) \approx x_0(t) + \varepsilon x_1(t),$$

with error $O(\varepsilon^2)$ for fixed t, is clearly not uniform on $0 < t \leqslant 1$ and breaks down completely at $t = 0$.

As in the last example, write the near-identity transformation

$$t = T(\varepsilon, \tau) = \tau + \varepsilon T_1(\tau) + \varepsilon^2 T_2(\tau) + \dots. \qquad (6.24)$$

Then

$$\frac{e^{-t}}{t^2} = \frac{e^{-\tau}}{\tau^2}\left\{1 - \varepsilon T_1(\tau)\left(\frac{2}{\tau} + 1\right)\right\} + O(\varepsilon^2) \qquad (6.25)$$

($O(\varepsilon^2)$ referring to fixed τ, not fixed t) and

$$\int_1^t e^{-u}\left(\frac{2}{u^4} + \frac{1}{u^3}\right)du = \int_1^\tau e^{-u}\left(\frac{2}{u^4} + \frac{1}{u^3}\right)du + \int_\tau^{\tau + \varepsilon T_1(\tau) + \dots} e^{-u}\left(\frac{2}{u^4} + \frac{1}{u^3}\right)du$$

$$= \int_1^\tau e^{-u}\left(\frac{2}{u^4} + \frac{1}{u^3}\right)du + O(\varepsilon). \qquad (6.26)$$

From (6.22), (6.23), (6.25), and (6.26) we find, to $O(\varepsilon^2)$ (referring to fixed τ),

$$x = \frac{e^{-\tau}}{\tau^2}\left\{1+\varepsilon\left[\int_1^\tau e^{-u}\left(\frac{2}{u^4}+\frac{1}{u^3}\right)du - T_1(\tau)\left(\frac{2}{\tau}+1\right)\right]\right\}. \tag{6.27}$$

The best thing we can do is to choose T_1 to eliminate the worst singularity for small τ arising from the integral. This is $-\frac{2}{3}\tau^{-3}$; so

$$T_1(\tau) = -\frac{1}{3\tau^2} \tag{6.28}$$

eliminates this singularity (of course, we are left with one of order τ^{-2}).

Finally, from (6.27), (6.24), and (6.28)

$$x = e^{-\tau}/\tau^2 \tag{6.29a}$$

together with

$$t = \tau - \frac{\varepsilon}{3\tau^2} \tag{6.29b}$$

gives a uniform approximation for $0 \leqslant t \leqslant 1$.

Note that there is now no infinity predicted at $t = 0$: from (6.29b), $t = 0$ corresponds to $\tau = (\varepsilon/3)^{1/3}$, and $x(0)$ therefore has the value approximately $(\varepsilon/3)^{-2/3}$. The error in (6.29a, b) is $O(\varepsilon^2)$ for fixed τ, but only $O(\varepsilon^{1/3})$ for fixed t.

6.3. Lighthill's method

We reconsider the equation

$$(\varepsilon x + t)\dot{x} + (2+t)x = 0, \tag{6.30a}$$

$$x(\varepsilon, 1) = e^{-1}, \tag{6.30b}$$

on $0 \leqslant t \leqslant 1$, using the direct attack suggested in the previous section. We substitute

$$x = X(\varepsilon, \tau) = X_0(\tau) + \varepsilon X_1(\tau) + \dots \tag{6.31a}$$

and

$$t = T(\varepsilon, \tau) = \tau + \varepsilon T_1(\tau) + \dots \tag{6.31b}$$

into the differential equation and boundary condition.

First, we do not expect that $t = 1$ will correspond with $\tau = 1$. Suppose that $t = 1$ corresponds to $\tau = \tau^*(\varepsilon)$. Then we must solve (6.31b) for τ^*; that is

$$1 = \tau^* + \varepsilon T_1(\tau^*) + \dots. \tag{6.32}$$

Then (6.30b), (6.31a) give the transformed boundary condition

$$e^{-1} = X_0(\tau^*) + \varepsilon X_1(\tau^*) + \dots. \tag{6.33}$$

To solve (6.32), assume τ^* is close to 1 for ε small and has the expansion
$$\tau^* = 1 + \varepsilon\tau_1 + \dots, \tag{6.34}$$
where $\tau_1, \dots$ are constants. Equation (6.32) becomes (writing $T_1(\tau^*) = T_1(1) + \varepsilon\tau_1 T_1'(1) + \dots$)
$$1 = 1 + \varepsilon(\tau_1 + T_1(1)) + \dots$$
so $\tau_1 = -T_1(1)$, and from (6.34), the boundary $t = 1$ corresponds to $\tau = \tau^*$, where
$$\tau^* = 1 - \varepsilon T_1(1) + \dots.$$
Therefore, by expanding $X_0(\tau^*)$ and $X_1(\tau^*)$ in (6.33), the boundary condition (6.33) becomes
$$\mathrm{e}^{-1} = X_0(1) + \varepsilon(X_1(1) - X_0'(1)T_1(1)) + \dots. \tag{6.35}$$

Next, the derivative in (6.30a) is transformed by
$$\frac{\mathrm{d}x}{\mathrm{d}t} = \frac{\mathrm{d}X}{\mathrm{d}\tau} \bigg/ \frac{\mathrm{d}T}{\mathrm{d}\tau} = \frac{X_0' + \varepsilon X_1' + \dots}{1 + \varepsilon T_1' + \dots} = X_0' + \varepsilon(X_1' - X_0'T_1') + \dots. \tag{6.36}$$

Thus eqn (6.30a) becomes, to order ε,
$$(\varepsilon X_0 + \tau + \varepsilon T_1)(X_0' + \varepsilon\{X_1' - X_0'T_1'\}) + (2 + \tau + \varepsilon T_1)(X_0 + \varepsilon X_1) = 0. \tag{6.37}$$

Equations (6.35) and (6.37) give
$$\tau X_0' + (2 + \tau)X_0 = 0, \quad X_0(1) = \mathrm{e}^{-1}; \tag{6.38a}$$
$$\tau X_1' + (2 + \tau)X_1 = -T_1(X_0' + X_0) + \tau X_0'T_1' - X_0 X_0',$$
$$X_1(1) = X_0'(1)T_1(1). \tag{6.38b}$$

From (6.38a) the zero-order approximation is
$$X_0(\tau) = \mathrm{e}^{-\tau}/\tau^2. \tag{6.39}$$

Then (6.38b) becomes
$$\tau X_1' + (2 + \tau)X_1 = \frac{2\mathrm{e}^{-\tau}}{\tau^3} T_1 - \mathrm{e}^{-\tau}\left(\frac{2}{\tau^2} - \frac{1}{\tau}\right)T_1' + \mathrm{e}^{-2\tau}\left(\frac{2}{\tau^5} - \frac{1}{\tau^4}\right), \tag{6.40}$$
with initial condition
$$X_1(1) = -3\mathrm{e}^{-1} T_1(1).$$

We now have a free choice of $T_1(\tau)$: it could be chosen to make the right-hand side of (6.40) zero, for example; this would lead to a solution for (6.40) of the type (6.39) again, but is in any case impracticable. We shall choose T_1 to nullify the worst visible singularity on the right of (6.40), which is of order $1/\tau^5$. We attempt this by writing $e^{-\tau}, e^{-2\tau} \approx 1$ for small τ and solving

$$\frac{2}{\tau^3} T_1 - \frac{2}{\tau^2} T_1' + \frac{2}{\tau^5} = 0.$$

The simplest solution is

$$T_1(\tau) = -1/3\tau^2 \qquad (6.41)$$

(compare (6.28)). We have therefore achieved the same result as in the last section; though, in this example, at considerably more effort. Note that throughout the argument we have regarded the equation as a member of a family of equations with parameter ε, as in Chapter 5.

6.4. Multiple time scales (two-timing)

This method can be used to extend the interval of validity of the simple perturbation method. We shall illustrate it with the linear equation

$$\ddot{x} + x = \varepsilon x, \quad t \geqslant 0. \qquad (6.42)$$

This has the general solution which it is convenient to write in the form (cf. eqn (6.4))

$$x(\varepsilon, t) = a \cos\{(1-\varepsilon)^{1/2}t + \alpha\} \qquad (6.43)$$

where a, α are arbitrary. We can expand this in powers of ε:

$$x(\varepsilon, t) = x_0(t) + \varepsilon x_1(t) + \ldots + \varepsilon^{n-1}x_{n-1}(t) + R_n(\varepsilon, t),$$

where R_n is the remainder after n terms. (This is also the expansion given by the ordinary perturbation method.) To three terms we have the obviously non-uniform approximation

$$x(\varepsilon, t) \approx a \cos(t + \alpha) + \tfrac{1}{2}\varepsilon a t \sin(t + \alpha)$$
$$+ \tfrac{1}{8}\varepsilon^2 a\{t \sin(t + \alpha) - t^2 \cos(t + \alpha)\}. \qquad (6.44)$$

If any fixed number of terms is taken the approximation eventually starts to fail as t increases since the remainder will cease to be small compared with the terms retained. For example, in (6.44) the terms retained consist of bounded sines and cosines, together with the combinations

$$\varepsilon t, \quad \varepsilon^2 t, \quad \varepsilon^2 t^2,$$

and $R_3 = O(\varepsilon^3 t^3)$. Therefore, so long as $t = O(1)$, the terms in (6.44) are successively $O(1), O(\varepsilon), O(\varepsilon^2)$, and the remainder is $O(\varepsilon^3)$. However, when t becomes comparable with $1/\varepsilon$, say

$$t = O(\varepsilon^{-1}),$$

the terms are all of order 1 and so is the remainder. The approximation then fails. The failure will occur for $t = O(\varepsilon^{-1})$ no matter how many terms are taken. This is the situation described in Section 6.1.

Now, other series representations, still retaining the general power series form, but satisfactory over a larger range of t, are possible. For instance, if we pick out the first two terms in the Taylor series for $(1 - \varepsilon)^{1/2}$ by writing

$$(1-\varepsilon)^{1/2}t \equiv (1-\tfrac{1}{2}\varepsilon)t + \{(1-\varepsilon)^{1/2} - (1-\tfrac{1}{2}\varepsilon)\}t, \tag{6.45}$$

and correspondingly expand (6.43) over the first couple of terms, we find that

$$x(\varepsilon, t) = a\cos\left[(t-\tfrac{1}{2}\varepsilon t) + \alpha\right] + \tfrac{1}{8}\varepsilon^2 at \sin\left[(t-\tfrac{1}{2}\varepsilon t) + \alpha\right] + O(\varepsilon^4 t^2). \tag{6.46}$$

In general, the remainder after n terms is of $O(\varepsilon^{2n} t^n)$. This expansion is well-behaved for t even as large as

$$t = O(\varepsilon^{-1})$$

the successive terms being $O(1), O(\varepsilon), \dots, O(\varepsilon^{n-1})$ and the remainder $O(\varepsilon^n)$. To obtain an expansion valid for t as large as $O(\varepsilon^{-2})$ we should have to write

$$(1-\varepsilon)^{1/2}t \equiv (1-\tfrac{1}{2}\varepsilon - \tfrac{1}{8}\varepsilon^2)t + O(\varepsilon^3 t),$$

recast (6.45) and (6.46) accordingly, and so on for further extensions of the range of t (Nayfeh 1973).

The procedure is possible when we already know the exact solution; the problem is to achieve an expansion of this type starting only with a differential equation. In the following process we aim at attaining an approximation with an error of $O(\varepsilon^2)$, at the cost of accepting a limitation on the range of t in non-periodic cases:

$$t = O(\varepsilon^{-1}). \tag{6.47}$$

When $x(\varepsilon, t)$ is periodic, however, the approximation will be uniform for $t \geqslant 0$.

Introduce a new time-like variable η ('slow time') defined by

$$\eta = \varepsilon t. \tag{6.48}$$

Then for $t = O(\varepsilon^{-1})$,

$$\eta = O(1). \tag{6.49}$$

The expansion (6.46) has the form

$$a \cos (t - \tfrac{1}{2}\eta + \alpha) + \tfrac{1}{8}\varepsilon a\eta \sin (t - \tfrac{1}{2}\eta + \alpha) + O(\varepsilon^2), \tag{6.50}$$

in which the function $x(\varepsilon, t)$ begins to emerge in the form

$$x(\varepsilon, t) = X(\varepsilon, t, \eta). \tag{6.51}$$

Now t and η are not in fact independent, being connected by (6.48), but X is certainly *a solution of some equation in which t and η figure as independent variables*. We obtain the required equation in the following way.

When $\eta = \varepsilon t$

$$\frac{\mathrm{d}x}{\mathrm{d}t} = \frac{\mathrm{d}}{\mathrm{d}t} X(\varepsilon, t, \varepsilon t) = \frac{\partial X}{\partial t} + \varepsilon \frac{\partial X}{\partial \eta}, \tag{6.52}$$

and

$$\frac{\mathrm{d}^2 x}{\mathrm{d}t^2} = \frac{\partial^2 X}{\partial t^2} + 2\varepsilon \frac{\partial^2 X}{\partial \eta \partial t} + \varepsilon^2 \frac{\partial^2 X}{\partial \eta^2}. \tag{6.53}$$

Since x satisfies (6.42), X must satisfy

$$\frac{\partial^2 X}{\partial t^2} + 2\varepsilon \frac{\partial^2 X}{\partial t \partial \eta} + \varepsilon^2 \frac{\partial^2 X}{\partial \eta^2} + X = \varepsilon X. \tag{6.54}$$

We now carry out a straightforward perturbation expansion by writing

$$X(\varepsilon, t, \eta) = X_0(t, \eta) + \varepsilon X_1(t, \eta) + \dots, \tag{6.55}$$

expecting the remainder, or error, after the nth term to be $O(\varepsilon^n)$. Matching the powers of ε in (6.54) we obtain

$$\frac{\partial^2 X_0}{\partial t^2} + X_0 = 0, \tag{6.56a}$$

$$\frac{\partial^2 X_1}{\partial t^2} + X_1 = X_0 - 2 \frac{\partial^2 X_0}{\partial t \partial \eta}, \tag{6.56b}$$

$$\frac{\partial^2 X_2}{\partial t^2} + X_2 = X_1 - 2 \frac{\partial^2 X_1}{\partial t \partial \eta} - \frac{\partial^2 X_0}{\partial \eta^2}, \tag{6.56c}$$

and so on.

The general solution of (6.56a) is

$$X_0(t,\eta) = A_0(\eta)\,\mathrm{e}^{it} + \bar{A}_0(\eta)\,\mathrm{e}^{-it}, \tag{6.57}$$

where A_0 is an arbitrary function. When (6.56a) is substituted into (6.56b) we have

$$\frac{\partial^2 X_1}{\partial t^2} + X_1 = \mathrm{e}^{it}(A_0 - 2\mathrm{i}A_0') + \text{complex conjugate.} \tag{6.58}$$

The secular terms must be removed; since otherwise X_1 will be $O(t)$, and εX_1 of order εt, instead of $\varepsilon^2 t$. Therefore we require

$$A_0 - 2\mathrm{i}\frac{\mathrm{d}A_0}{\mathrm{d}\eta} = 0$$

so that

$$A_0(\eta) = A\,\mathrm{e}^{-\frac{1}{2}\mathrm{i}\eta} \tag{6.59}$$

where A is any constant. From (6.57) we obtain

$$X_0(t,\eta) = A\,\mathrm{e}^{\mathrm{i}(t-\frac{1}{2}\eta)} + \text{complex conjugate} = a\cos(t - \tfrac{1}{2}\eta + \alpha), \tag{6.60}$$

where $A = \frac{1}{2}a\,\mathrm{e}^{\mathrm{i}\alpha}$. Thus we have recovered the first term of (6.50).

The right-hand side of (6.56b) is now zero, so

$$X_1(t,\eta) = A_1(\eta)\,\mathrm{e}^{it} + \bar{A}_1(\eta)\,\mathrm{e}^{-it} \tag{6.61}$$

is its general solution. Putting (6.60) and (6.61) into (6.56c) we find

$$\frac{\partial^2 X_2}{\partial t^2} + X_2 = (A_1 - 2\mathrm{i}A_1' + \tfrac{1}{4}A\,\mathrm{e}^{-\frac{1}{2}\mathrm{i}\eta})\,\mathrm{e}^{it} + \text{complex conjugate.} \tag{6.62}$$

Again it is necessary to remove the secular terms by equating the bracket to zero. We find that

$$X_1(t,\eta) = -\tfrac{1}{8}A\eta\mathrm{i}\mathrm{e}^{\mathrm{i}(t-\frac{1}{2}\eta)} + \text{complex conjugate}$$

$$= \tfrac{1}{8}a\eta\sin(t - \tfrac{1}{2}\eta + \alpha), \tag{6.63}$$

and we have recovered the second term of (6.50).

Example 6.3   *Find solutions of Rayleigh's equation*

$$\ddot{x} + \varepsilon(\tfrac{1}{3}\dot{x}^3 - \dot{x}) + x = 0, \tag{6.64}$$

correct to $O(\varepsilon)$ so long as $t = O(\varepsilon^{-1})$.

We are not concerned only with the limit cycle. Suppose that x has the expansion

$$x(\varepsilon, t) = X(\varepsilon, t, \eta) = X_0(t, \eta) + \varepsilon X_1(t, \eta) + O(\varepsilon^2) \tag{6.65}$$

at least for

$$t = O(\varepsilon^{-1}). \tag{6.66}$$

By substituting (6.52) and (6.53) into (6.64), retaining terms up to order ε, and matching the coefficients of ε we obtain

$$\frac{\partial^2 X_0}{\partial t^2} + X_0 = 0, \tag{6.67}$$

$$\frac{\partial^2 X_1}{\partial t^2} + X_1 = \frac{\partial X_0}{\partial t} - \frac{1}{3}\left(\frac{\partial X_0}{\partial t}\right)^3 - 2\frac{\partial^2 X_0}{\partial t \partial \eta}. \tag{6.68}$$

Equation (6.67) has the general solution

$$X_0(\eta, t) = A_0(\eta)\,\mathrm{e}^{\mathrm{i}t} + \bar{A}_0(\eta)\,\mathrm{e}^{-\mathrm{i}t}. \tag{6.69}$$

Equation (6.68) becomes

$$\frac{\partial^2 X_1}{\partial t^2} + X_1 = \mathrm{i}(A_0 - A_0^2\bar{A}_0 - 2A_0')\,\mathrm{e}^{\mathrm{i}t} - \tfrac{1}{3}A_0^3\,\mathrm{e}^{3\mathrm{i}t}$$

$$+ \text{complex conjugate.} \tag{6.70}$$

The secular terms arise from the presence of $\mathrm{e}^{\mathrm{i}t}, \mathrm{e}^{-\mathrm{i}t}$ in (6.70), and vanish when

$$A_0' - \tfrac{1}{2}A_0 + \tfrac{1}{2}A_0^2\bar{A}_0 = 0. \tag{6.71}$$

The conjugate equation gives nothing new. To solve (6.71), write

$$A_0(\eta) = \rho(\eta)\exp(\mathrm{i}\alpha(\eta)), \tag{6.72}$$

where ρ and α are real functions. We obtain

$$\rho' - \tfrac{1}{2}\rho + \tfrac{1}{2}\rho^3 = 0,$$

$$\alpha' = 0;$$

from which

$$\begin{aligned}\rho(\eta) &= (1 + A\,\mathrm{e}^{-\eta})^{-1/2}, \quad A \text{ real,} \\ \alpha(\eta) &= \text{real constant.}\end{aligned} \tag{6.73}$$

From (6.72) and (6.73)

$$A_0(\eta) = (1 + A\,\mathrm{e}^{-\eta})^{-1/2}\,\mathrm{e}^{\mathrm{i}\alpha},$$

where A, α are real.

Equation (6.69) becomes

$$X_0(t, \eta) = 2\mathrm{Re}\{A_0(\eta)\,\mathrm{e}^{\mathrm{i}t}\} = 2(1 + A\,\mathrm{e}^{-\eta})^{-1/2}\cos(t + \alpha);$$

or

$$X_0(t, \eta) = 2(1 + A\,\mathrm{e}^{-\varepsilon t})^{-1/2}\cos(t + \alpha),$$

after writing $\eta = \varepsilon t$. From the initial conditions

$$x(0) = a, \quad \dot{x}(0) = 0$$

we obtain

$$a = 2(1+A)^{-1/2}\cos\alpha, \quad 0 = -2(1+A)^{-1/2}\sin\alpha + O(\varepsilon).$$

Therefore, to order ε, $\alpha = 0$ and $A = -1 + 4/a^2$. Finally

$$x(t) = 2[1 - (1 - 4/a^2)e^{-\varepsilon t}]^{-1/2}\cos t \qquad (6.74)$$

to order ε, at least when $t = O(\varepsilon^{-1})$. The limit cycle is given by $a = 2$.

6.5. Matching approximations on an interval

In this section we shall mainly be concerned with boundary-value problems, and we conform with the literature by using y as the dependent and x as the independent variable. For fuller information on the methods of this section the reader is referred to Nayfeh (1973) and O'Malley (1974).

As in Section 6.1, we shall illustrate the problem by looking at the approximation to a particular function y:

$$y(\varepsilon, x) = e^{-\frac{1}{2}x} - e^{\frac{1}{2}x}e^{-2x/\varepsilon}, \quad 0 \leqslant x \leqslant 1, \quad \varepsilon \geqslant 0. \qquad (6.75)$$

(The particular form chosen appears in the solution to a differential equation later on.)

Note that, for $x > 0$,

$$\lim_{\varepsilon \to 0+} (e^{-2x/\varepsilon}/\varepsilon^n) = 0$$

for every positive n: the function tends to zero very rapidly as $\varepsilon \to 0+$. Therefore, for every *fixed* $x > 0$

$$y(\varepsilon, x) \approx e^{-\frac{1}{2}x} \qquad (6.76)$$

with error $O(\varepsilon^n)$ for every $n > 0$. But by looking back at (6.75) it can be seen that as x takes smaller and smaller values, smaller and smaller values of ε are required before (6.76) becomes an acceptable approximation. It fails altogether at $x = 0$, where (6.75) gives zero and (6.76) gives 1. Therefore (6.76) is not a uniform approximation on $0 \leqslant x \leqslant 1$, and another form is needed near $x = 0$. Figure 6.3 shows the nature of the approximation for some particular values of ε; as ε decreases the interval of good fit becomes extended, but it is always poor near $x = 0$. The approximation begins to fail when x is comparable with ε in magnitude. The region of failure is called a *boundary layer* (from a hydrodynamical analogy), having in this case *thickness of order ε*. The function (6.76) is called the *outer approximation* to y.

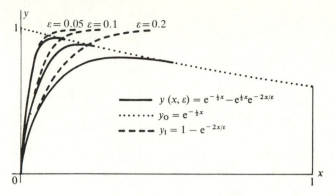

FIG. 6.3. Inner approximation, y_I, and outer approximation, y_O, to $y(\varepsilon, x)$ (eqn 6.75)

To get an approximation near $x = 0$ it is no use trying to work with x fixed: this is covered by (6.76). We therefore consider x tending to zero with ε by putting, say,

$$x(\varepsilon) = \xi\varepsilon, \tag{6.77}$$

where ξ may take any value. ξ is called a *stretched variable*: from one point of view we are magnifying the boundary layer to thickness $O(1)$. Then

$$y(\varepsilon, x) \approx 1 - e^{-2\xi}. \tag{6.78}$$

The error is $O(\varepsilon)$ for every fixed ξ, though naturally it works better when ξ is not too large. We can express this idea alternatively by saying

$$y(\varepsilon, x) \approx 1 - e^{-2x/\varepsilon} \tag{6.79}$$

with error $O(\varepsilon)$ so long as $x = O(\varepsilon)$ (which includes the case $x = o(\varepsilon)$ too). This is rather like the situation described for $t > 0$ at the beginning of Section 6.1. The approximation (6.79) is shown in Fig. 6.3. The function (6.79) is called the *inner approximation* to y.

If our information about y consisted only of the approximations (6.76) and (6.79), together with the associated error estimates, it would be pure guesswork, given a value of x and a value of ε, to decide which to use, since an error of low *order* in ε is not necessarily small for any *given* small value of ε. We can see what the regions of validity look like in a general way by experimenting and displaying the results in the plane of ε, x. The regions in which (6.76) and (6.79) give errors of less than 0.05 and 0.01 are shown in Fig. 6.4.

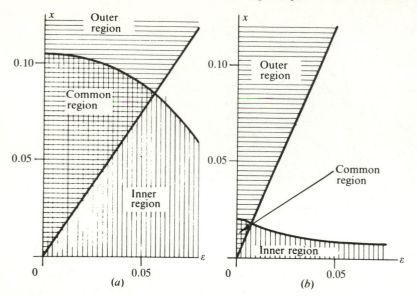

FIG. 6.4. Regions of validity for the approximations (6.76) (outer approximation) and (6.79) (inner approximation) with error E: (a) $E = 0.05$; (b) $E = 0.01$

The figures show that there is a region on the ε, x plane where (6.76) and (6.79) both give a small error. We therefore expect that there may be cases 'between' the cases $x = \text{constant}$ and $x = O(\varepsilon)$, in which both approximations have a small remainder. For example, if

$$x = \eta\sqrt{\varepsilon}, \quad \eta \text{ constant}, \tag{6.80}$$

$y(\varepsilon, x)$ becomes

$$e^{-\frac{1}{2}\eta\sqrt{\varepsilon}} - e^{\frac{1}{2}\eta\sqrt{\varepsilon}} e^{-2\eta/\sqrt{\varepsilon}} = 1 + O(\sqrt{\varepsilon}); \tag{6.81}$$

also (6.76) becomes

$$e^{-\frac{1}{2}\eta\sqrt{\varepsilon}} = 1 + O(\sqrt{\varepsilon}) \tag{6.82}$$

and (6.79) becomes

$$1 - e^{-2\eta/\sqrt{\varepsilon}} = 1 + O(\sqrt{\varepsilon}) \tag{6.83}$$

(in fact, $1 + o(\sqrt{\varepsilon})$). Thus the original function and both approximations have an error tending to zero with ε when $x = \eta\sqrt{\varepsilon}, \eta$ fixed. We say that the functions 'match' to $o(1)$ in the 'overlap region'.

Figure 6.5 indicates the progress of a point $(\varepsilon, \eta\sqrt{\varepsilon}) = (\varepsilon, x)$ as it moves into regions where the two approximations (6.76) and (6.79) have in common an error diminishing to zero with ε. It is desirable to show that there are no 'gaps' in the postulated 'common region'. Instead of (6.80),

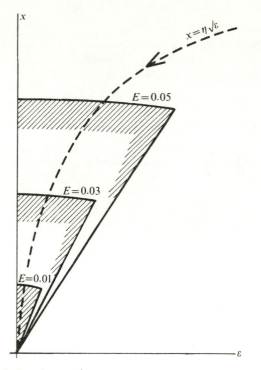

FIG. 6.5. As $\varepsilon \to 0$, the point $(\varepsilon, \eta\sqrt{\varepsilon})$ (η constant) always lies in the 'overlap region' for the approximations (6.76), (6.79). Error E

therefore, consider the more general case

$$x(\varepsilon) = \zeta\psi(\varepsilon),$$

where ζ is any constant and ψ tends to zero, but more slowly than ε, so that

$$\lim_{\varepsilon \to 0} \varepsilon/\psi(\varepsilon) = 0.$$

Then y, from (6.75), becomes

$$y(\varepsilon, x) = e^{-\frac{1}{2}\zeta\psi(\varepsilon)} - e^{\frac{1}{2}\zeta\psi(\varepsilon)}e^{-2\zeta\psi(\varepsilon)/\varepsilon} = 1 + o(1).$$

The outer approximation (6.76) gives

$$e^{-\frac{1}{2}x} = e^{-\frac{1}{2}\zeta\psi(\varepsilon)} = 1 + o(1),$$

and the inner approximation (6.79) gives

$$1 - e^{-2x/\varepsilon} = 1 - e^{-2\zeta\psi(\varepsilon)/\varepsilon} = 1 + o(1).$$

These are in agreement to $o(1)$.

The following example shows how the assumption of a 'common region' is used in connection with differential equations to establish an unknown constant. In this case we have put

$$\psi(\varepsilon) = \varepsilon^{1-\delta}, \quad 0 < \delta < 1,$$

to give a certain generality.

Example 6.4 *A function y has the two approximations on $0 \leqslant x \leqslant 1$: an inner approximation*

$$y(\varepsilon, x) \approx A + (1 - A)e^{-x/\varepsilon} = y_1, \text{ say}, \tag{6.84}$$

with error $O(\varepsilon)$ when $x = O(\varepsilon)$, and an outer approximation

$$y(\varepsilon, x) \approx e^{1-x} = y_O, \text{ say}, \tag{6.85}$$

with error $O(\varepsilon)$ for x constant. Find the value of A.

Assuming that both approximations are valid (though possibly with a larger error) for

$$x = \eta\varepsilon^{1-\delta}, \quad 0 < \delta < 1, \quad \eta \text{ constant},$$

we must have at least

$$\lim_{\substack{\varepsilon \to 0 \\ (\eta \text{ constant})}} \left[A + (1 - A)e^{-\eta/\varepsilon^\delta} \right] = \lim_{\substack{\varepsilon \to 0 \\ (\eta \text{ constant})}} e^{1-\eta\varepsilon^{1-\delta}}$$

Therefore $A = e^1$.

Example 6.5 *From (6.84) and (6.85) make up a single approximate expression applying uniformly on $0 \leqslant x \leqslant 1$.*

Write, (with $A = e^1$),

$$y_1 + y_O = e^1 + (1 - e^1)e^{-x/\varepsilon} + e^{1-x}$$

When x is constant and $\varepsilon \to 0$, we obtain

$$e^1 + e^{1-x}$$

instead of e^{1-x} required by (6.85); and when $x = \xi\varepsilon$ and $\varepsilon \to 0$ (ξ constant), we obtain

$$2e^1 + (1 - e^1)e^{-\xi}$$

instead of $e^1 + (1 - e^1)e^{-\xi}$ required by (6.84). Therefore the required uniform expansion is given by

$$y_I + y_O - e^1 = (1 - e^1)e^{-x/\varepsilon} + e^{1-x}.$$

6.6. A matching technique for differential equations

In the following problems the solution has different approximations in different (x, ε) regions as described in Section 6.5. Characteristic of these is *the presence of ε as the coefficient of the highest derivative appearing in the equation.* The term in the highest derivative is, then, negligible except where the derivative itself is correspondingly large. Thus, over most of the range the equation is effectively of lower order, and can satisfy fewer boundary conditions. However, in certain intervals (boundary layers) the derivative may be large enough for the term to be significant. In such intervals, y will change very rapidly, and may be caused to satisfy another boundary condition.

Example 6.6 (An initial-value problem) *Find a first-order approximation to the solution of*

$$\varepsilon\frac{dy}{dx} + y = x, \quad x > 0, \quad \varepsilon > 0, \tag{6.86}$$

subject to

$$y(\varepsilon, 0) = 1. \tag{6.87}$$

The obvious first step is to put $\varepsilon = 0$ into (6.86): this gives the 'outer approximation'

$$y \approx y_O = x. \tag{6.88}$$

The error is $O(\varepsilon)$, as can be seen by viewing it as the first step in a perturbation process. It is not clear where this holds, but assume we have numerical or other indications that it is right except near $x = 0$, where there is a boundary layer. We do not know how thick this is, so we allow some freedom in choosing a new, stretched variable, ξ, by writing

$$x = \xi\phi(\varepsilon) \tag{6.89}$$

where

$$\lim_{\varepsilon \to 0} \phi(\varepsilon) = 0.$$

Equation (6.86) becomes

$$\frac{\varepsilon}{\phi(\varepsilon)}\frac{dy}{d\xi} + y = \xi\phi(\varepsilon). \tag{6.90}$$

The choice of $\phi(\varepsilon)$ which retains $dy/d\xi$ as a leading term in order of magnitude, and which gives an equation which can describe a boundary layer is

$$\phi(\varepsilon) = \varepsilon; \tag{6.91}$$

and (6.90) becomes

$$\frac{dy}{d\xi} + y = \varepsilon\xi. \tag{6.92}$$

The first approximation to the solution fitting the initial condition (6.87) (the inner approximation), is given by

$$y \approx y_1 = e^{-\xi} \tag{6.93}$$

with error $O(\varepsilon)$ as before. This may be interpreted

$$y_1 = e^{-x/\varepsilon}, \quad x = O(\varepsilon). \tag{6.94}$$

There is no scope for 'matching' y_0 and y_1 here: either they are approximations to the same solution or they are not; there are no arbitrary constants left to be settled. We note, however, that in an 'intermediate region' such as

$$x = \eta\sqrt{\varepsilon}$$

for η constant, both approximations do agree to order $\sqrt{\varepsilon}$.

To construct a uniform approximation, start with the form $y_1 + y_0$ as in Example 6.5

$$y_1 + y_0 = e^{-x/\varepsilon} + x. \tag{6.95}$$

This agrees to order ε with eqns (6.88) and (6.93) for $x = $ constant and $x = \xi\varepsilon$ respectively, so that eqn (6.95) is already a uniform approximation. (The reader should compare the exact solution

$$y(\varepsilon, x) = x - \varepsilon + e^{-x/\varepsilon} + \varepsilon e^{-x/\varepsilon}.)$$

Example 6.7 (A boundary-value problem)   *Find a first approximation to the equation*

$$\varepsilon\frac{d^2y}{dx^2} + 2\frac{dy}{dx} + y = 0, \quad 0 < x < 1 \tag{6.96}$$

subject to

$$y(0) = 0, \quad y(1) = 1. \tag{6.97}$$

Putting $\varepsilon = 0$ in (6.96) (or finding the first term in an ordinary perturbation process), gives the differential equation for the outer approximation $y_0(\varepsilon, x)$:

$$2\frac{dy_0}{dx} + y_0 = 0. \tag{6.98}$$

Since this is first order, only one boundary condition can be satisfied. We shall assume (if necessary by showing that the contrary assumption leads to failure of the method) that (6.98) holds approximately at $x = 1$. We require $y_0(\varepsilon, 1) = 1$, so

$$y \approx y_0(x, \varepsilon) = e^{1/2} e^{-\frac{1}{2}x} \tag{6.99}$$

with error $O(\varepsilon)$. The non-uniformity is not self-evident, but certainly the boundary condition at $x = 0$ is not satisfied. Assuming that the condition is attained by a sudden change in the nature of the solution near $x = 0$, introduce a new, stretched, variable ξ, where

$$x = \xi\phi(\varepsilon), \tag{6.100}$$

ξ fixed, where

$$\lim_{\varepsilon \to 0} \phi(\varepsilon) = 0. \tag{6.101}$$

Equation (6.96) becomes

$$\frac{\varepsilon}{\phi^2(\varepsilon)} \frac{d^2y}{d\xi^2} + \frac{2}{\phi(\varepsilon)} \frac{dy}{d\xi} + y = 0.$$

The choice

$$\phi(\varepsilon) = \varepsilon \tag{6.102}$$

yields the equation

$$\frac{d^2y}{d\xi^2} + 2\frac{dy}{d\xi} + \varepsilon y = 0 . \tag{6.103}$$

which simplifies to the equation for the inner approximation y_1:

$$\frac{d^2y_1}{d\xi^2} + 2\frac{dy_1}{d\xi} = 0 \tag{6.104}$$

with boundary condition

$$y_1(\varepsilon, 0) = 0. \tag{6.105}$$

Therefore

$$y \approx y_1(\varepsilon, x) = A(1 - e^{-2\xi}) \tag{6.106}$$

where ξ is a constant, the error being $O(\varepsilon)$.

The value of A must be determined by the condition that (6.99) and (6.106) should both approximate to the same function (the solution): if there is an 'overlapping region' of approximation to some order, the approximations themselves must agree to this order, and there is only one choice of A making this agreement possible.

Allow the behaviour of x to be given by

$$x = \eta\psi(\varepsilon) \tag{6.107}$$

where η is any constant, and

$$\lim_{\varepsilon \to 0} \psi(\varepsilon) = 0 \tag{6.108}$$

but where ψ does not tend to zero so fast as ϕ, so that

$$\lim_{\varepsilon \to 0} \phi(\varepsilon)/\psi(\varepsilon) = \lim_{\varepsilon \to 0} \varepsilon/\psi(\varepsilon) = 0. \qquad (6.109)$$

(Previously we put $\psi(\varepsilon) = \sqrt{\varepsilon}$, which satisfies these conditions.) Then

$$y_0(\varepsilon, x) = e^{1/2} e^{-\frac{1}{2}\eta\psi(\varepsilon)} = e^{\frac{1}{2}} + O(\psi) \qquad (6.110)$$

by (6.108), and

$$y_1(\varepsilon, x) = A(1 - e^{-2\eta\psi/\varepsilon}) = A + o(1) \qquad (6.111)$$

by (6.109). Agreement is only possible if $A = e^{1/2}$, so finally

and
$$y \approx y_0(\varepsilon, x) = e^{\frac{1}{2}} e^{-\frac{1}{2}x} \quad \text{for} \quad x = O(1) \qquad (6.112)$$
$$y \approx y_1(\varepsilon, x) = e^{1/2}(1 - e^{-2x/\varepsilon}) \quad \text{for} \quad x = O(\varepsilon). \qquad (6.113)$$

To find an approximation uniform on $0 \leqslant x \leqslant 1$, form

$$y_0 + y_1 = e^{1/2} e^{-\frac{1}{2}x} + e^{1/2}(1 - e^{-2x/\varepsilon}).$$

Then for $x = O(1)$ this becomes $e^{1/2} e^{-\frac{1}{2}x} + e^{1/2}$ as $\varepsilon \to 0$; that is it has the unwanted term $e^{1/2}$ in it. Similarly, when $x = \xi\varepsilon$ it becomes $e^{1/2} + e^{1/2}(1 - e^{-2\xi})$ as $\varepsilon \to 0$, which again contains an unwanted $e^{1/2}$. Therefore

$$y \approx y_0 + y_1 - e^{1/2} = e^{1/2}(e^{-\frac{1}{2}x} - e^{-2x/\varepsilon}) \qquad (6.114)$$

is a uniform approximation on $0 \leqslant x \leqslant 1$.

The reader should compare the exact solution

$$y(\varepsilon, x) = (e^{\lambda_1 x} - e^{\lambda_2 x})/(e^{\lambda_1} - e^{\lambda_2}) \qquad (6.115)$$

where λ_1, λ_2 are the roots of $\varepsilon\lambda^2 + 2\lambda + 1 = 0$: the expansions of the roots are

$$\lambda_1 = -\tfrac{1}{2} - \tfrac{1}{8}\varepsilon - \dots,$$
$$\lambda_2 = -\frac{2}{\varepsilon} + \tfrac{1}{2} + \tfrac{1}{8}\varepsilon + \dots.$$

The expansion and matching process may be extended to take in terms beyond the first-order terms we have considered here. Details are given by Nayfeh (1973).

Exercises

1. Work through the details of Example 6.1 to obtain an approximate solution of $\ddot{x} + x = \varepsilon x^3$, with $x(\varepsilon, 0) = 1$, $\dot{x}(\varepsilon, 0) = 0$, with error $O(\varepsilon^3)$ uniformly on $t \geqslant 0$.

Obtain the coefficients of $\cos t$ and $\cos 3t$ to order ε^3 for the approximate pendulum equation.

2. How does the period obtained by the method of Exercise 1 compare with that derived in Exercise 34, Chapter 1?

3. Apply the method of Exercise 1 to the equation

$$\ddot{x} + x = \varepsilon x^3 + \varepsilon^2 \alpha x^5$$

with $x(\varepsilon, 0) = 1$, $\dot{x}(\varepsilon, 0) = 0$. Obtain the period to order ε^3, and confirm that the period is correct for the pendulum when the right-hand side is the first two terms in the expansion of $x - \sin x$. (Compare the result of Exercise 33, Chapter 1. To obtain the required equations simply add the appropriate term to the right-hand side of the equation for x_2 in Example 6.1.)

4. Use a substitution to show that the case considered in Exercise 1 covers all boundary conditions $x(\varepsilon, 0) = a$, $\dot{x}(\varepsilon, 0) = 0$.

5. The equation for the relativistic perturbation of a planetary orbit is

$$\frac{d^2 u}{d\theta^2} + u = k(1 + \varepsilon u^2)$$

(see Exercise 21, Chapter 5). Apply the coordinate perturbation technique to eliminate the secular term in $u_1(\theta)$ in the expansion $u(\varepsilon, \theta) = u_0(\theta) + \varepsilon u_1(\theta) + \ldots$, with $\theta = \phi + \varepsilon T_1(\phi) + \ldots$. Confirm that the perihelion of the orbit advances by approximately $2\pi\varepsilon k^2$ in each planetary year.

6. Apply the multiple-scale method to van der Pol's equation $\ddot{x} + \varepsilon(x^2 - 1)\dot{x} + x = 0$. Show that, for $x(0) = a$, $\dot{x}(0) = 0$, and for $t = O(\varepsilon^{-1})$,

$$x = 2\{1 + (4/a^2 - 1)e^{-\varepsilon t}\}^{-1/2} \cos t.$$

7. Apply the multiple-scale method to the equation $\ddot{x} + x - \varepsilon x^3 = 0$, with initial conditions $x(0) = a$, $\dot{x}(0) = 0$. Show that, for $t = O(\varepsilon^{-1})$,

$$x(t) = a \cos \{t(1 - \tfrac{3}{8}\varepsilon a^2)\}.$$

8. Obtain the exact solution of Example 6.7, and show that it has the first approximation equal to that obtained by the matching method.

9. Consider the problem

$$\varepsilon y'' + y' + y = 0, \quad y(\varepsilon, 0) = 0, \quad y(\varepsilon, 1) = 1,$$

on $0 \leqslant x \leqslant 1$, where ε is small and positive.
  (i) Obtain the outer approximation

$$y(\varepsilon, x) \approx y_0 = e^{1-x}, \quad x \text{ fixed}, \quad \varepsilon \to 0+\,;$$

and the inner approximation

$$y(\varepsilon, x) \approx y_1 = C(1 - e^{-x/\varepsilon}), \quad x = O(\varepsilon), \quad \varepsilon \to 0+,$$

where C is a constant.

(ii) Obtain the value of C by matching y_O and y_i in the intermediate region.

(iii) Construct from y_O and y_i a first approximation to the solution which is uniform on $0 \leqslant x \leqslant 1$.

10. Repeat the procedure of Exercise 9 for the problem

$$\varepsilon y'' + y' + xy = 0, \quad y(0) = 0, \quad y(1) = 1$$

on $0 \leqslant x \leqslant 1$.

11. Explain why the procedure of Exercise 9 fails for the van der Pol equation $\varepsilon \ddot{x} + (x^2 - 1)\dot{x} + x = 0, x(\varepsilon, 0) = 2, x(\varepsilon, 1) = 1$.

12. By using the method of multiple scales, with variables x and $\xi = x/\varepsilon$, obtain a first approximation uniformly valid on $0 \leqslant x \leqslant 1$ to the solution of

$$\varepsilon y'' + y' + xy = 0, \quad y(\varepsilon, 0) = 0, \quad y(\varepsilon, 1) = 1,$$

on $0 \leqslant x \leqslant 1$, with $\varepsilon > 0$. Show that the result agrees to order ε with that of Exercise 10.

13. The steady flow of a conducting liquid between insulated parallel plates under the influence of a transverse magnetic field satisfies

$$w'' + Mh' = -1, \quad h'' + Mw' = 0, \quad w = (\pm 1) = h(\pm 1) = 0,$$

where, in dimensionless form, w is the fluid velocity, h the induced magnetic field, and M is the Hartmann number. By putting $p = w + h$ and $q = w - h$, find the exact solution. Alternatively, solve the equations as a singular perturbation problem for large M.

14. Obtain an approximation, to order ε and for $t = O(\varepsilon^{-1})$, to the solutions of $\ddot{x} + 2\varepsilon\dot{x} + x = 0$, by using the method of multiple scales with the variables t and $\eta = \varepsilon t$.

15. Use the method of multiple scales to obtain a uniform approximation to the solutions of the equation $\ddot{x} + \omega^2 x + \varepsilon x^3 = 0$, in the form

$$x(\varepsilon, t) \approx a_0 \cos \left[\{\omega_0 + (3\varepsilon a_0^2/8\omega_0)\}t + \alpha \right],$$

α constant. Explain why the approximation is uniform, and not merely valid for $t = O(\varepsilon^{-1})$.

16. Use the coordinate perturbation technique to obtain the first approximation

$$x = \tau^{-1}, \quad t = \tau + \tfrac{1}{2}\varepsilon\tau(1 - \tau^{-2})$$

to the solution of

$$(t + \varepsilon x)\dot{x} + x = 0, \quad x(\varepsilon, 1) = 1, \quad 0 \leqslant x \leqslant 1.$$

Confirm that the approximation is, in fact, the exact solution, and that an alternative approximation

$$x = \tau^{-1} + \tfrac{1}{2}\varepsilon\tau^{-1}, \quad t = \tau - \tfrac{1}{2}\varepsilon\tau^{-1}$$

is correct to order ε, for fixed τ.

17. Apply the method of multiple scales, with variables t and $\eta = \varepsilon t$, to van der Pol's equation $\ddot{x} + \varepsilon(x^2 - 1)\dot{x} + x = 0$. Show that, for $t = O(\varepsilon^{-1})$,

$$x(\varepsilon, t) = \frac{2a_0^{\frac{1}{2}} e^{\frac{1}{2}\varepsilon t}}{\sqrt{(1 + a_0\, e^{\varepsilon t})}} \cos{(t + \alpha)} + O(\varepsilon),$$

where a_0 is the initial radius.

18. Use the method of matched approximations to obtain a uniform approximation to the solution of

$$\varepsilon\left(y'' + \frac{2}{x}y'\right) - y = 0, \quad y(\varepsilon, 0) = 0, \quad y'(\varepsilon, 1) = 1,$$

($\varepsilon > 0$) on $0 \leqslant x \leqslant 1$. Show that there is a boundary layer of thickness $O(\varepsilon^{1/2})$ near $x = 1$, by putting $1 - x = \xi\phi(\varepsilon)$.

19. Use the method of matched approximations to obtain a uniform approximation to the solution of the problem

$$\varepsilon(y'' + y') - y = 0, \quad y(\varepsilon, 0) = 1, \quad y(\varepsilon, 1) = 1,$$

($\varepsilon > 0$) given that there are boundary layers at $x = 0$ and $x = 1$. Show that both boundary layers have thickness $O(\varepsilon^{1/2})$. Compare with the exact solution.

20. Obtain a first approximation, uniformly valid on $0 \leqslant x \leqslant 1$, to the equation

$$\varepsilon y'' + \frac{1}{1+x}y' + \varepsilon y = 0,$$

with $y(\varepsilon, 0) = 0$, $y(\varepsilon, 1) = 1$.

21. Apply the Lighthill technique to obtain a uniform approximation to the solution of

$$(t + \varepsilon x)\dot{x} + x = 0, \quad x(\varepsilon, 1) = 1, \quad 0 \leqslant x \leqslant 1.$$

(Compare Exercise 16.)

22. Obtain a first approximation, uniform on $0 \leqslant x \leqslant 1$, to the solution of

$$\varepsilon y' + y = x, \quad y(\varepsilon, 0) = 1,$$

using inner and outer approximations. Compare the exact solution and explain geometrically why the outer approximation is independent of the boundary condition.

23. Use the method of multiple scales with variables t and $\eta = \varepsilon t$ to show that, to a first approximation, the response of the van der Pol equation to a 'soft' forcing term described by

$$\ddot{x} + x = \varepsilon\{(1-x^2)\dot{x} + \gamma \cos \omega t\}, \quad \varepsilon > 0,$$

is the same as the unforced response. (Consider $|\omega|$ not near to 1; that is, the non-resonant case.)

24. Repeat Exercise 23 for

$$\ddot{x} + x = \varepsilon(1-x^2)\dot{x} + \Gamma \cos \omega t, \quad \varepsilon > 0,$$

where $\Gamma = O(1)$ and $|\omega|$ is not near 1. Show that

$$x(\varepsilon, t) = \frac{\Gamma}{1-\omega^2} \cos \omega t + O(\varepsilon), \quad \Gamma^2 \geqslant 2(1-\omega^2)^2;$$

and that for $\Gamma^2 < 2(1-\omega^2)^2$

$$x(\varepsilon, t) = 2\{1 - \Gamma^2/2(1-\omega^2)^2\}^{1/2} \cos t + \frac{\Gamma}{1-\omega^2} \cos \omega t + O(\varepsilon).$$

25. Apply the matching technique to the damped pendulum equation

$$\varepsilon\ddot{x} + \dot{x} + \sin x = 0, \quad x(\varepsilon, 0) = 1, \quad \dot{x}(\varepsilon, 0) = 1,$$

for ε small and positive. Show that the inner and outer approximations are given by

$$x_I = 1, \quad x_O = 2 \tan^{-1}\{e^{-t} \tan \tfrac{1}{2}\}.$$

(The pendulum has strong damping and strong restoring action, but the damping dominates.)

26. The equation for a tidal bore on a shallow stream is

$$\varepsilon\frac{d^2\eta}{d\xi^2} - \frac{d\eta}{d\xi} - \eta + \eta^2 = 0;$$

where (in appropriate dimensions) η is the height of the free surface, and $\xi = x - ct$, where c is the wave speed. For $0 < \varepsilon \ll 1$, find the equilibrium points for the equation and classify them according to their linear approximations. Apply the coordinate perturbation method to the equation for the phase paths,

$$\varepsilon\frac{dw}{d\eta} = \frac{w^2 + \eta - \eta^2}{w},$$

and show that

$$w = -\zeta + \zeta^2 + O(\varepsilon^2), \quad \eta = \zeta - \varepsilon(-\zeta + \zeta^2) + O(\varepsilon^2).$$

Confirm that, to this degree of approximation, a separatrix from the origin reaches the other equilibrium point. Interpret the result in terms of the shape of the bore.

27. Following Exercise 35, Chapter 2, on the motion of a satellite, show that the equation can be written in the form

$$\frac{d^2 z}{d\tau^2} = -\frac{1-\mu}{z^2} + \frac{\mu}{(1-z)^2},$$

where

$$\tau = \gamma^{1/2}(m_1 + m_2)^{1/2} a^{-3/2} t, \quad \mu = m_2/(m_1 + m_2), \quad z = x/a.$$

Show that, on the path with zero energy,

$$\tfrac{1}{2} z'^2 = \frac{1-\mu}{z} + \frac{\mu}{1-z}.$$

Construct the first two terms of a perturbation solution of the form

$$\tau(\mu, z) = T_0(z) + \mu T_1(z) + \ldots, \quad \tau(\mu, 0) = 0,$$

valid for small μ, and indicate for what values of z it is valid.

28. Replace the time, t, in van der Pol's equation by the 'slow' variable $\tau = \varepsilon t$, so that $\ddot{x} + \varepsilon(x^2 - 1)\dot{x} + x = 0$ becomes

$$\varepsilon^2 x'' + \varepsilon^2 (x^2 - 1)x' + x = 0$$

where the dashes represent derivatives with respect to τ.

Attempt an asymptotic solution

$$x \sim u_0(\varepsilon, t, \tau) + \varepsilon u_1(\varepsilon, t, \tau) + \varepsilon^2 u_2(\varepsilon, t, \tau) + \ldots$$

for $\varepsilon \to 0$. Show that u_0 must satisfy $(\partial^2 u_0/\partial t^2) + u_0 = 0$. Write the solution in the form

$$u_0 = a(\varepsilon, \tau)\cos(t + b(\varepsilon, \tau))$$

and assume expansions of the form

$$a(\varepsilon, \tau) \sim a_0(\tau) + \varepsilon a_1(\tau) + \ldots, \quad b(\varepsilon, \tau) \sim b_0(\tau) + \varepsilon b_1(\tau) + \ldots.$$

Show that u_1 must satisfy

$$u_{1_{tt}} + u_1 = (2a_0' - a_0 + \tfrac{1}{4} a_0^3)\sin(t + B) + 2a_0 b_0' \cos(t + B) + \tfrac{1}{4} a_0^3 \sin 3(t + B),$$

and that the secular terms vanish if

$$2a_0' - a_0 + \tfrac{1}{4} a_0^3 = 0, \quad b_0' = 0.$$

Assume the initial conditions $x(\varepsilon, 0) = 2$, $\dot{x}(\varepsilon, 0) = 0$, and deduce that

$$x \sim 2\cos(t - \tfrac{1}{16}\varepsilon^2 t) + \tfrac{3}{4}\varepsilon \sin(t - \tfrac{1}{16}\varepsilon^2 t) - \tfrac{1}{4}\varepsilon \sin(3t - \tfrac{3}{16}\varepsilon^2 t).$$

7 Forced oscillations: harmonic and subharmonic response, stability, entrainment

WE CONSIDER second-order differential equations of the general form $\ddot{x} + f(x, \dot{x}) = F \cos \omega t$, which have a periodic forcing term. When the equation is linear the structure of its solutions is very simple. There are two parts combined additively. One part, the 'free oscillation' is a linear combination of the solutions of the homogeneous equation and involves the initial conditions. The second part, the 'forced solution', is proportional to F and is independent of initial conditions. When damping is present the free oscillation dies away in time. This independence of the free and the forced components, and the ultimate independence of initial conditions when there is damping allows a very restricted range of phenomena: typically the only things to look at are the amplitude and phase of the ultimate response, the rate of decay of the free oscillation, and the possibility of resonance. When the equation is nonlinear, on the other hand, there is no such simple separation possible, and the resulting interaction between the free and forced terms, especially when a self-excited motion is possible, and the enduring importance of the initial conditions in some cases, generates a range of entirely new phenomena. The present chapter is concerned with using the method of harmonic balance (Chapter 4) and the perturbation method (Chapter 5) to obtain approximate solutions which clearly show these new features.

7.1. General forced periodic solutions

Consider, for definiteness, Duffing's equation

$$\ddot{x} + k\dot{x} + \alpha x + \beta x^3 = \Gamma \cos \omega t, \tag{7.1}$$

of which various special cases have appeared in earlier chapters. Suppose that $x(t)$ is a periodic solution with period $2\pi/\lambda$. Then $x(t)$ can be represented by a Fourier series for all t:

$$x(t) = a_0 + a_1 \cos \lambda t + b_1 \sin \lambda t + a_2 \cos 2\lambda t + \ldots. \tag{7.2}$$

If this series is substituted into (7.1) the nonlinear term x^3 is periodic, and so generates another Fourier series. When the contributions are assembled eqn (7.1) takes the form

$$A_0 + A_1 \cos \lambda t + B_1 \sin \lambda t + A_2 \cos 2\lambda t + \ldots = \Gamma \cos \omega t \qquad (7.3)$$

for all t, the coefficients being functions of $a_0, a_1, b_1, a_2 \ldots$. Matching the two sides gives, in principle, an infinite set of equations for $a_0, a_1, a_2, \ldots$ and enables λ to be determined. For example, the obvious matching

$$\lambda = \omega$$

and

$$A_1 = \Gamma, \quad A_0 = B_1 = A_2 = \ldots = 0$$

resembles the harmonic balance method of Section 4.4; though in the harmonic balance method a drastically truncated series (7.2), consisting of only the single term $a_1 \cos \lambda t$, is employed. The above matching leads to a harmonic (or autoperiodic) response to frequency ω.

A less obvious matching can sometimes be achieved when

$$\lambda = \omega/n \quad (n \text{ an integer});$$

$$A_n = \Gamma;$$

$$A_i = 0 \ (i \neq n); \quad B_i = 0 \ (\text{all } i).$$

If solutions of this set exist, the system will deliver responses of period $2\pi n/\omega$ and frequency ω/n. These are called *subharmonics* of order $\frac{1}{2}, \frac{1}{3}, \ldots$. Not all of these actually occur (see Section 7.7 and Exercises 16 and 17).

It is important to realize where the terms in (7.3) come from: there is 'throwback' from terms of high order in (7.2) to contribute to terms of low order in (7.3). For example, in the expansion of x^3:

$$x^3(t) = (a_0 + \ldots + a_{11} \cos 11\omega t + \ldots + a_{21} \cos 21\omega t + \ldots)^3,$$

one of the terms is

$$3a_{11}^2 a_{21} \cos^2 11\omega t \cos 21\omega t = 3a_{11}^2 a_{21} (\tfrac{1}{2} \cos 21\omega t + \tfrac{1}{4} \cos 43\omega t + \tfrac{1}{4} \cos \omega t),$$

which includes a term in $\cos \omega t$. Normally we assume that terms in (7.2) above a certain small order are negligible, that is, that the coefficients are small. Hopefully, then, the combined throwback as measured by the modification of coefficients by high-order terms will be small.

An alternative way of looking at the effect of a nonlinear term is as a feedback. Consider the undamped form of (7.1) (with $k = 0$) written as

$$\mathscr{L}(x) \equiv \left(\frac{\mathrm{d}^2}{\mathrm{d}t^2} + \alpha\right)x = -\beta x^3 + \Gamma \cos \omega t.$$

We can represent this equation by the block diagram shown in Fig. 7.1.

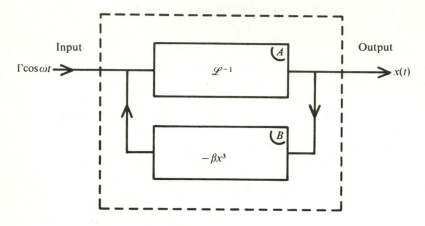

FIG. 7.1. Feedback diagram for $\ddot{x} + \alpha x + \beta x^3 = \Gamma \cos \omega t$

Regard $\Gamma \cos \omega t$ as the input to the system and $x(t)$ as the output. The box A represents the operator $\mathscr{L}^{-1}$ which solves the equation $\ddot{x} + \alpha x = f(t)$ for a given input f and for assigned initial conditions. Here, the input to A is equal to the current value of $-\beta x^3 + \Gamma \cos \omega t$. Its output is $x(t)$. Suppose the output from A is assumed to be simple, containing only a few harmonic components. Then B generates a shower of harmonics of higher and possibly of lower orders which are fed back into A. The higher harmonics are the most attenuated on passing through A (roughly like n^{-2} where n is the order). It is therefore to be expected that a satisfactory consistency between the inputs might be obtained by a representation of in terms only of the lowest harmonics present. The low-order approximation should be most adequate when the lowest harmonic present is *amplified* by $\mathscr{L}^{-1}$; that is, when $\omega^2 \simeq n^2\alpha$, being a state of near-resonance.

7.2. Harmonic solutions, transients, and stability for Duffing's equation

As a direct illustration of the method described in Section 7.1, consider the undamped Duffing equation (7.1) in the form

$$\ddot{x} + x + \beta x^3 = \Gamma \cos \omega t, \tag{7.4}$$

where $\Gamma > 0$. As an approximation to the solution we use the truncated Fourier series

$$x(t) = a \cos \omega t + b \sin \omega t. \tag{7.5}$$

This form allows for a possible phase difference between the forcing term and the solution. The omission of the constant term involves fore-knowledge that the only relevant solutions of reasonably small amplitude will give this constant as zero. Then

$$\ddot{x}(t) = -a\omega^2 \cos \omega t - b\omega^2 \sin \omega t \tag{7.6}$$

and

$$
\begin{aligned}
x^3(t) &= a^3 \cos^3 \omega t + 3a^2 b \cos^2 \omega t \sin \omega t + 3ab^2 \cos \omega t \sin^2 \omega t + b^3 \sin^3 \omega t \\
&= \tfrac{3}{4} a (a^2 + b^2) \cos \omega t + \tfrac{3}{4} b (a^2 + b^2) \sin \omega t \\
&\quad + \tfrac{1}{4} a (a^2 - 3b^2) \cos 3\omega t + \tfrac{1}{4} b (3a^2 - b^2) \sin 3\omega t. \tag{7.7}
\end{aligned}
$$

We disregard the terms in $\cos 3\omega t$, $\sin 3\omega t$ on the grounds that, regarded as feedback input to the differential equation, the attenuation will be large compared with that of terms having the fundamental frequency. When (7.5), (7.6), (7.7) are substituted into (7.4) and the coefficients of $\cos \omega t$, $\sin \omega t$ are matched we obtain

$$b\{(\omega^2 - 1) - \tfrac{3}{4}\beta(a^2 + b^2)\} = 0, \tag{7.8}$$

$$a\{(\omega^2 - 1) - \tfrac{3}{4}\beta(a^2 + b^2)\} = -\Gamma. \tag{7.9}$$

The only admissible solution of (7.8) and (7.9) requires $b = 0$, and the corresponding values of a are the solutions of (7.9):

$$\tfrac{3}{4}\beta a^3 - (\omega^2 - 1)a - \Gamma = 0. \tag{7.10}$$

The roots are given by the intersections, in the plane of a and z, of the curves

$$z = -\Gamma, \quad z = -\tfrac{3}{4}\beta a^3 + (\omega^2 - 1)a.$$

From Fig. 7.2 it can be seen that, for $\beta < 0$, there are three solutions for Γ small and one for Γ larger when $\omega^2 < 1$, and that when $\omega^2 > 1$ there is one solution only. This reproduces the results of Sections 5.4 and 5.5. The oscillations are in phase with the forcing term when the critical value of a is positive and out of phase by a half cycle when a is negative.

FIG. 7.2. Graph of $z = -\frac{3}{4}\beta a^3 + (\omega^2 - 1)a$ for $\beta < 0$ and (a) $\omega^2 < 1$; (b) $\omega^2 > 1$

We do not yet know the stability of these various oscillations and it is necessary to decide this matter since an unstable oscillation will not occur in practice. To investigate stability we shall look at the 'transient' conditions, which lead to or away from periodic states, by supposing the coefficients a and b to be slowly-varying functions of time, at any rate near to periodic states. Assume that

$$x(t) = a(t)\cos \omega t + b(t)\sin \omega t, \tag{7.11}$$

where a and b are *slowly-varying amplitudes* (compared with $\cos \omega t$ and $\sin \omega t$).

Then

$$\dot{x}(t) = (\dot{a} + \omega b)\cos \omega t + (-\omega a + \dot{b})\sin \omega t, \tag{7.12}$$

and (neglecting $\ddot{a}, \ddot{b}$)

$$\ddot{x}(t) = (-\omega^2 a + 2\omega\dot{b})\cos \omega t + (-2\omega\dot{a} - \omega^2 b)\sin \omega t. \tag{7.13}$$

Also, as in (7.7),

$$x^3(t) = \frac{3}{4}a(a^2 + b^2)\cos \omega t + \frac{3}{4}b(a^2 + b^2)\sin \omega t$$

$$+ \text{harmonics in } \cos 3\omega t, \sin 3\omega t. \tag{7.14}$$

As before we shall ignore the terms in $\cos 3\omega t, \sin 3\omega t$.

When (7.12), (7.13), (7.14) are substituted into the differential equation

(7.4), and the terms are rearranged, we have

$$[2\omega\dot{b} - a\{(\omega^2 - 1) - \tfrac{3}{4}\beta(a^2 + b^2)\}]\cos\omega t$$
$$+ [-2\omega\dot{a} - b\{(\omega^2 - 1) - \tfrac{3}{4}\beta(a^2 + b^2)\}]\sin\omega t = \Gamma\cos\omega t. \quad (7.15)$$

Appealing again to the supposition that a and b are slowly-varying we may approximately match the coefficients of $\cos\omega t$ and $\sin\omega t$, giving the *autonomous system*

$$\dot{a} = -\frac{1}{2\omega}b\{(\omega^2 - 1) - \tfrac{3}{4}\beta(a^2 + b^2)\} \equiv A(a, b), \quad \text{say}; \quad (7.16)$$

$$\dot{b} = \frac{1}{2\omega}a\{(\omega^2 - 1) - \tfrac{3}{4}\beta(a^2 + b^2)\} + \frac{\Gamma}{2\omega} \equiv B(a, b), \quad \text{say.} \quad (7.17)$$

Initial conditions are given in terms of those for the original equation, (7.4), by

$$a(0) = x(0), \quad b(0) = \dot{x}(0)/\omega. \quad (7.18)$$

The phase plane for a, b in the system above is called the *van der Pol plane*. The equilibrium points, given by $A(a, b) = B(a, b) = 0$, represent the steady periodic solutions already obtained. The other paths correspond to solutions of (7.4) which are non-periodic in general. The phase diagram computed for a particular case is shown in Fig. 7.3. The point $(a_3, 0)$ is a saddle and represents an unstable oscillation, and $(a_1, 0), (a_2, 0)$ are centres.

The equilibrium points may be analysed algebraically as follows. Consider a case when there are three equilibrium points

$$\omega^2 < 1, \quad \beta < 0 \quad (7.19)$$

and let a_0 represent any one of the values $a = a_1, a_2$, or a_3 of Fig. 7.2(a). Putting

$$a = a_0 + \xi, \quad (7.20)$$

the local linear approximation to (7.16), (7.17) is

$$\dot{\xi} = A_1(a_0, 0)\xi + A_2(a_0, 0)b,$$
$$\dot{b} = B_1(a_0, 0)\xi + B_2(a_0, 0)\xi, \quad (7.21)$$

where $A_1(a, b) = \partial A(a, b)/\partial a$, and so on. It is easy to confirm that

$$A_1(a_0, 0) = B_2(a_0, 0) = 0,$$

$$A_2(a_0, 0) = -\frac{\omega^2 - 1}{2\omega} + \frac{3\beta a_0^2}{8\omega} = -\frac{z_0}{a_0},$$

FIG. 7.3. Paths in the van der Pol plane for the undamped pendulum, eqns (7.16), (7.17);
$\Gamma = 0.005$, $\omega = 0.975$, $\beta = -0.167$ showing the equilibrium points

(where, in Fig. 7.2(a), z_0 is the ordinate at a_0), and that

$$B_1(a_0, 0) = s_0$$

(where s_0 is the slope of the curve in Fig. 7.2(a) at a_0). Therefore (7.21) can be written

$$\dot{\xi} = -\frac{z_0}{a_0}b, \quad \dot{b} = s_0\xi.$$

By considering the signs of a_0, z_0, s_0 in Fig. 7.2(a) it can be seen that $(a_1, 0)$, $(a_2, 0)$ are centres for the linear system (7.21), and that $(a_3, 0)$ is a saddle. The saddle will never be observed exactly, though if a state can be set up near enough to this point, $\dot{a}$ and $\dot{b}$ will be very small and a nearly periodic motion at the forcing frequency may linger long enough to be observed.

We may introduce a damping term into (7.4) so that it becomes

$$\ddot{x} + k\dot{x} + x + \beta x^3 = \Gamma \cos \omega t, \quad k > 0. \tag{7.22}$$

The equilibrium points (Exercise 10) are given by

$$b\{\omega^2 - 1 - \tfrac{3}{4}\beta(a^2 + b^2)\} + k\omega a = 0,$$
$$a\{\omega^2 - 1 - \tfrac{3}{4}\beta(a^2 + b^2)\} - k\omega b = -\Gamma;$$

so that

$$r^2\{k^2\omega^2 + (\omega^2 - 1 - \tfrac{3}{4}\beta r^2)^2\} = \Gamma^2, \quad r = \sqrt{(a^2 + b^2)}. \tag{7.23}$$

and the closed paths of Fig. 7.3 become spirals. In this case the theory predicts that one of two stable periodic states is approached from any initial state (we shall later show, however, that from some initial states we arrive at subharmonics). A calculated example is shown in Fig. 7.4.

FIG. 7.4. Paths in the van der Pol plane for the damped pendulum; $\Gamma = 0.005$, $\omega = 0.975$, $k = 0.005$, $\beta = -0.167$

Typical response curves are shown in Fig. 5.3; the governing equations for the equilibrium points, and the conclusions about stability, being the same as in Chapter 5.

Referring back to Fig. 7.3, notice that the existence of closed curves, which implies periodicity of $a(t)$ and $b(t)$, does not in general imply periodicity of $x(t)$, for since $x(t) = a(t)\cos \omega t + b(t)\sin \omega t$ the motion is periodic only if the period of $a(t)$ and $b(t)$ is a rational multiple of $2\pi/\omega$.

Though this may occur infinitely often in a family of closed paths, the theory is too rough to decide which periodic motions, possibly of very long period, are actually present. Normally the closed paths represent special cases of 'almost periodic motion'.

If P is a representative point on a given closed path with polar representation $r(t), \phi(t)$, then (7.5) can be written

$$x(t) = r(t)\cos(\omega t - \phi(t)).$$

Let $2\pi/\Omega$ be the period of $a(t), b(t)$, and hence of $r(t), \phi(t)$, on the path. Then since a and b are slowly varying, $\Omega \ll \omega$. When the path does not encircle the origin (Fig. 7.5(a)) the phase ϕ is restored after each

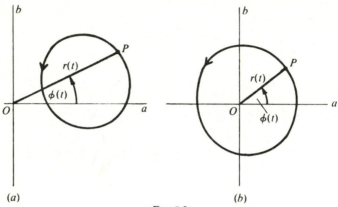

(a)   (b)

FIG. 7.5.

complete circuit. The effect is of an oscillation with frequency ω modulated by a slowly varying periodic amplitude. The phenomenon of 'beats' (Fig. 7.6) appears.

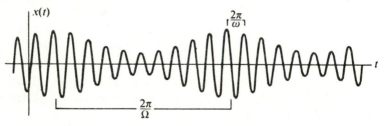

FIG. 7.6. Illustrating beats

If the path encircles the origin as in Fig. 7.5(b) the phase increases by 2π in every circuit, which effectively modulates the approximate 'frequency' of the oscillation from ω to $\omega + \Omega$.

7.3. The jump phenomenon

Equation (7.22) for the damped Duffing equation has periodic solutions which are approximately of the form $a \cos \omega t + b \sin \omega t$, where the amplitudes r of the oscillations are given by

$$r^2\{k^2\omega^2 + (\omega^2 - 1 - \tfrac{3}{4}\beta r^2)^2\} = \Gamma^2, \quad r = \sqrt{(a^2 + b^2)}, \quad \beta < 0.$$

Suppose that $\beta < 0$ as in the case of the pendulum, and put $\rho = r\sqrt{(-\beta)}$, $\gamma = \Gamma\sqrt{(-\beta)}$. Then the equation for the amplitudes becomes

$$g(\omega, \rho) \equiv \rho^2\{k^2\omega^2 + (\omega^2 - 1 + \tfrac{3}{4}\rho^2)^2\} = \gamma^2. \tag{7.24}$$

Assume that k is a parameter having a given value. Then the graph of (7.24) is a surface in (γ, ω, ρ)-space which displays the response amplitude ρ in terms of the forcing amplitude γ and forcing frequency ω. As we have observed in the preceding section there exist either one or three amplitudes, depending on the parameter values; and the way in which these arise is shown by the sketch of the γ, ω, ρ-surface shown in Fig. 7.7.

The surface has a 'fold' originating at a certain point in the field—inside the fold there are three possible amplitudes of oscillation and outside it there is one. Clearly, if γ or ω is allowed to vary in such a way that the system passes from one of these regions to another then a significant change in the general character of the system takes place, and the boundary of the fold as projected on to the (γ, ω)-plane is therefore of importance. It is in fact a cusp, whose equation is obtained by eliminating ρ between (7.24) and the condition for vertical tangency $(\partial g / \partial \rho) = 0$. The latter equation reduces to a quadratic equation in ρ^2

$$27\rho^4 + 48\rho^2(\omega^2 - 1) + 16k^2\omega^2 + 16(\omega^2 - 1)^2 = 0. \tag{7.25}$$

The projection can easily be plotted by calculating ρ from (7.25) for a given ω, $((\omega^2 - 1)^2 \geqslant 3k^2\omega^2$ is required for real, positive roots); and then obtaining the corresponding value of γ from (7.24). The cusp originates from the point where

$$\omega^2 = \tfrac{1}{2}[2 + 3k^2 - \sqrt{\{(2 + 3k^2)^2 - 4\}}], \tag{7.26}$$

from which can be calculated the corresponding γ

$$\gamma = \sqrt{\{32\sqrt{3}\omega^3 k^3 / 27\}}.$$

The cusp is shown in Fig. 7.8 for the case where $k = 1.2$.

Fig. 7.7. Response amplitude ρ versus parameters γ, ω for the damped pendulum (k fixed) showing the projection of the fold on the γ, ω plane

The theory of Section 7.2 predicts that for values of γ, ω where there exists only a single response then this response is stable; and that where there are three responses the oscillations of greatest and least amplitude are stable and the other one is unstable. Thus in Fig. 7.7 the oscillations corresponding to the two outermost sections of the surface are stable, and the part between which joins these two is unstable—that is, in practical circumstances the amplitude will not remain on this part of the sheet but will pass, in the manner of Fig. 7.4, to the greater or lesser amplitude. This characteristic leads to a so-called *jump phenomenon*, or to a particular type of 'catastrophe'. (See Poston and Stewart 1978 for an extensive account of catastrophes; further discussion of such phenomena is given in Chapter 12.) To illustrate the consequences of this distribution of stabilities, consider what happens in an experiment in which γ is held constant, and the applied frequency is very slowly varied in such a way that it crosses the boundary of the cusp of Fig. 7.8 at some point. A cross-section of the fold for this value of γ is shown in Fig. 7.9, with the stable and unstable sections indicated.

FIG. 7.8. This is a graph of the 'vertical' tangents to the surface defined by (7.24) in the (γ, ω, ρ)-space projected on to the (γ, ω)-plane. The folded part of the surface occurs near the cusp

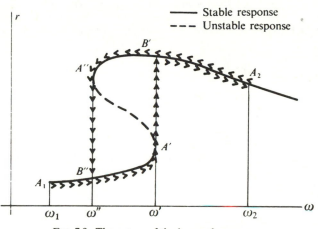

FIG. 7.9. The nature of the jump phenomenon

Begin the experiment at applied frequency $\omega = \omega_1$, steadily increase to ω_2, then bring it back again to ω_1. Starting at A_1, the response moves to A' at frequency ω'. As ω increases beyond ω', the oscillation will, after some irregular motion, settle down to the condition represented by B'; that is to say, the amplitude 'jumps' at the critical frequency ω'. After this it follows the smooth curve from B' to A_2. On the way back the response point moves along the upper curve as far as A''. Here it must drop to B'' on the lower curve and then back to A_1.

Clearly a jump of a similar kind will occur whenever the parameters ω and Γ change in such a way as to pass into or out of the projection of the fold in Fig. 7.7. This so-called *jump phenomenon* has been observed in many nonlinear physical systems.

7.4. Harmonic oscillations, stability, and transients for the forced van der Pol equation

The equation considered is

$$\ddot{x} + \varepsilon(x^2 - 1)\dot{x} + x = \Gamma \cos \omega t, \tag{7.27}$$

where $\varepsilon > 0$. In the absence of a forcing term there is a single, self-excited oscillation (justified in Section 4.1 but not proved until Section 11.3). We look for responses approximately of the form

$$x(t) = a(t)\cos \omega t + b(t)\sin \omega t \tag{7.28}$$

where a, b are slowly varying functions. Then, neglecting $\ddot{a}$, $\ddot{b}$,

$$\dot{x}(t) = (\dot{a} + \omega b)\cos \omega t + (-\omega a + \dot{b})\sin \omega t, \tag{7.29a}$$

$$\ddot{x}(t) = (-\omega^2 a + 2\omega \dot{b})\cos \omega t + (-2\omega \dot{a} - \omega^2 b)\sin \omega t. \tag{7.29b}$$

After some algebra,

$$(x^2 - 1)\dot{x} = \{(\tfrac{3}{4}a^2 + \tfrac{1}{4}b^2 - 1)\dot{a} + \tfrac{1}{2}ab\dot{b} - \omega b(1 - \tfrac{1}{4}a^2 - \tfrac{1}{4}b^2)\} \cos \omega t$$
$$+ \{\tfrac{1}{2}ab\dot{a} + (\tfrac{1}{4}a^2 + \tfrac{3}{4}b^2 - 1)\dot{b} + \omega a(1 - \tfrac{1}{4}a^2 - \tfrac{1}{4}b^2)\} \sin \omega t$$
$$+ \text{higher harmonics.} \tag{7.30}$$

Finally substitute (7.28) to (7.30) into (7.27), ignoring the higher harmonics. As in Section 7.2, the equation is satisfied approximately if the coefficients of $\cos \omega t$, $\sin \omega t$ are equated to zero. We obtain

$$(2\omega - \tfrac{1}{2}\varepsilon ab)\dot{a} + \varepsilon(1 - \tfrac{1}{4}a^2 - \tfrac{3}{4}b^2)\dot{b} = \varepsilon \omega a(1 - \tfrac{1}{4}r^2) - (\omega^2 - 1)b,$$
$$\tag{7.31a}$$

$$-\varepsilon(1 - \tfrac{3}{4}a^2 - \tfrac{1}{4}b^2)\dot{a} + (2\omega + \tfrac{1}{2}\varepsilon ab)\dot{b} = (\omega^2 - 1)a$$
$$+ \varepsilon \omega b(1 - \tfrac{1}{4}r^2) + \Gamma, \tag{7.31b}$$

where

$$r = \sqrt{(a^2 + b^2)}.$$

The equilibrium points occur when $\dot{a} = \dot{b} = 0$, that is, when the right-hand sides are zero. We can reduce the parameters $(\omega, \Gamma, \varepsilon)$ to two by multiplying through by $1/\varepsilon \omega$ and putting

$$v = (\omega^2 - 1)/\varepsilon \omega, \quad \gamma = \Gamma/\varepsilon \omega; \tag{7.32}$$

(the quantity v is a measure of the 'detuning'). The equations for the equilibrium points become

$$a(1 - \tfrac{1}{4}r^2) - vb = 0, \tag{7.33a}$$

$$va + b(1 - \tfrac{1}{4}r^2) = -\gamma. \tag{7.33b}$$

By squaring and adding these equations we obtain

$$r^2\{v^2 + (1 - \tfrac{1}{4}r^2)^2\} = \gamma^2, \tag{7.34}$$

and when this is solved the roots a, b can be recovered from (7.33a, b). Equation (7.34) may have either 1 or 3 real roots (since $r > 0$) depending on the parameter values v and γ (or $\omega, \Gamma, \varepsilon$). This dependence can be conveniently represented on a single figure, the 'response diagram', Fig. 7.10(a).

There are two questions which can be settled by a study of (7.31): that of the stability of the periodic solutions just found and, connected with this, the changes in the general behaviour of the system when one of the parameters ω or Γ varies. We shall simplify equations (7.31) by supposing that ε is small enough for $\varepsilon\dot{a}$, $\varepsilon\dot{b}$ to be of negligible effect. Then (7.31) becomes

$$\dot{a} = \tfrac{1}{2}\varepsilon(1 - \tfrac{1}{4}r^2)a - \frac{\omega^2 - 1}{2\omega}b \equiv A(a, b), \quad \text{say}, \tag{7.35a}$$

$$\dot{b} = \frac{\omega^2 - 1}{2\omega}a + \tfrac{1}{2}\varepsilon(1 - \tfrac{1}{4}r^2)b + \frac{\Gamma}{2\omega} \equiv B(a, b), \quad \text{say}. \tag{7.35b}$$

Now consider the stability of an equilibrium point at $a = a_0, b = b_0$. In the neighbourhood of the point put

$$a = a_0 + \xi, \quad b = b_0 + \eta. \tag{7.36}$$

The corresponding linearized equations are

$$\dot{\xi} = A_1(a_0, b_0)\xi + A_2(a_0, b_0)\eta,$$
$$\dot{\eta} = B_1(a_0, b_0)\xi + B_2(a_0, b_0)\eta.$$

The point is a stable equilibrium point (Fig. 2.8) if

$$\det\begin{pmatrix} A_1 & A_2 \\ B_1 & B_2 \end{pmatrix} = A_1 B_2 - A_2 B_1 > 0, \tag{7.37}$$

and

$$\text{tr}\begin{pmatrix} A_1 & A_2 \\ B_1 & B_2 \end{pmatrix} = A_1 + B_2 \leqslant 0. \tag{7.38}$$

From (7.35) and (7.36),

$$A_1 = \tfrac{1}{2}\varepsilon(1 - \tfrac{3}{4}a_0^2 - \tfrac{1}{4}b_0^2), \quad A_2 = -\tfrac{1}{4}\varepsilon a_0 b_0 - \frac{\omega^2 - 1}{2\omega},$$

$$B_1 = -\tfrac{1}{4}\varepsilon a_0 b_0 + \frac{\omega^2 - 1}{2\omega}, \quad B_2 = \tfrac{1}{2}\varepsilon(1 - \tfrac{1}{4}a_0^2 - \tfrac{3}{4}b_0^2).$$

Therefore, after simplification, we have

$$A_1 B_2 - A_2 B_1 = \tfrac{1}{4}\varepsilon^2(\tfrac{3}{16}r_0^4 - r_0^2 + 1) + \left(\frac{\omega^2 - 1}{2\omega}\right)^2, \tag{7.39}$$

$$A_1 + B_2 = \tfrac{1}{2}\varepsilon(2 - r_0^2), \tag{7.40}$$

where $r_0^2 = \sqrt{(a_0^2 + b_0^2)}$.

In terms of the parameter v, (7.32), the conditions for stability, (7.37) and (7.38), become

$$\tfrac{3}{16}r_0^4 - r_0^2 + 1 + v^2 > 0, \tag{7.41}$$

$$2 - r_0^2 \leqslant 0. \tag{7.42}$$

The response curves derived from (7.34), and the stability regions given by (7.41) and (7.42), are exhibited in Fig. 7.10, together with some further detail about the nature of the equilibrium points in various regions. It can be seen that the stable and unstable regions are correctly given by the argument in Section 5.4, (vi), that the responses are stable when $d\Gamma/dr > 0$ and unstable when $d\Gamma/dr < 0$.

If we attempt to force an oscillation corresponding to an equilibrium point lying in the unstable region of Fig. 7.10(*a*) the system will drift away from this into another state: into a stable equilibrium point if one is available, or into a limit cycle, which corresponds to an almost-periodic motion. A guide to the transition states is provided by the phase paths of (7.35). Although we are not directly concerned with the detail of the phase plane it is interesting to see the variety of patterns that can arise according to the parameter values concerned, and three typical cases are shown in Figs. 7.12, 7.13, and 7.14.

We can show more directly how the settling-down takes place from a given initial condition to a periodic oscillation by a phase-plane representation of (7.27) given by

$$\dot{x} = y, \quad \dot{y} = -\varepsilon(x^2 - 1)y - x + \Gamma \cos \omega t. \tag{7.43}$$

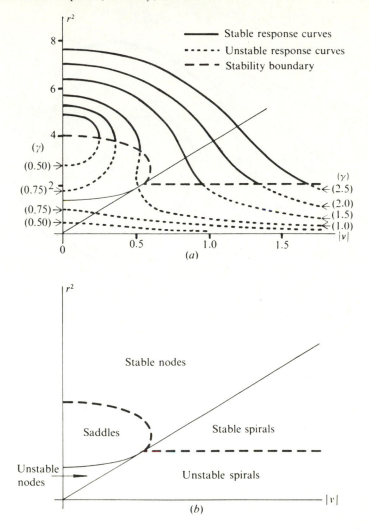

FIG. 7.10. Response curves and the stability region for the forced van der Pol equation (see eqns (7.32), (7.33), (7.41), (7.42)). — — — — stability boundary

A periodic solution shows up as a limiting closed path (Fig. 7.11(a)). Since the system (7.43) is not autonomous, the paths may cross. The corresponding path in the van der Pol plane is shown in Fig. 7.11(b).

FIG. 7.11. (a) Solution of the forced van der Pol equation in the $x, \dot{x}$ plane. (b) The corresponding path in the van der Pol plane: ● equilibrium point.

7.5. Frequency entrainment for the van der Pol equation

The description offered by the treatment of Sections 7.3 and 7.4 is incomplete because of the approximate representation used. For example, no subharmonics (of period $2\pi n/\omega, n > 1$) can be revealed by the representation (7.28) if a and b are slowly-varying. Since sub-harmonics do exist they must be arrived at from some region of initial conditions $x(0)$, $\dot{x}(0)$; (see Section 7.8), but such regions are not identifiable on the van der Pol plane for a, b. References to 'all paths' and 'any initial condition' in the remarks below must therefore be taken as merely a broad descriptive wording.

Referring to Fig. 7.10, the phenomena to be expected for different ranges of v, γ (or ω, Γ, ε, eqn (7.32)) fall into three main types.

(I) When $\gamma < 0.77$ about, and v is appropriately small enough, there are three equilibrium points. One is a stable node, and the other two are unstable. Starting from any initial condition, the corresponding path will in practice run into the stable node. Figures 7.12(a) and (b) illustrate this case.

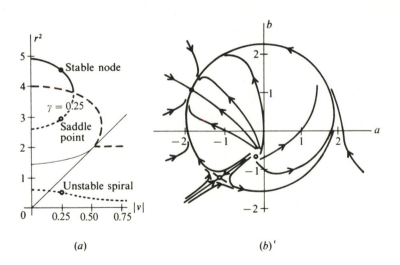

FIG. 7.12. (a) Response region for case I. (b) Corresponding paths in the van der Pol plane ($\gamma = 0.25$, $v = 0.25$)

FIG. 7.13. (*a*) Response region for case II. (*b*) Corresponding paths in the van der Pol plane
($\gamma = 2.25$, $v = 1.25$)

(II) For any v, and γ large enough, there is a single stable point which all
paths approach (Fig. 7.14).

(III) For any v, and γ small enough, the only equilibrium point is an
unstable one. It can be shown (Exercise 11.15) that all paths approach a
limit cycle (Fig. 7.14). There are no stable harmonic solutions in this case.

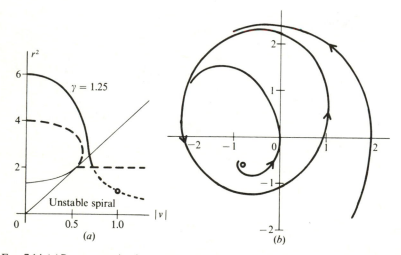

FIG. 7.14. (*a*) Response region for case III. (*b*) Corresponding paths in the van der Pol plane
showing limit cycle ($\gamma = 1.25$, $v = 1.0$)

In cases I and II the final state is periodic with the period of the forcing function and the natural oscillation of the system appears to be completely suppressed despite its self-sustaining nature. This condition is arrived at from arbitrary initial states (at any rate, so far as this approximate theory tells us). The system is said to be *entrained* at the frequency of the forcing function. Moreover, this phenomenon does not depend critically on exact parameter values: if any of the parameters $\omega, \Gamma, \varepsilon$ of the system fluctuate a little, the system will continue to be entrained at the prevailing frequency.

FIG. 7.15. Harmonic entrainment under varying frequency input: van der Pol equation

To illustrate the sudden passage from case III to case I conditions, consider an experiment in which the applied amplitude remains constant at Γ_0 but the frequency varies slowly. We treat ω close to unity and Γ_0 small as representing the physically interesting case. Figure 7.15 is a rendering of Fig. 7.10 in terms of ω and Γ instead of v and γ for $\varepsilon = 0.1$.

Suppose the experiment starts with $\omega = \omega_0$ where there is no stable periodic oscillation, and then ω is then reduced slowly. At $\omega = \omega_1$ the stable branch is encountered, and in the space of a few cycles a periodic oscillation of frequency ω_1 is established. This effect is allied to the jump phenomenon of Section 7.3. Figure 7.16 shows the occurrence of entrainment in a particular case. The phenomenon of entrainment is related to that of *synchronization* (Minorsky 1962), in which two coupled systems having slightly different natural frequencies may fall into a common frequency of oscillation.

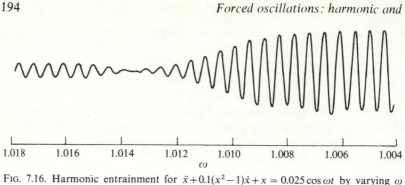

1.018   1.016   1.014   1.012   1.010   1.008   1.006   1.004
ω

FIG. 7.16. Harmonic entrainment for $\ddot{x}+0.1(x^2-1)\dot{x}+x = 0.025\cos\omega t$ by varying ω
slowly from 1.018 down to 1.004

7.6. Comparison of the theory with computations

The approximations discussed in the previous sections appear to be very rough. It is therefore of interest to compare, for a special case, the theoretical predictions summarized in Fig. 7.10 with exact calculations. In Fig. 7.17 the data for Fig. 7.10 is replotted for the case $\varepsilon = 1$ in terms of input frequency ω, input amplitude Γ, and output amplitude r.

Also plotted are the corresponding curves derived by direct numerical

FIG. 7.17. Theoretical (eqn (7.33)) and computed response curves for the forced van der Pol equation with $\varepsilon = 1$, eqn (7.25). Values of Γ are shown on curves

solution of (7.23), giving the amplitudes of the stable periodic oscillations to which the solutions settle down after a sufficient time. These computed response curves terminate near the stability boundary, where no steady oscillations could be found.

In the hatched region the computed amplitudes cannot be compared directly with the theory since there is a heavy admixture of third, fifth, and higher harmonics, ignored by the theory. A typical steady oscillation in this region is shown in Fig. 7.18, together with its harmonic analysis. The amplitude of the first harmonic is as consistent with prediction as are the other curves of Fig. 7.17.

FIG. 7.18. Computed harmonic response for the forced van der Pol equation with $\varepsilon = 1$, $\Gamma = 1\cdot5$, $\omega = 0\cdot2$

Order of harmonic	1	2	3	4	5
Amplitude (from computed curve)	1.59	0.04	0.79	0.06	0.91

7.7. Subharmonics of Duffing's equation by perturbation

In considering periodic solutions of differential equations with a forcing term we have so far looked only for solutions having the period of the forcing term. But even a linear equation may have a periodic solution with a different principal period. For example, all solutions of

$$\ddot{x} + \frac{1}{n^2}x = \Gamma \cos t$$

are of the form

$$x(t) = a\cos\frac{1}{n}t + b\sin\frac{1}{n}t - \frac{\Gamma}{1-n^{-2}}\cos t.$$

If n is an integer the period is $2\pi n$ instead of 2π. The response is said to be a *subharmonic of order* $1/n$. Figure 7.19 shows a case when $n = 3$.

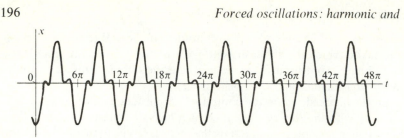

FIG 7.19. A subharmonic of order $\frac{1}{3}$ for $\ddot{x} + \frac{1}{9}x = \cos t$

For the linear equation this periodic motion appears to be merely an anomalous case of the usual almost-periodic motion, depending on a precise relation between the forcing and natural frequencies. Also, any damping will cause its disappearance. When the equation is nonlinear, however, the generation of alien harmonics by the nonlinear terms may cause a stable subharmonic to appear for a range of the parameters, and in particular for a range of applied frequencies. Also, the forcing amplitude plays a part in generating and sustaining the subharmonic even in the presence of damping. Thus there will exist the tolerance of slightly varying conditions necessary for the consistent appearance of a subharmonic and its use in physical systems. The possibility of such a response was pointed out in Section 7.1.

Assuming that a subharmonic exists, its form can be established by using the perturbation method. Consider Duffing's equation without damping

$$\ddot{x} + \alpha x + \beta x^3 = \Gamma \cos \omega t, \tag{7.44}$$

$\alpha, \Gamma, \omega > 0$. Take β to be the 'small parameter' of the perturbation method, so that we regard (7.44) as a member of a family of equations for which β lies in an interval including $\beta = 0$. The values of ω for which the subharmonics occur are unknown. To simplify the subsequent algebra we put

$$\omega t = \tau \tag{7.45}$$

and look for periodic solutions of the transformed equation

$$\omega^2 x'' + \alpha x + \beta x^3 = \Gamma \cos \tau, \tag{7.46}$$

where the derivatives are with respect to τ. There are no subharmonics of order 1/2 (also for the damped case as in Exercise 16), and we shall look for those of order 1/3. The solution of (7.44) will then have period $6\pi/\omega$ and the solution of (7.46) will have period 6π.

Now write, in (7.46),

$$x(\tau) = x_0(\tau) + \beta x_1(\tau) + \dots, \tag{7.47}$$

$$\omega = \omega_0 + \beta \omega_1 + \dots. \tag{7.48}$$

The condition that x has period 6π for all β implies that $x_0, x_1, \dots$ have period 6π. After substitution in (7.46) we obtain

$$\omega_0^2 x_0'' + \alpha x_0 = \Gamma \cos \tau, \tag{7.49a}$$

$$\omega_0^2 x_1'' + \alpha x_1 = -2\omega_0 \omega_1 x_0'' - x_0^3. \tag{7.49b}$$

The periodicity condition applied to (7.49a) gives $\alpha/\omega_0^2 = 1/9$, or

$$\omega_0 = 3\sqrt{\alpha}. \tag{7.50}$$

The $\frac{1}{3}$ subharmonic is therefore stimulated by applied frequencies in the neighbourhood of three times the natural frequency of the linearized equation ($\beta = 0$), as expected. Then

$$x_0(\tau) = a_{1/3} \cos \tfrac{1}{3}\tau + b_{1/3} \sin \tfrac{1}{3}\tau - \frac{\Gamma}{8\alpha} \cos \tau \tag{7.51}$$

where $a_{1/3}, b_{1/3}$ are constants, to be settled at the next stage. The solution therefore bifurcates from a solution of the linearized version of (7.46). For (7.49b), we have

$$x_0''(\tau) = -\tfrac{1}{9} a_{1/3} \cos \tfrac{1}{3}\tau - \tfrac{1}{9} b_{1/3} \sin \tfrac{1}{3}\tau + \frac{\Gamma}{8\alpha} \cos \tau \tag{7.52}$$

and

$$x_0^3(\tau) = \frac{3}{4} \left\{ a_{1/3} \left(a_{1/3}^2 + b_{1/3}^2 + 2\left(\frac{\Gamma}{8\alpha}\right)^2 \right) - \frac{\Gamma}{8\alpha} (a_{1/3}^2 - b_{1/3}^2) \right\} \cos \tfrac{1}{3}\tau$$

$$+ \frac{3}{4} \left\{ b_{1/3} \left(a_{1/3}^2 + b_{1/3}^2 + 2\left(\frac{\Gamma}{8\alpha}\right)^2 \right) + \frac{\Gamma}{4\alpha} a_{1/3} b_{1/3} \right\} \sin \tfrac{1}{3}\tau$$

$$+ \text{terms in } \cos \tau, \sin \tau \text{ and higher harmonics}. \tag{7.53}$$

In order that $x_1(\tau)$ should be periodic it is necessary that resonance should not occur in (7.49b). The coefficients of $\cos \tfrac{1}{3}\tau, \sin \tfrac{1}{3}\tau$ on the right must therefore be zero. Clearly this condition is insufficient to determine all three of $a_{1/3}, b_{1/3}$, and ω_1. Further stages will still give no condition to determine ω_1 so we shall consider it to be arbitrary, and determine $a_{1/3}$ and $b_{1/3}$ in terms of ω_0, ω_1 and so (to order β^2) in terms of ω. It is convenient to write

$$2\omega_0 \omega_1 \approx (\omega^2 - 9\alpha)/\beta \tag{7.54}$$

this being correct with an error of order β, so that the arbitrary status of ω is clearly displayed. (Remember, however, that ω must remain near ω_0, or $3\sqrt{\alpha}$, by (7.50).) Then the conditions that the coefficients of $\cos\frac{1}{3}\tau, \sin\frac{1}{3}\tau$ on the right of (7.49b) should be zero are

$$a_{1/3}\left(a_{1/3}^2 + b_{1/3}^2 + \frac{\Gamma}{32\alpha^2} - \frac{4}{27}\frac{(\omega^2-9\alpha)}{\beta}\right) - \frac{\Gamma}{8\alpha}(a_{1/3}^2 - b_{1/3}^2) = 0,$$

$$(7.55a)$$

$$b_{1/3}\left(a_{1/3}^2 + b_{1/3}^2 + \frac{\Gamma}{32\alpha^2} - \frac{4}{27}\frac{(\omega^2-9\alpha)}{\beta}\right) + \frac{\Gamma}{4\alpha}a_{1/3}b_{1/3} = 0.$$

$$(7.55b)$$

The process of solution can be simplified in the following way. One solution of (7.55b) is given by

$$b_{1/3} = 0, \qquad (7.56)$$

and then (7.55a) becomes

$$a_{1/3}^2 - \frac{\Gamma}{8\alpha}a_{1/3} + \left(\frac{\Gamma^2}{32\alpha^2} - \frac{4}{27}\frac{(\omega^2-9\alpha)}{\beta}\right) = 0 \qquad (7.57)$$

(rejecting the trivial case $a_{1/3} = 0$). Thus there are either two real solutions of (7.57), or none. Now consider the case when $b_{1/3} \neq 0$. Solve (7.55b) for $b_{1/3}$ and substitute for $b_{1/3}^2$ in (7.55a). We obtain

$$b_{1/3} = \pm\sqrt{3}a_{1/3} \qquad (7.58)$$

and (compare (7.57))

$$(-2a_{1/3})^2 - \frac{\Gamma}{8\alpha}(-2a_{1/3}) + \left(\frac{\Gamma}{32\alpha^2} - \frac{4}{27}\frac{(\omega^2-9\alpha)}{\beta}\right) = 0. \quad (7.59)$$

If the roots are displayed on the $a_{1/3}, b_{1/3}$ plane, those arising from (7.58) and (7.59) are obtained from the first pair by rotations through an angle $\pm\frac{2}{3}\pi$ (Fig. 7.20). Another interpretation is obtained by noting that the differential equation (7.46) is invariant under the change of variable $\tau \to \tau \pm 2\pi$. If $b_{1/3} = 0$ and $a_{1/3}$ represent one solution of (7.56) and (7.57), and $x^*(\tau)$ is the corresponding solution of (7.46), then

$$x^*(\tau \pm 2\pi) = a_{1/3}\cos\left(\tfrac{1}{3}\tau \pm \tfrac{2}{3}\pi\right) - \frac{\Gamma}{8\alpha}\cos\tau$$

$$= -\tfrac{1}{2}a_{1/3}\cos\tfrac{1}{3}\tau \pm \frac{\sqrt{3}}{2}a_{1/3}\sin\tfrac{1}{3}\tau - \frac{\Gamma}{8\alpha}\cos\tau$$

are also solutions. The new coefficients are clearly the solutions of (7.58),

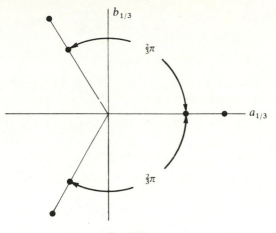

FIG. 7.20.

(7.59). The stability properties of corresponding pairs of equilibrium points are the same.

Figure 7.21 shows the subharmonic for a special case. Note that the subharmonic state will only be entered when the initial conditions are suitable.

——— Subharmonic $\cdots\cdots\cdots$ $\frac{1}{8}\Gamma\cos\omega t$

FIG. 7.21. Subharmonic order $\frac{1}{3}$, for the pendulum equation $\ddot{x}+x-\frac{1}{6}x^3 = 1.5\cos 2.85t$

We should expect that the perturbation method would be satisfactory only if $\alpha = O(1), |\beta| \ll 1$ and $(\omega^2 - 9\alpha)/\beta \approx 2\omega_0\omega_1$ is of order 1. However, put

$$y = x/C \qquad (7.60)$$

where C is constant, into (7.44). This becomes

$$y'' + c_1 y + c_3 y^3 = \gamma \cos \tau \qquad (7.61)$$

where

$$c_1 = \alpha/\omega^2, \quad c_3 = \beta C^2/\omega^2, \quad \gamma = \Gamma/C\omega^2. \qquad (7.62)$$

Now choose

$$C^2 = (\omega^2 - 9\alpha)/\beta \tag{7.63}$$

so that

$$9c_1 + c_3 = 1. \tag{7.64}$$

The perturbation method applied to (7.61) in the same way as to (7.46) (and obviously leading to the same result), requires at most

$$|c_3| \ll 1, \quad (1 - 9c_1)/c_3 = O(1). \tag{7.65}$$

But from (7.64)

$$(1 - 9c_1)/c_3 = 1. \tag{7.66}$$

So the second condition in (7.65) is satisfied, and $|c_3| \ll 1$ requires only

$$|c_3| = \left| \beta \left(\frac{\omega^2 - 9\alpha}{\beta} \right) \frac{1}{\omega^2} \right| \ll 1,$$

or

$$\alpha \simeq \tfrac{1}{9}\omega^2. \tag{7.67}$$

In other words, so long as ω is near enough to $3\sqrt{\alpha}$ (a near-resonant state for the subharmonic), the calculations are valid without the restriction on β. The usefulness of the approximate form (7.51) is even wider than this, and the reader is referred to Hayashi (1964) for numerical comparisons.

7.8. Stability and transients for subharmonics of Duffing's equation

Some of the subharmonics obtained in Section 7.7 may not be stable, and so will not appear in practice. When there are stable subharmonics, the question arises of how to stimulate them. We know that there are stable solutions having the period of the forcing term; there may also be subharmonics of order other than $1/3$ (so far as we know at the moment). Which state of oscillation is ultimately adopted by a system depends on the initial conditions. The following method, due to Mandelstam and Papalexi, is similar to the use of the van der Pol plane in Section 7.4. It enables the question of stability to be settled and gives an idea of the domain of initial conditions leading ultimately to a subharmonic of order $1/3$.

The method involves accepting the form (7.51) as a sufficiently good approximation to the subharmonic, and that the solutions for the 'transient' states of

$$\ddot{x} + \alpha x + \beta x^3 = \Gamma \cos \omega t \tag{7.68}$$

are approximately of the form

$$x(t) = a_{1/3}(t)\cos\tfrac{1}{3}\omega t + b_{1/3}(t)\sin\tfrac{1}{3}\omega t - \frac{\Gamma}{8\alpha}\cos\omega t, \qquad (7.69)$$

where $a_{1/3}$ and $b_{1/3}$ are slowly-varying functions in the sense of Section 7.4. Further justification for the use of the form (7.69) will be found in Hayashi (1964) and McLachlan (1956).

From (7.69)

$$\dot{x}(t) = (\dot{a}_{1/3} + \tfrac{1}{3}\omega b_{1/3})\cos\tfrac{1}{3}\omega t + (-\tfrac{1}{3}\omega a_{1/3} + \dot{b}_{1/3})\sin\tfrac{1}{3}\omega t$$
$$+ (\omega\Gamma/8\alpha)\sin\omega t,$$

and

$$\ddot{x}(t) = (-\tfrac{1}{9}\omega^2 a_{1/3} + \tfrac{2}{3}\omega\dot{b}_{1/3})\cos\tfrac{1}{3}\omega t$$
$$+ (-\tfrac{2}{3}\omega\dot{a}_{1/3} - \tfrac{1}{9}\omega^2 b_{1/3})\sin\tfrac{1}{3}\omega t - (\omega^2\Gamma/8\alpha)\cos\omega t, \qquad (7.70)$$

where $\ddot{a}_{1/3}$, $\ddot{b}_{1/3}$ have been neglected. Equations (7.69) and (7.70) are substituted into (7.68). When the terms are assembled, neglect all harmonics of higher order than $\tfrac{1}{3}$, and balance the coefficients of $\cos\tfrac{1}{3}\omega t, \sin\tfrac{1}{3}\omega t$. This leads to

$$\dot{a}_{1/3} = \frac{9\beta}{8\omega}\left\{ b_{1/3}\left(a_{1/3}^2 + b_{1/3}^2 + \frac{\Gamma^2}{32\alpha^2} - \frac{4}{27}\frac{(\omega^2 - 9\alpha)}{\beta}\right)\right.$$
$$\left. + \frac{\Gamma}{4\alpha}a_{1/3}b_{1/3}\right\} \equiv A(a_{1/3}, b_{1/3}), \qquad (7.71a)$$

$$\dot{b}_{1/3} = \frac{9\beta}{8\omega}\left\{ -a_{1/3}\left(a_{1/3}^2 + b_{1/3}^2 + \frac{\Gamma^2}{32\alpha^2} - \frac{4}{27}\frac{(\omega^2 - 9\alpha)}{\beta}\right)\right.$$
$$\left. + \frac{\Gamma}{8\alpha}(a_{1/3}^2 - b_{1/3}^2)\right\} \equiv B(a_{1/3}, b_{1/3}). \qquad (7.71b)$$

The phase paths of these autonomous equations, representing 'transients' for $x(t)$, can be displayed on a van der Pol plane of $a_{1/3}, b_{1/3}$ (for example see Fig. 7.22). Stable solutions for the form (7.69) correspond stable equilibrium points of the system (7.71), and unstable solutions to unstable equilibrium points. By comparing with (7.55) we see that the equilibrium points are the same as those found earlier by the perturbation method. The point

$$a_{1/3} = b_{1/3} = 0$$

is also an equilibrium point and this may be taken to acknowledge the possibility that from some range of initial conditions the harmonic, rather than a subharmonic, oscillation is approached, namely, the solution

$$x(t) = -\frac{\Gamma}{8\alpha}\cos\omega t.$$

This, as expected, is approximately equal to the forced solution of the linearized equation, for the exact solution of $\ddot{x} + \alpha x = \Gamma\cos\omega t$ is

$$\frac{\Gamma}{\alpha - \omega^2}\cos\omega t \simeq -\frac{\Gamma}{8\alpha}\cos\omega t$$

so long as $\omega^2 \approx 9\alpha$ (eqn (7.67)).

We may determine the stability of the equilibrium points as follows. It is only necessary to consider the pair for which $b_{1/3} = 0$, by the argument on p. 198. Let $a_{1/3}^*, b_{1/3}^* = 0$, be such a point, and consider the neighbourhood of this point by writing

$$a_{1/3} = a_{1/3}^* + \xi, \quad b_{1/3} = b_{1/3}^* + \eta = \eta$$

with ξ, η small. Then the system (7.71) becomes

$$\left.\begin{array}{l} \dot{\xi} = A_1(a_{1/3}^*, 0)\xi + A_2(a_{1/3}^*, 0)\eta, \\ \dot{\eta} = B_1(a_{1/3}^*, 0)\xi + B_2(a_{1/3}^*, 0)\eta. \end{array}\right\} \tag{7.72}$$

Writing

$$\gamma = \frac{\Gamma}{8\alpha}, \quad v = \frac{\Gamma^2}{32\alpha^2} - \frac{4}{27}\frac{(\omega^2 - 9\alpha)}{\beta}, \tag{7.73}$$

we find that the coefficients in (7.72), with $b_{1/3}^* = 0$, are given by

$$A_1(a_{1/3}^*, 0) = 0,$$

$$A_2(a_{1/3}^*, 0) = \frac{9\beta}{8\omega}(a_{1/3}^{*2} + 2\gamma a_{1/3}^* + v) = \frac{9\beta}{8\omega}3\gamma a_{1/3}^*$$

(from (7.57)), $\tag{7.74}$

$$B_1(a_{1/3}^*, 0) = \frac{9\beta}{8\omega}(-3a_{1/3}^{*2} + 2\gamma a_{1/3}^* - v) = \frac{9\beta}{8\omega}(-\gamma a_{1/3}^* + 2v),$$

$$B_2(a_{1/3}^*, 0) = 0.$$

The equilibrium points are therefore either centres (in the linear approximation) or saddles. The origin can similarly be shown to be a centre. A typical layout for the $a_{1/3}, b_{1/3}$ plane, using the data of Fig. 7.21, is shown in Fig. 7.22.

FIG. 7.22. Van der Pol plane for subharmonic order $\frac{1}{3}$; data as in Fig. 7.21. S_1, S_2, S_3, stable; U_1, U_2, U_3, unstable.

When damping is taken into account the closed paths become spirals. Figure 7.23(*a*) shows a typical pattern (see also Hayashi 1964). The shaded areas are *domains of attraction* for the equilibrium points to which they relate. It can be seen that in this case the subharmonics occur over a fairly narrow range of initial conditions. Figure 7.23(*b*) shows a typical subharmonic.

It may be inferred that when the parameters of the system (such as the applied amplitude or frequency) are varied slowly, the system will suddenly become entrained at a subharmonic if the domain of attraction of a non-zero equilibrium point, appropriate to the instantaneous value of the changing parameter, is encountered. Figures such as Fig. 7.22 have to be treated with caution so far as the details are concerned. The nature and positions of the equilibrium points are likely to be nearly correct, but the phase paths are not necessarily very close to the correct ones.

Subharmonics provide a means for 'gearing down' from a high-frequency input by stages to a low-frequency output. In the quartz crystal clock the basic timing mechanism is the oscillating crystal. Its electrical output is passed through a succession of circuits producing submultiples of the input frequency until a frequency appropriate for a

FIG. 7.23. (*a*) Van der Pol plane; subharmonics order $\frac{1}{3}$ of

$$\ddot{x} + 0.002\dot{x} + x - \tfrac{1}{6}x^3 = 1.5\cos 2.85t.$$

S_1, S_2, S_3, stable; U_1, U_2, U_3, unstable.   (*b*) Subharmonic near S_1.

mechanical drive is reached. Subharmonic entrainment is essential for
this process since small variations in the electronic system must be
accommodated to maintain exact timing.

Exercises

1. Show that the solutions of eqns (7.16) and (7.17), for the undamped Duffing equation in the van der Pol plane, are given exactly by

$$r^2\{(\omega^2 - 1) - \tfrac{3}{8}\beta r^2\} + 2\Gamma a = \text{constant}, \quad r = \sqrt{(a^2 + b^2)}.$$

Show that these approximate to circles when r is large. Estimate the period on such a path of $a(t), b(t)$.

2. Express eqns (7.16) and (7.17) in polar coordinates. Deduce the approximate period of $a(t)$ and $b(t)$ for large r. Find the approximate equations for these distant paths. Show how frequency modulation occurs, by deriving an expression for $x(t)$.

3. Consider the equation

$$\ddot{x} + \text{sgn}\,(x) = \Gamma \cos \omega t.$$

Assume solutions of the form $x = a \cos \omega t + b \sin \omega t$. Show that solutions of period $2\pi/\omega$ exist when $|\Gamma| \leqslant 4/\pi^2\omega^2$. Show also that

$$a(4 - \pi\omega^2|a|) = \pi\Gamma|a|, \quad b = 0.$$

$$\left(\text{sgn}\,\{x(t)\} = \frac{4a}{\pi\sqrt{(a^2 + b^2)}} \cos \omega t + \frac{4b}{\pi\sqrt{(a^2 + b^2)}} \sin \omega t + \text{higher harmonics.} \right)$$

4. Show that solutions, period 2π, of the equation $\ddot{x} + x^3 = \Gamma \cos t$ are given approximately by $x = a \cos t$, where a is a solution of $3a^3 - 4a = 4\Gamma$.

5. Show that solutions, period 2π, of $\ddot{x} + k\dot{x} + x + x^3 = \Gamma \cos t$ are given approximately by $x = a \cos t + b \sin t$, where

$$ka - \tfrac{3}{4}br^2 = 0, \quad kb + \tfrac{3}{4}ar^2 = \Gamma, \quad r = \sqrt{(a^2 + b^2)}.$$

Deduce that the response curves are given by

$$r^2(k^2 + \tfrac{9}{16}r^4) = \Gamma^2.$$

6. Obtain approximate solutions, period $2\pi/\omega$, of the equation

$$\ddot{x} + \alpha x + \beta x^2 = \Gamma \cos \omega t,$$

by assuming the form $x = c + a \cos \omega t$, and deducing equations for c and a.

Show that if β is small, $\Gamma = O(\beta)$, and $\omega^2 - \alpha = O(\beta)$, then there is a solution with $c \approx -\beta a^2/2\alpha$, $a \approx \Gamma/(\alpha - \omega^2)$.

7. Consider the equation $\ddot{x} + x^3 = \Gamma \cos t$. Substitute $x = a \cos t + b \sin t$, and obtain the solution $x = a \cos t$, where $\tfrac{3}{4}a^3 - a = \Gamma$.

Now fit x^3, by a least squares procedure, by a straight line of the form px where p is constant on $-A \leqslant x \leqslant A$, so that

$$\int_{-A}^{A} (x^3 - px)^2 \, dx$$

is a minimum with respect to p. Deduce that this linear approximation to the restoring force is compatible with an oscillation, period 2π, of amplitude A, provided $\frac{3}{5}A^3 - A = \Gamma$.
Compare a with A when $\Gamma = 0.1$.

8. Consider the equation

$$\ddot{x} + 0.16x^2 = 1 + 0.2\cos t.$$

By linearizing the restoring force about the equilibrium points of the unforced system, show that there are two modes of oscillation period 2π, given by

$$x \approx 2.5 - \cos t, \quad x \approx -2.5 - 0.11\cos t.$$

Find to what extent the predicted modes differ when a substitution of the form $x = c + a\cos t + b\sin t$ is used instead.

9. By examining the transients in the linearized equations obtained in the first part of Exercise 8, show that the two solutions, period 2π, obtained are respectively stable and unstable.

10. Show that the equations giving the equilibrium points in the van der Pol plane for solutions period $2\pi/\omega$ for the forced, damped pendulum equation

$$\ddot{x} + k\dot{x} + x - \tfrac{1}{6}x^3 = \Gamma\cos\omega t, \quad k > 0$$

are

$$k\omega a + b\{\omega^2 - 1 + \tfrac{1}{8}(a^2 + b^2)\} = 0,$$

$$-k\omega b + a\{\omega^2 - 1 + \tfrac{1}{8}(a^2 + b^2)\} = -\Gamma.$$

Deduce that

$$r^2(\omega^2 - 1 + \tfrac{1}{8}r^2)^2 + \omega^2 k^2 r^2 = \Gamma^2, \quad \omega k r^2 = \Gamma b,$$

where $r = \sqrt{(a^2 + b^2)}$. Show that there is a single equilibrium point when $\omega^2 > 1$.

11. For the equation $\ddot{x} + x - \tfrac{1}{6}x^3 = \Gamma\cos\omega t$, show that the solutions of period $2\pi/\omega$ are stable whenever $\omega^2 > 1$; and when $\omega^2 < 1$ and $|\Gamma| > \frac{2}{3}\sqrt{(\frac{8}{3})}(1 - \omega^2)^{3/2}$.

12. For the equation $\ddot{x} + \alpha x + \beta x^2 = \Gamma\cos t$, substitute $x = c(t) + a(t)\cos t + b(t)\sin t$, and show that, neglecting $\ddot{a}$ and $\ddot{b}$,

$$\dot{a} = \tfrac{1}{2}b(\alpha - 1 + 2\beta c),$$

$$\dot{b} = -\tfrac{1}{2}a(\alpha - 1 + 2\beta c) + \Gamma,$$

$$\ddot{c} = -\alpha c - \beta\{c^2 + \tfrac{1}{2}(a^2 + b^2)\}.$$

Deduce that if $|\Gamma|$ is large there are no solutions of period 2π, and that if $\alpha < 1$ and Γ is sufficiently small there are two solutions of period 2π.

13. Substitute $x = c(t) + a(t)\cos t + b(t)\sin t$ into the equation

$$\ddot{x} + \alpha x^2 = 1 + \Gamma \cos t$$

(compare Exercise 8), and show that if $\ddot{a}$ and $\ddot{b}$ are neglected, then

$$2\dot{a} = b(2\alpha c - 1), \quad 2\dot{b} = a(1 - 2\alpha c) + \Gamma, \quad \ddot{c} + \alpha(c^2 + \tfrac{1}{2}a^2 + \tfrac{1}{2}b^2) = 1.$$

Use a graphical argument to show that there are two equilibrium points when $\alpha < \tfrac{1}{4}$. Explain in general terms why, unlike $\ddot{a}$ and $\ddot{b}$, $\ddot{c}$ cannot be suppressed.

14. In the Duffing equation

$$\ddot{x} + k\dot{x} + x - \tfrac{1}{6}x^3 = \Gamma \cos \omega t,$$

substitute $x = a(t)\cos \omega t + b(t)\sin \omega t$ to investigate the solutions of period $2\pi/\omega$. Assume that a and b are slowly-varying and that $k\dot{a}, k\dot{b}$ can be neglected. Show that the paths in the van der Pol plane are given by

$$\dot{a} = -\frac{b}{2\omega}\{\omega^2 - 1 + \tfrac{1}{8}(a^2 + b^2)\} - \tfrac{1}{2}ka,$$

$$\dot{b} = \frac{a}{2\omega}\{\omega^2 - 1 + \tfrac{1}{8}(a^2 + b^2)\} - \tfrac{1}{2}kb + \frac{\Gamma}{2\omega}.$$

Find the linear approximation in the neighbourhood of the equilibrium point when $\omega^2 > 1$, and show that it is a stable node or spiral when $k > 0$.

15. For the equation $\ddot{x} + \alpha x + \beta x^3 = \Gamma \cos \omega t$, show that the restoring force $\alpha x + \beta x^3$ is represented in the linear least-squares approximation on $-A \leqslant x \leqslant A$ by $(\alpha + \tfrac{3}{5}\beta A^2)x$. Obtain the general solution of the approximating equation corresponding to a solution of amplitude A. Deduce that there may be a subharmonic of order $\tfrac{1}{3}$ if $\alpha + \tfrac{3}{5}\beta A^2 = \tfrac{1}{9}\omega^2$ has a real solution A. Compare eqn (7.57) for the case when $\Gamma/8\alpha$ is small. Deduce that when $\alpha \approx \tfrac{1}{9}\omega^2$ (close to subharmonic resonance), the subharmonic has the approximate form

$$A\cos(\tfrac{1}{3}\omega t + \phi) - \frac{\Gamma}{8\alpha}\cos \omega t$$

where ϕ is a constant.

(The interpretation is that when $\Gamma/8\alpha$ is small enough for the oscillation to lie in $[-A, A]$, A can be adjusted so that the slope of the straight-line fit on $[-A, A]$ is appropriate to the generation of a natural oscillation which is a subharmonic. The phase cannot be determined by this method.)

Show that the amplitude predicted for the equation $\ddot{x} + 0.15x - 0.1x^3 = 0.1\cos t$ is $A = 0.085$. Compare this with the solution of (7.57).

16. Use the perturbation method to show that

$$\ddot{x} + k\dot{x} + \alpha x + \beta x^3 = \Gamma \cos \omega t$$

has no subharmonic of order $\frac{1}{2}$ when β is small and $k = O(\beta)$. (Assume that

$$(a\cos\tfrac{1}{2}\tau + b\sin\tfrac{1}{2}\tau + c\cos\tau)^3 = \tfrac{3}{4}c(a^2 - b^2) + \tfrac{3}{4}a(a^2 + b^2 + 2c^2)\cos\tfrac{1}{2}\tau$$
$$+ \tfrac{3}{4}b(a^2 + b^2 + 2c^2)\sin\tfrac{1}{2}\tau + \text{higher harmonics.})$$

17. Use the perturbation method to show that

$$\ddot{x} + k\dot{x} + \alpha x + \beta x^3 = \Gamma\cos\omega t$$

has no subharmonic of order other than $\frac{1}{3}$ when β is small and $k = O(\beta)$.

18. Look for subharmonics of order $\frac{1}{2}$ for the equation

$$\ddot{x} + \varepsilon(x^2 - 1)\dot{x} + x = \Gamma\cos\omega t$$

by substituting

$$x = a\cos\tfrac{1}{2}\omega t + b\sin\tfrac{1}{2}\omega t.$$

Show that there are no non-zero equilibrium points unless $|\omega| = 2$ and $\frac{2}{9}\Gamma^2 - 4$ < 0, in which case the circle $r = \sqrt{(4 - \frac{2}{9}\Gamma^2)}$ consists of equilibrium points. (The algebra can be simplified by writing $x^2\dot{x} = \frac{1}{3}\mathrm{d}(x^3)/\mathrm{d}t$, and using the hint in Exercise 16.)

19. Extend the analysis of the equation

$$\ddot{x} + \varepsilon(x^2 - 1)\dot{x} + x = \Gamma\cos\omega t$$

in Exercise 18 by assuming that

$$x = a(t)\cos\tfrac{1}{2}\omega t + b(t)\sin\tfrac{1}{2}\omega t - \tfrac{1}{3}\Gamma\cos\omega t,$$

where a and b are slowly-varying. Show that when $\ddot{a}, \ddot{b}, \varepsilon\dot{a}, \varepsilon\dot{b}$, are neglected,

$$\omega\dot{a} = (1 - \tfrac{1}{4}\omega^2)b - \tfrac{1}{8}\varepsilon\omega a(a^2 + b^2 + \tfrac{2}{9}\Gamma^2 - 4),$$
$$\omega\dot{b} = -(1 - \tfrac{1}{4}\omega^2)a - \tfrac{1}{8}\varepsilon\omega b(a^2 + b^2 + \tfrac{2}{9}\Gamma^2 - 4),$$

on the van der Pol plane for the subharmonic.

By using $\rho = a^2 + b^2$ and ϕ the polar angle on the plane show that

$$\dot{\rho} = -\tfrac{1}{4}\varepsilon\rho(\rho + K), \quad \dot{\phi} = (1 - \tfrac{1}{4}\omega^2)/\omega; \quad K = \tfrac{2}{9}\Gamma^2 - 4.$$

Deduce that
(i) When $\omega \neq 2$ and $K \geqslant 0$, all paths spiral in to the origin, which is the only equilibrium point (so no subharmonic exists).
(ii) When $\omega = 2$ and $K \geqslant 0$, all paths are radial straight lines entering the origin (so there is no subharmonic).
(iii) When $\omega \neq 2$ and $K < 0$, all paths spiral on to a limit cycle, which is a circle, radius $\sqrt{(-K)}$ and centre the origin (so x is not periodic).
(iv) When $\omega = 2$ and $K < 0$, the circle centre the origin and radius $\sqrt{(-K)}$ consists entirely of equilibrium points, and all paths are radial straight lines approaching these points (each such point represents a subharmonic).

(Since subharmonics are expected only in case (iv), and for a critical value of ω, entrainment cannot occur. For practical purposes, even if the theory were exact we could never expect to observe the subharmonic, though solutions near to it may occur.)

20. Given eqns (7.34), (7.41), and (7.42) for the response curves and the stability boundaries for van der Pol's equation (Fig. 7.10), eliminate r^2 to show that the boundary of the entrainment region in the γ, v-plane is given by

$$\gamma^2 = 8\{1+9v^2 - (1-3v^2)^{3/2}\}/27.$$

Show that, for small v, $\gamma \approx \pm 2v$.

21. Consider the equation

$$\ddot{x} + \varepsilon(x^2 + \dot{x}^2 - 1)\dot{x} + x = \Gamma \cos \omega t.$$

To obtain solutions of period $2\pi/\omega$, substitute

$$x = a(t)\cos \omega t + b(t)\sin \omega t$$

and deduce that, if $\ddot{a}$, $\ddot{b}$, $\varepsilon\dot{a}$, $\varepsilon\dot{b}$ can be neglected, then

$$\dot{a} = \tfrac{1}{2}\varepsilon\{a - vb - \tfrac{1}{4}\mu a(a^2+b^2)\},$$
$$\dot{b} = \tfrac{1}{2}\varepsilon\{va + b - \tfrac{1}{4}\mu b(a^2+b^2)\} + \tfrac{1}{2}\varepsilon\gamma,$$

where

$$\mu = 1+3\omega^2, \quad v = (\omega^2-1)/\varepsilon\omega, \quad \text{and} \quad \gamma = \Gamma/\varepsilon\omega.$$

Show that the stability boundaries are given by

$$1 + v^2 - \mu r^2 + \tfrac{3}{16}\mu^2 r^4 = 0, \quad 2 - \mu r^2 = 0.$$

22. Show that the equation

$$\ddot{x}(1 - x\dot{x}) + (\dot{x}^2 - 1)\dot{x} + x = 0$$

has an *exact* periodic solution $x = \cos t$. Show that the corresponding forced equation:

$$\ddot{x}(1 - x\dot{x}) + (\dot{x}^2 - 1)\dot{x} + x = \Gamma \cos \omega t$$

has an *exact* solution of the form $a \cos \omega t + b \sin \omega t$, where

$$a(1 - \omega^2) - \omega b + \omega^3 b(a^2+b^2) = \Gamma,$$
$$b(1 - \omega^2) + \omega a - \omega^3 a(a^2+b^2) = 0.$$

Deduce that the amplitude $r = \sqrt{(a^2+b^2)}$ satisfies

$$r^2\{(1-\omega^2)^2 + \omega^2(1 - r^2\omega^2)^2\} = \Gamma^2.$$

23. The frequency–amplitude relation for the damped forced pendulum is (eqn (7.23), with $\beta = -1/6$)

$$r^2\{k^2\omega^2 + (\omega^2 - 1 + \tfrac{1}{8}r^2)^2\} = \Gamma^2.$$

Show that the vertex of the cusp bounding the fold in Fig. 7.7 occurs where

$$\omega^2 = \tfrac{1}{2}[2 + 3k^2 - \sqrt{(9k^4 + 12k^2)}].$$

Find the corresponding value for Γ^2.

24. (Combination tones.) Consider the equation

$$\ddot{x} + \alpha x + \beta x^2 = \Gamma_1 \cos \omega_1 t + \Gamma_2 \cos \omega_2 t, \quad \alpha > 0, \quad |\beta| \ll 1,$$

where the forcing term contains two distinct frequencies ω_1 and ω_2. To find an approximation to the response, construct the iterative process leading to the sequence of approximations $x^0(t), x^1(t), \ldots$, and starting with

$$\ddot{x}^0 + \alpha x^0 = \Gamma_1 \cos \omega_1 t + \Gamma_2 \cos \omega_2 t,$$

$$\ddot{x}^1 + \alpha x^1 = \Gamma_1 \cos \omega_1 t + \Gamma_2 \cos \omega_2 t - \beta(x^0)^2,$$

show that a particular solution is given approximately by

$$x(t) = -\frac{\beta}{2\alpha}(a^2 + b^2) + a \cos \omega_1 t + b \cos \omega_2 t$$

$$+ \frac{\beta a^2}{2(4\omega_1^2 - \alpha)} \cos 2\omega_1 t + \frac{\beta b^2}{2(4\omega_2^2 - \alpha)} \cos 2\omega_2 t$$

$$+ \frac{\beta ab}{(\omega_1 + \omega_2)^2 - \alpha} \cos(\omega_1 + \omega_2)t + \frac{\beta ab}{(\omega_1 - \omega_2)^2 - \alpha} \cos(\omega_1 - \omega_2)t,$$

where

$$a \approx \Gamma_1/(\alpha - \omega_1^2), \quad b \approx \Gamma_2/(\alpha - \omega_2^2).$$

(The presence of 'sum and difference tones' with frequencies $\omega_1 \pm \omega_2$ can be detected in sound resonators having suitable nonlinear characteristics, or as an auditory illusion attributed to the nonlinear detection mechanism in the ear (McLachlan 1956). The iterative method of solution can be adapted to simpler forced oscillation problems involving a single input frequency.)

25. Apply the method of Exercise 24 to the Duffing equation

$$\ddot{x} + \alpha x + \beta x^3 = \Gamma_1 \cos \omega_1 t + \Gamma_2 \cos \omega_2 t.$$

26. Investigate the resonant solutions of Duffing's equation in the form

$$\ddot{x} + x + \varepsilon^3 x^3 = \cos t, \quad |\varepsilon| \ll 1,$$

by the method of multiple scales (Section 6.4) using a slow time $\eta = \varepsilon t$ and a solution of the form

$$x(\varepsilon, t) = \frac{1}{\varepsilon} X(\varepsilon, t, \eta) = \frac{1}{\varepsilon} \sum_{n=0}^{\infty} \varepsilon^n X_n(t, \eta).$$

Show that $X_0 = a_0(\eta) \cos t + b_0(\eta) \sin t$, where

$$8a_0' + 3b_0(a_0^2 + b_0^2) = 0, \quad 8b_0' - 3a_0(a_0^2 + b_0^2) = 4.$$

(This example illustrates that even a small nonlinear term may inhibit the growth of resonant solutions.)

27. Repeat the multiple scale procedure of the previous exercise for the equation

$$\ddot{x} + x + \varepsilon^3 x^2 = \cos t, \quad |\varepsilon| \ll 1,$$

which has an unsymmetrical, quadratic departure from linearity. Use a slow time $\eta = \varepsilon^2 t$ and an expansion $x(\varepsilon, t) = \varepsilon^{-2} \sum_{n=0}^{\infty} \varepsilon^n X_n(t, \eta)$.

8 Stability

THE WORD 'stability' is used freely, as a descriptive term, in earlier chapters. In Chapter 2 *equilibrium points* are classified as stable or unstable according to their appearance in the phase diagram. Roughly speaking, if every initial state close enough to equilibrium leads to states which continue permanently to be close, then the equilibrium is stable. Some equilibrium states may be thought of as being more stable than others; for example, all initial states near a stable node lead ultimately into the node, but in the case of a centre there always remains a residual oscillation after disturbance of the equilibrium. *Limit cycles* are classified as stable or unstable according to whether or not 'nearby' paths 'spiral' into the limit cycle. In Chapter 7 we considered the stability of *forced periodic oscillations*, reducing the question to that of the stability of the corresponding equilibrium points on a van der Pol plane.

These cases are formally different, as well as being imprecisely defined. Moreover, they are not exhaustive: the definition of stability chosen must be that appropriate to the types of phenomena we want to distinguish between. The question to be answered is usually of this kind. If a system is in some way disturbed, will its subsequent behaviour differ from its undisturbed behaviour by an acceptably small amount? In practice, physical and other systems are always subject to small, unpredictable influences: to variable switch-on conditions, to maladjustment, to variation in physical properties, and the like. If such variations produce large changes in the operating conditions the system is probably unusable, and its normal operating condition would be described as unstable. Even if the effect of small disturbances does not grow, the system may not be 'stable enough' for its intended purpose: for proper working it might be necessary for the system to re-approach the normal operating condition, rather than to sustain a permanent superimposed deviation, however small. We might even require it to approach its normal state without ultimately suffering any time delay.

Some of these possibilities are treated in the following chapter. We are concerned with *regular systems* throughout (see Appendix A). The treatment is not restricted merely to second-order systems.

The general nth-order autonomous system can be written as

$$\dot{\mathbf{x}} = \mathbf{X}(\mathbf{x})$$

where $\mathbf{x}(t) = (x_1(t), x_2(t), \ldots, x_n(t))$ is a vector with n components $x_1(t)$, $x_2(t), \ldots, x_n(t)$. A solution $\mathbf{x} = \mathbf{x}(t)$ defines a phase path in the $(x_1, x_2, \ldots, x_n)$ phase space, and a set of phase paths defines a phase diagram for the system. A point $\mathbf{x}_0$ is an equilibrium point if the constant vector $\mathbf{x}(t) = \mathbf{x}_0$ is a solution for all t. As described in Chapter 2 for the case $n = 2$ the system can be linearized in the neighbourhood of $\mathbf{x}_0$. The local behaviour can then be interpreted from the eigenvalues of the linearized system. Obviously for $n \geqslant 3$ the classification of linear equations in terms of their eigenvalues becomes extensive. The general nth-order non-autonomous system can be represented by

$$\dot{\mathbf{x}} = \mathbf{X}(\mathbf{x}, t).$$

8.1. Poincaré stability (stability of paths)

This is a relatively undemanding criterion, which applies to autonomous systems. We shall treat it intuitively, in the two-dimensional case only. It agrees with the tentative classification of equilibrium points as stable or unstable in Chapter 2, and of limit cycles in Section 3.4.

We identify a phase path (or an equilibrium point) of the system $\dot{\mathbf{x}} = \mathbf{X}(\mathbf{x})$ representing a solution $\mathbf{x}^*(t)$ whose stability is in question, called the *standard path*. We are usually interested in this path only from a particular point $\mathbf{a}^*$ onwards: then we have a *positive half-path* $\mathscr{H}^*$ (or semi-path) *with initial point* $\mathbf{a}^*$ (Fig. 8.1). The solution which $\mathscr{H}^*$ represents is

$$\mathbf{x}^*(t), \quad t \geqslant t_0,$$

where

$$\mathbf{x}^*(t_0) = \mathbf{a}^*,$$

and since the system is autonomous, all choices of t_0 lead to the same $\mathscr{H}^*$.

The solution $\mathbf{x}^*(t)$, $t \geqslant t_0$ is *Poincaré stable* (orbitally stable), if all sufficiently small disturbances of the initial value $\mathbf{a}^*$ lead to half-paths remaining uniformly at a small distance from $\mathscr{H}^*$. To form an analytical definition it is necessary to reverse this statement. Firstly, choose $\varepsilon > 0$ arbitrarily and construct a strip of 'radius' ε about $\mathscr{H}^*$ (Fig. 8.2). Then for stability it must be shown that *all* paths starting within *some* distance δ of $\mathbf{a}^*$ (where $\delta \leqslant \varepsilon$ necessarily), remain *permanently* within the strip. Such a condition must hold for every ε: in general, the smaller ε, the smaller δ must be.

The formal definition is as follows.

DEFINITION 8.1 (*Poincaré, or orbital stability*). *Let $\mathscr{H}^*$ be the half-path for the solution $\mathbf{x}^*(t)$ of $\dot{\mathbf{x}} = \mathbf{X}(\mathbf{x})$ which starts at $\mathbf{a}^*$ at $t = t_0$. Suppose that for every $\varepsilon > 0$ there exists $\delta(\varepsilon) > 0$ such that if $\mathscr{H}$ is the half-path starting at $\mathbf{a}$, then*

$$|\mathbf{a} - \mathbf{a}^*| < \delta \Rightarrow \sup_{\mathbf{x} \in \mathscr{H}} \text{dist}\,(\mathbf{x}, \mathscr{H}^*) < \varepsilon.$$

Then $\mathscr{H}^*$ (or the corresponding solution) is said to be Poincaré stable. Otherwise $\mathscr{H}^*$ is unstable.

Here, the distance between a point $\mathbf{x}$ and a curve $\mathscr{C}$ is defined by

$$\text{dist}\,(\mathbf{x}, \mathscr{C}) = \inf_{\mathbf{y} \in \mathscr{C}} |\mathbf{x} - \mathbf{y}|\,;$$

(a)   (b)   (c)

FIG. 8.1. Typical half-paths $\mathscr{H}^*$. In (*b*), the half-path is a closed curve, repeated indefinitely

FIG. 8.2.

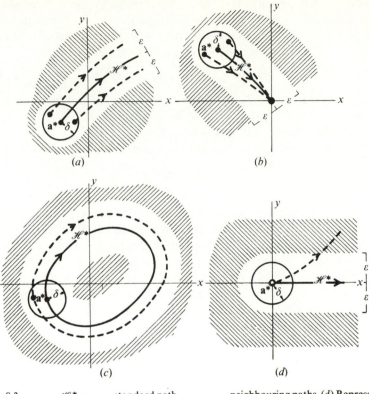

FIG. 8.3. ———— $\mathscr{H}*$ ———— standard path, ———— neighbouring paths. (*d*) Represents an unstable case

that is, it is the least distance in the ordinary sense (which is also a perpendicular distance). The maximum permitted separation of $\mathscr{H}$ from $\mathscr{H}*$ is limited, in two dimensions, by the following construction: two discs of diameter ε are rolled along either side of $\mathscr{H}*$ to sweep out a strip; for stability, there must exist a $\delta > 0$ such that every $\mathscr{H}$ with starting point $\mathbf{a}$, $|\mathbf{a} - \mathbf{a}*| < \delta$, lies permanently in the strip (Fig. 8.2).

Figure 8.3 illustrates schematically some cases of stability and instability. (*a*), (*b*), and (*c*) show stable cases. Case (*d*) represents the system

$$\dot{x} = x, \quad \dot{y} = 2y$$

with solutions

$$x = A e^{t}, \quad y = B e^{2t}$$

where A and B are constants. The paths are given by $y = Cx^2$ for various C. Consider, for example, the Poincaré stability of the half-path $y = 0$, $x > 0$. Choose $\varepsilon > 0$ and consider any neighbouring path starting at $\mathbf{a} = (x_0, y_0)$. Parametrically, the corresponding half-path is

$$x(t) = x_0 e^{t - t_0}, \quad y(t) = y_0 e^{t - t_0}$$

when the initial time is t_0. $|y(t)|$ becomes equal to, and then exceeds, ε (at $t = t_0 + \log_e(\varepsilon/|y_0|)$), no matter how close to the origin (x_0, y_0) is chosen. The solution is therefore unstable.

An example of a Poincaré stable solution is any one of the periodic solutions of the system

$$\dot{x} = y, \quad \dot{y} = -x,$$

whose phase diagram consists of circles, centre the origin. The positive half-paths consist of these circles, traversed endlessly. It is clear from Fig. 8.4 that every circular path is Poincaré stable; for if the path considered is on the circle of radius $(x_0^2 + y_0^2)^{1/2}$, starting at (x_0, y_0), the strip of arbitrary given width 2ε is that bounded by two more circles, and every half-path starting within a distance $\delta = \varepsilon$ of (x_0, y_0) lies permanently between these circles.

The same will be true of paths representing periodic solutions, and forming part of a centre, when these are not circular. However, closed

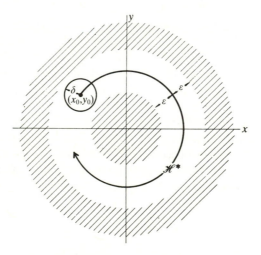

Fig. 8.4. Stability of the system $\dot{x} = y, \dot{y} = -x$

FIG. 8.5. Limit cycles: (*a*) is Poincaré stable; (*b*) and (*c*) are not

paths are not in general Poincaré stable. In Fig. 8.5 idealized limit cycles
are shown; the first is Poincaré stable and the other two are not.

Consider the system

$$\dot{x} = y, \quad \dot{y} = x,$$

the paths being given by $x^2 - y^2 = C$ where C is the parameter of the
family. The phase diagram shows a saddle point (Fig. 8.6).

On the basis of the previous discussion it can be seen that the half-
paths starting typically at A and C, and all the curved half-paths, are
Poincaré stable. Those starting typically at B and D are unstable
despite the fact that they approach the origin, since every nearby path
eventually goes to infinity.

Consider now the special half-paths which are *equilibrium points*,
representing constant solutions. That corresponding to a saddle (for
example, Fig. 8.6) is unstable, since there is no circle with centre at the
equilibrium point, of however small radius, such that every half-path
starting in it remains within an arbitrary preassigned distance ε of the
equilibrium point.

Those types of equilibrium point which we earlier designated as being
stable—the centre, the stable node, and the stable spiral—are indeed
Poincaré stable according to the formal definition (8.1), but the proof is
difficult. In the case of a centre, however, a simple geometrical argument
is convincing. Suppose the centre to be the origin (Fig. 8.7). Then
Definition 8.1 requires that if, given any $\varepsilon > 0$, a circle is drawn, $\mathscr{C}_\varepsilon$,
centre the origin and radius ε, then a similar circle $\mathscr{C}_\delta$, of radius δ can be
found such that no half-path starting in $\mathscr{C}_\delta$ will ever reach $\mathscr{C}_\varepsilon$. The
construction is clear from the Figure: first a path $\mathscr{P}$ is chosen wholly
inside $\mathscr{C}_\varepsilon$, then $\mathscr{C}_\delta$ is drawn inside $\mathscr{P}$.

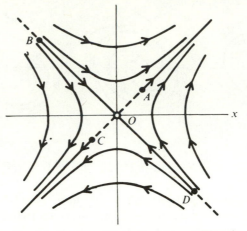

FIG. 8.6. Stability of the paths near a saddle point

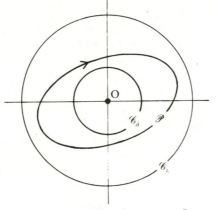

FIG. 8.7. Stability of a centre at O

8.2. Paths and solution curves

Poincaré stability is concerned with autonomous systems; another concept is required for non-autonomous systems. Even when a system is autonomous, however, a more sensitive criterion may be needed.

Consider, for example, the system

$$\dot{x} = y, \quad \dot{y} = -\sin x,$$

representing the motion of a pendulum. Near the origin the phase paths have the form shown in Fig. 8.8. Consider a solution $\mathbf{x}^*(t)$, $t \geqslant t_0$, with $\mathbf{x}^*(t_0) = \mathbf{a}^*$, its half-path being the heavy line starting at P; and a second solution $\mathbf{x}(t)$, $t \geqslant t_0$, $\mathbf{x}(t_0) = \mathbf{a}$, its half-path starting at Q.

$\mathbf{x}^*$ is Poincaré stable: if Q is close to P the corresponding paths remain close. However, *the representative points do not remain close*. It is known (Exercise 33, Chapter 1) that the period of $\mathbf{x}(t)$ is greater than that of $\mathbf{x}^*(t)$. Therefore the representative point Q' lags increasingly behind P': no matter how close Q and P are at time t_0, after a sufficient time Q' will be as much as half a cycle behind P' and the difference between the *solutions* at this time will therefore be large, a condition which recurs at regular intervals. If the pendulum (or other nonlinear oscillating system) were simply used to take a system through a sequence of states at more or less predetermined intervals this would be unimportant, but for

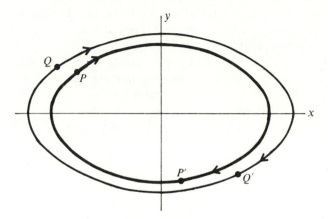

FIG. 8.8. Paths for the pendulum are Poincaré stable, but the representative points do not stay in step

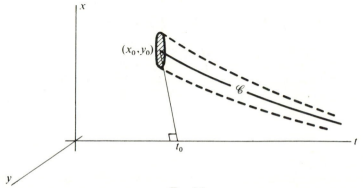

FIG. 8.9.

timekeeping the progressively increasing error resulting from a disturbance would make the uncontrolled pendulum useless.

In two dimensions another representation is desirable which displays
solutions rather than *phase paths*, and therefore allows us to keep track of
time variation, as in Fig. 8.9. For the system $\dot{x} = X(x, y, t), \dot{y} = Y(x, y, t)$,
the axes are x, y, t. A solution $x(t), y(t)$ appears as a curve in x, y, t space.
The heavy line $\mathscr{C}$ represents a solution on $t \geqslant t_0$ satisfying the initial
conditions $x(t_0) = x_0, y(t_0) = y_0$. The bundle of curves surrounding it
represents solutions with initial conditions close to those satisfied by
$x(t), y(t)$. The solution curves do not intersect at ordinary points of the
system since this would violate the uniqueness theorem (Appendix A).

When the system is autonomous, the *phase paths are the projections of
the solution curves on the x, y plane.* We know that these meet only at
equilibrium points, which represent constant solutions which are
straight lines parallel to the t axis in Fig. 8.9.

The figure suggests that $\mathscr{C}$ depicts a stable solution in an extended
sense. *Curves which start* (at t_0) *close together are close together at every
subsequent time.* If the equation generating these solutions were autonomous, then the projections of the curves on the plane would give part
of a phase diagram whose paths are close: this suggests that Poincaré
stability is also implied.

Contrast the case of the type illustrated in Fig. 8.10. The solution
curves corresponding to neighbouring paths are helices with the same
rate of advance along the t axis, but with different pitch: these become
widely separated although the paths (projections) do not.

Fig. 8.10. Solution curves corresponding to paths about a centre

We shall formulate the definition of this new sort of stability in the
next section. We require a measure of the distance between solutions at

particular times; that is, of the distance between the vectors of solution values. In n dimensions, let

$$\mathbf{x}(t) = (x_1(t), x_2(t), \ldots, x_n(t))^t$$

be a column vector of n real or complex functions, normally a solution vector of some system of first-order equations of dimension n. Several measures of the 'size' of $\mathbf{x}$ (for fixed t), called *norms* of $\mathbf{x}$, are available: we use the following:

$$\|\mathbf{x}\| = \left(\sum_{i=1}^{n} |x_i|^2 \right)^{1/2}; \qquad (8.1)$$

which is the length of $\mathbf{x}$ when its components are real. We retain the notation '$\|.\|$' since most of the theory is unaffected when other norms are used.

8.3. Liapunov stability (stability of solutions)

Consider the regular system

$$\dot{\mathbf{x}} = \mathbf{X}(\mathbf{x}, t), \qquad (8.2)$$

or, in component form,

$$\dot{x}_i = X_i(x_1, \ldots, x_n, t), \quad i = 1, 2, \ldots, n.$$

The formal definition of the type of stability suggested in Section 8.2 is as follows.

DEFINITION 8.2 (*Liapunov Stability*)   *Let* $\mathbf{x}^*(t)$ *be a given real or complex solution of* (8.2). *Then*
(i) $\mathbf{x}^*(t)$ *is Liapunov stable on* $t \geq t_0$ *if, for any* $\varepsilon > 0$, *there exists* $\delta(\varepsilon, t_0)$ *such that*

$$\|\mathbf{x}(t_0) - \mathbf{x}^*(t_0)\| < \delta \Rightarrow \|\mathbf{x}(t) - \mathbf{x}^*(t)\| < \varepsilon, \qquad (8.3)$$

for all $t \geq t_0$, *where* $\mathbf{x}(t)$ *is any other solution.*
(ii) *It can be shown (Cesari 1971) that if* (i) *is satisfied for initial conditions at* t_0, *then a similar condition is satisfied when any* $t_1 > t_0$ *is substituted for* t_0: *that is, if* $\mathbf{x}^*(t)$ *is stable for* $t \geq t_0$, *it is stable for* $t \geq t_1$ *> t_0.*
(iii) *Otherwise* $\mathbf{x}^*(t)$ *is said to be unstable (in the Liapunov sense).*

It can be shown that when (8.2) is autonomous, Liapunov stability implies Poincaré stability.

FIG. 8.11.

Example 8.1 *Show that all solutions of $\dot{x} = -x$ (a one-dimensional system) are stable.*

We will consider real solutions only, with initial condition at t_0. All real solutions are given by $x(t) = x(t_0) e^{-(t-t_0)}$. Consider the stability of $x^*(t)$ starting at $(t_0, x^*(t_0))$ (Fig. 8.11). Choose $\varepsilon > 0$ arbitrarily and consider the strip $x^*(t) - \varepsilon < x < x^*(t) + \varepsilon$. For stability, we must show that all solutions starting at t_0 and sufficiently close to $x^*(t_0)$ lie inside this strip. Since all solutions 'close up' on $x^*(t)$ as t increases it is clear that the conditions of the definition are satisfied by the choice $\delta = \varepsilon$.

The following example shows that the corresponding n-dimensional problem leads to the same conclusion.

Example 8.2 *Show that all solutions of the n-dimensional system $\dot{\mathbf{x}} = -\mathbf{x}$ are stable for $t \geq 0$.*

The general solution is given by

$$\mathbf{x}(t) = \mathbf{x}(0) e^{-t}.$$

Consider the stability of $\mathbf{x}^*(t)$ for $t > 0$, where

$$\mathbf{x}^*(t) = \mathbf{x}^*(0) e^{-t}.$$

We have

$$\|\mathbf{x}(t) - \mathbf{x}^*(t)\| \leq \|\mathbf{x}(0) - \mathbf{x}^*(0)\| e^{-t}$$
$$\leq \|\mathbf{x}(0) - \mathbf{x}^*(0)\|, \quad t \geq 0.$$

Therefore, given any $\varepsilon > 0$,

$$\|\mathbf{x}(0) - \mathbf{x}^*(0)\| < \varepsilon \Rightarrow \|\mathbf{x}(t) - \mathbf{x}^*(t)\| < \varepsilon, \quad t > 0.$$

Thus $\delta = \varepsilon$ in Definition 8.2. (The process can be modified to hold for any t_0. However, the system is autonomous, so this is automatic.)

The implication of (ii) in Definition 8.2 is as follows. (i) taken alone implies that arbitrarily small enough changes in the initial condition at $t = t_0$ lead to uniformly small departures from the tested solution on $t_0 \leqslant t < \infty$. Such behaviour could be described as stability with respect only to variations in initial conditions at t_0. Armed with (ii), however, it can be said that the solution is stable with respect to variation in initial conditions at all times t_1 where $t_1 \geqslant t_0$, and that the stability is therefore a property of the solution as a whole. (ii) holds by virtue of (i) under very general smoothness conditions that we do not discuss here.

If (i) is satisfied, and therefore (ii), we may still want to be assured that a particular solution does not become 'less stable' as time goes on. It is possible that a system's sensitivity to disturbance might increase indefinitely with time although it remains technically stable, the symptom being that $\delta(\varepsilon, t_0)$ decreases to zero as t_0 increases.

To see that this is possible, consider the family of curves with parameter c:

$$x = f(c, t) = c\,e^{(c^2-1)t}/t, \quad t > 0. \tag{8.4}$$

These are certainly solutions of *some* first-order equation. The equation is of no particular interest: the criterion for stability can be applied to the members of (8.4) without reference to the equation.

In Fig. 8.12, the curve separating those which tend to infinity ($c > 1$) from those tending to zero ($c < 1$) is the curve $f(1, t) = t^{-1}$, which also tends to zero. Now consider the stability of the 'solution' having $c = 0$: $f(0, t) \equiv 0$. It is apparent from the diagram that this function is stable for

FIG. 8.12. $x(t) = c\,e^{(c^2-1)t}/t$

all $t_0 > 0$. However, it can afford no greater disturbance in 'initial condition' at time t_0 than $|f(1, t_0)| = 1/t_0$: a larger disturbance might encounter a curve which tends to infinity with t. In order that the disturbed and undisturbed curves should, for $t \geq t_0$, be permanently within a preassigned strip $-\varepsilon < x < \varepsilon$, starting values $x(t_0)$ must lie in the interval

$$\|x(t_0)\| \leqslant \min(\varepsilon, |f(1, t_0)|) = \min(\varepsilon, 1/t_0).$$

Therefore

$$\delta \leqslant \min(\varepsilon, 1/t_0)$$

in Definition 8.2, and δ tends to zero as $t_0 \to \infty$.

The stability is said to be *non-uniform* in this case. The definition of uniform stability is as follows.

DEFINITION 8.3 (*Uniform stability*)   *If a solution is stable for $t \geq t_0$ and the δ of Definition 8.2 is independent of t_0, the solution is uniformly stable on $t \geq t_0$.*

It is clear that any stable solutions of an *autonomous* system must be uniformly stable, since the system is invariant with respect to time translation.

A third desirable property is asymptotic stability. The system $\dot{\mathbf{x}} = \mathbf{0}$ has the constant solutions $\mathbf{x}(t) = \mathbf{x}(t_0)$ for given t_0. These are all stable on $t \geq t_0$ for any t_0 (and also uniformly stable); however, a disturbed solution shows no tendency to return to the original solution: it remains a constant distance away. On the other hand, solutions of the type examined in Figs. 8.11 and 8.12 do re-approach the undisturbed solution after being disturbed; thus the system tends to return to its original operating curve. Such solutions are said to be *asymptotically stable* (but uniformly and nonuniformly respectively in the cases of Figs. 8.11 and 8.12.

DEFINITION 8.4 (*Asymptotic stability*)   *Let $\mathbf{x}^*$ be a stable (or uniformly stable) solution for $t \geq t_0$. If additionally there exists $\eta(t_0) > 0$ such that*

$$\|\mathbf{x}(t_0) - \mathbf{x}^*(t_0)\| < \eta \Rightarrow \lim_{t \to \infty} \|\mathbf{x}(t) - \mathbf{x}^*(t)\| = 0, \tag{8.5}$$

then the solution is said to be asymptotically stable (or uniformly and asymptotically stable).

In the rest of this chapter we prove theorems enabling general statements to be made about the stability of solutions of certain classes of equations. Such theorems are necessary since in general it is not possible to find solutions of nonlinear systems explicitly.

8.4. Stability of linear systems

We consider first the stability (in the Liapunov sense) of the constant solutions (equilibrium points) of the two-dimensional, constant-coefficient systems classified in Chapter 2. Without loss of generality we can place the equilibrium point at the origin, since this only involves modifying the solutions by additive constants. We have then

$$\dot{x}_1 = ax_1 + bx_2, \quad \dot{x}_2 = cx_1 + dx_2 \tag{8.6}$$

in the present notation.

The stability properties of the constant solutions $\mathbf{x}^*(t) = \mathbf{0}, t \geq t_0$ can sometimes be read off from the phase diagram: for example a saddle is unstable. Consider also the case of a centre. Figure 8.13(a) shows the phase plane and 8.13(b) the solution space: the solution whose stability is being considered lies along the t axis, and another typical solution, corresponding to a closed path, is shown.

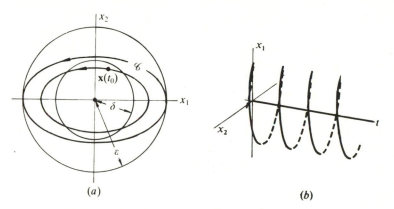

(a)   (b)

Fig. 8.13. Phase diagram and solution space near a centre

Choose a t_0, and an arbitrary $\varepsilon > 0$, and define a domain by

$$\|\mathbf{x}\| < \varepsilon \tag{8.7}$$

as indicated in Fig. 8.13. This disc contains at least one closed path

$\mathscr{C}$. Now construct a domain of initial values, near $\mathbf{x}^*(t_0) \equiv \mathbf{0}$:

$$\|\mathbf{x}(t_0) - \mathbf{x}^*(t_0)\| = \|\mathbf{x}(t_0)\| < \delta, \tag{8.8}$$

by taking δ small enough for this domain to lie entirely within $\mathscr{C}$. Then the solution with any initial condition $\mathbf{x}(t_0)$, subject to (8.8), lies permanently inside $\mathscr{C}$ and therefore satisfies (8.7) for all $t \geqslant t_0$. Since ε is arbitrary the condition (8.3) for stability is satisfied. The zero solution is uniformly stable, since given ε, δ is independent of t_0; this is also a consequence of the autonomous nature of (8.6). It is not asymptotically stable since the solutions do not approach $\mathbf{x} = \mathbf{0}$ as $t \to \infty$.

The other types of equilibrium point are more difficult, and we merely summarize the results here:

(i) *centre*: uniformly stable,
(ii) *stable node*: uniformly and asymptotically stable,
(iii) *stable spiral*: uniformly and asymptotically stable,
(iv) *saddle point*: unstable.

(8.9)

It must be understood that the stability of the *equilibrium points only* (the constant solutions), and not that of the solutions represented by neighbouring paths, is being considered here. For example, although a centre is stable, in the nonlinear case the paths surrounding it will generally be unstable due to amplitude–frequency dependence.

The stability of a non-constant solution of a linear system (and, in fact, of a general system: see Chapter 9) can be reduced to the question of the stability of the zero solution of a related system, as follows.

Consider the nth-order forced linear system which is not autonomous,

$$\dot{\mathbf{x}} = A(t)\mathbf{x} + \mathbf{f}(t) \tag{8.10}$$

where $A(t)$ is an $n \times n$ matrix. We wish to investigate the stability of a solution $\mathbf{x}^*(t)$. Let $\mathbf{x}(t)$ represent any other solution, and define $\boldsymbol{\xi}(t)$ by

$$\boldsymbol{\xi}(t) = \mathbf{x}(t) - \mathbf{x}^*(t). \tag{8.11}$$

Then $\boldsymbol{\xi}(t)$ tracks the difference between the 'test' solution and a solution having a different initial value. The initial condition for $\boldsymbol{\xi}$ is

$$\boldsymbol{\xi}(t_0) = \mathbf{x}(t_0) - \mathbf{x}^*(t_0). \tag{8.12}$$

Also, $\boldsymbol{\xi}$ satisfies the homogeneous equation derived from (8.10):

$$\boldsymbol{\xi} = A(t)\boldsymbol{\xi}. \tag{8.13}$$

By comparison of (8.11), (8.12), and (8.13) with (8.3), it can be seen that the stability of $\mathbf{x}^*(t)$ is equivalent to the stability of the zero solution of (8.13). $\xi(t)$ is called a *perturbation* of the solution $\mathbf{x}^*(t)$.

Since this new formulation of the problem is independent of the solution of (8.10) initially chosen, we can make the following statement:

THEOREM 8.1 *All solutions of the regular linear system* $\dot{\mathbf{x}} = A(t)\mathbf{x} + \mathbf{f}(t)$ *have the same stability property (unstable, stable, uniformly stable, asymptotically stable, uniformly and asymptotically stable), this being the same as that of the zero (or any other) solution of the homogeneous equation* $\dot{\xi} = A(t)\xi$.

Thus, in the *linear case* (only), the classifications (8.9) apply equally to the surrounding paths.

Example 8.3 *All the solutions of the equation* $\dot{x}_1 = x_2$, $\dot{x}_2 = -\omega^2 x_1 + f(t)$ *are uniformly stable, but not asymptotically stable (forced linear oscillator).*

Equation (9.19) becomes $\dot{\xi}_1 = \xi_2$, $\dot{\xi}_2 = -\omega^2 \xi_1$. The zero solution is a centre, which has the specified properties.

Example 8.4 *All the solutions of the equations* $\dot{x}_1 = x_2$, $\dot{x}_2 = -kx_2 - \omega^2 x_1$ *+f(t) are uniformly and asymptotically stable (damped, forced linear oscillator).*

Equation (8.13) becomes

$$\dot{\xi}_1 = \xi_2, \quad \dot{\xi}_2 = -\omega^2 \xi_1 - k\xi_2.$$

The zero solution corresponds to the equilibrium point $\xi_1 = \xi_2 = 0$ which (Chapter 2) is a 'stable' node or spiral (as named in that chapter). By (8.9), these are uniformly and asymptotically stable. This result corresponds to the damping out of free oscillations set up by the initial conditions: the amplitude and phase of the forced oscillation are independent of these.

8.5. Structure of the solutions of *n*-dimensional linear systems

The need to consider linear systems arises in connection with nonlinear systems which are in some sense close to linear. In such cases certain characteristics of their solutions, notably stability properties, may follow those of the approximating linear systems.

The general homogeneous, first-order linear system of n dimensions is

$$\dot{\mathbf{x}} = A(t)\mathbf{x} \tag{8.14}$$

where $A(t)$ is an $n \times n$ matrix whose elements $a_{ij}(t)$ are functions of time, and $\mathbf{x}(t)$ is a column vector of the n dependent variables. The explicit appearance of t shows that the system is non-autonomous. In component form,

$$\begin{pmatrix} \dot{x}_1 \\ \dot{x}_2 \\ \vdots \\ \dot{x}_n \end{pmatrix} = \begin{pmatrix} a_{11} & a_{12} & \cdots & a_{1n} \\ a_{21} & a_{22} & \cdots & a_{2n} \\ \vdots & \vdots & \vdots & \vdots \\ a_{n1} & a_{n2} & \cdots & a_{nn} \end{pmatrix} \begin{pmatrix} x_1 \\ x_2 \\ \vdots \\ x_n \end{pmatrix},$$

or, when expanded,

$$\dot{x}_i = \sum_{j=1}^{n} a_{ij}(t)x_j; \quad i,j = 1, 2, \ldots, n.$$

We shall assume that each $a_{ij}(t)$ is continuous on $-\infty < t < \infty$, so that the system is regular (Appendix A). In that case, given any initial conditions $\mathbf{x}(t_0) = \mathbf{x}_0$, there is a unique solution satisfying this condition. Moreover, it exists for all t, $-\infty < t < \infty$ (the same is not true in general of nonlinear equations).

If $\mathbf{x}_1(t)$, $\mathbf{x}_2(t), \ldots, \mathbf{x}_m(t)$ are real or complex solutions of (8.14), then so is $\alpha_1\mathbf{x}_1(t) + \alpha_2\mathbf{x}_2(t) + \ldots + \alpha_m\mathbf{x}_m(t)$ where $\alpha_1, \alpha_2, \ldots, \alpha_m$ are any constants, real or complex. A central question to be settled is the dimension of the solution space: the minimum number of solution functions which, with their linear combinations, generate every solution of (8.14).

DEFINITION 8.5   *Let $\boldsymbol{\psi}_1(t)$, $\boldsymbol{\psi}_2(t), \ldots, \boldsymbol{\psi}_m(t)$ be vector functions (real or complex), continuous on $-\infty < t < \infty$, none being identically zero. If there exist constants (real or complex) $\alpha_1, \alpha_2, \ldots, \alpha_m$, not all zero, such that*

$$\alpha_1\boldsymbol{\psi}_1(t) + \alpha_2\boldsymbol{\psi}_2(t) + \ldots + \alpha_m\boldsymbol{\psi}_m(t) = \mathbf{0}$$

for $-\infty < t < \infty$, the functions are linearly dependent. Otherwise they are linearly independent.

(As a consequence of the definition, note that the functions $(1, 1)$, (t, t) are linearly independent, although the constant vectors $(1, 1)$, (t_0, t_0) are linearly dependent for every t_0.)

Example 8.5   cos t *and* sin t *are linearly independent on* $-\infty < t < \infty$.

$$\alpha_1 \cos t + \alpha_2 \sin t = \sqrt{(\alpha_1^2 + \alpha_2^2)} \sin(t + \beta)$$

where $\tan \beta = \alpha_1/\alpha_2$. There is no choice of α_1, α_2, except for $\alpha_1 = \alpha_2 = 0$, which makes this function zero for all t.

Example 8.6   *The functions* cos t, sin t, 2 sin t *are linearly dependent.*

Since, for example,

$$0(\cos t) + 2(\sin t) - 1(2 \sin t) \equiv 0.$$

If functions are linearly dependent, at least one of them can be expressed as a linear combination of the others. From now on we shall be concerned with the linear dependence of solutions of (8.14). Given any set of n solution (column) vectors $\phi_1, \phi_2, \ldots, \phi_n$, real or complex, where ϕ_j has elements $\phi_{1j}(t), \phi_{2j}(t), \ldots, \phi_{nj}(t)$, we shall use the notation

$$\Phi(t) = (\phi_1, \phi_2, \ldots, \phi_n) = \begin{pmatrix} \phi_{11} & \phi_{12} & \cdots & \phi_{1n} \\ \phi_{21} & \phi_{22} & \cdots & \phi_{2n} \\ \vdots & \vdots & & \vdots \\ \phi_{n1} & \phi_{n2} & \cdots & \phi_{nn} \end{pmatrix}$$

for the matrix of these solutions. Then from (8.14),

$$\dot{\Phi}(t) = A(t)\Phi(t). \tag{8.15}$$

Remember that the constituent solutions may be real or complex; for example it is convenient to be able to choose a matrix of solutions for

$$\dot{x}_1 = x_2, \quad \dot{x}_2 = -x_1$$

of the form

$$\Phi(t) = \begin{pmatrix} e^{it} & e^{-it} \\ ie^{it} & -ie^{-it} \end{pmatrix}.$$

Correspondingly, the constants occurring in linear dependence, and the eigenvalues and eigenvectors, may all be complex. This possibility will not always be specially pointed out in the rest of this chapter.

THEOREM 8.2 *Any $n+1$ nontrivial solutions of* (8.14) *are linearly dependent.*

Proof Let the solutions be $\phi_1(t), \phi_2(t), \ldots, \phi_{n+1}(t)$. Let t_0 be any value of t. Then the $n+1$ *constant* vectors $\phi_1(t_0), \phi_2(t_0), \ldots, \phi_{n+1}(t_0)$ are linearly dependent (as are any set of $n+1$ constant vectors of length n); that is, there exist constants $\alpha_1, \alpha_2, \ldots, \alpha_{n+1}$, not all zero, such that

$$\sum_{j=1}^{n+1} \alpha_j \phi_j(t_0) = \mathbf{0}.$$

Let

$$\mathbf{x}(t) = \sum_{j=1}^{n+1} \alpha_j \phi_j(t).$$

Then $\mathbf{x}(t_0) = \mathbf{0}$, and $\mathbf{x}(t)$ is a solution of (1). Therefore, by the Uniqueness Theorem (Appendix A), $\mathbf{x}(t) = \mathbf{0}$ for all t: that is to say, the solutions $\phi_j(t), j = 1, 2, \ldots, n+1$ are linearly dependent.

THEOREM 8.3   *There exists a set of n linearly independent solutions of* (8.14)

Proof By the Existence Theorem (Appendix A), there is a set of n solutions $\psi_1(t), \psi_2(t), \ldots, \psi_n(t)$ whose matrix $\Psi(t)$ satisfies

$$\Psi(0) = I,$$

where I is the identity matrix. Since the columns of $\Psi(0): \psi_1(0), \ldots, \psi_n(0)$, are linearly independent, so are the solutions.

These two theorems settle the dimension of the solution space: every solution is a linear combination of the solutions ψ_j, $j = 1, 2, \ldots, n$ of Theorem 8.2; but since these solutions are themselves linearly independent, we cannot do without any of them. Instead of the special solutions $\psi_j(t)$ we may take any set of n linearly independent solutions as the basis, as is shown in Theorem 8.4.

THEOREM 8.4   *Let $\phi_1(t)$, $\phi_2(t)$, ..., $\phi_n(t)$ be any set of linearly independent solutions (real or complex) of (8.14). Then every solution is a linear combination of these solutions.*

Proof Let $\phi(t)$ be any non-trivial solution of (8.14). Then by Theorem 8.2, $\phi, \phi_1, \phi_2, \ldots, \phi_n$ are linearly dependent. The coefficient of ϕ in any linear relation of linear dependence is non-zero, or we should violate the requirement that the ϕ_j are linearly independent. Therefore ϕ is a linear combination of $\phi_1, \ldots, \phi_n$.

DEFINITION 8.6   *Let $\phi_1(t), \phi_2(t), \ldots, \phi_n(t)$ be n linearly independent solutions of (8.14). Then the matrix*

$$\Phi(t) = (\phi_1, \phi_2, \ldots, \phi_n) = \begin{pmatrix} \phi_{11} & \phi_{12} & \cdots & \phi_{1n} \\ \phi_{21} & \phi_{22} & \cdots & \phi_{2n} \\ \vdots & \vdots & & \vdots \\ \phi_{n1} & \phi_{n2} & \cdots & \phi_{nn} \end{pmatrix}$$

is called a fundamental matrix of the system (8.14).

Note that any two fundamental matrices Φ_1 and Φ_2 are related by

$$\Phi_2(t) = \Phi_1(t)C, \tag{8.16}$$

where C is a non-singular $n \times n$ matrix (since by Theorem 8.3, each column of Φ_2 is a linear combination of the columns of Φ_1, and vice versa).

THEOREM 8.5 *Given any $n \times n$ solution matrix $\Phi(t) = (\phi_1(t), \ldots, \phi_n(t))$, then either* (i) *for all t, $\det\{\Phi(t)\} = 0$, or* (ii) *for all t, $\det\{\Phi(t)\} \neq 0$. Case* (i) *occurs if and only if the solutions are linearly dependent, and case* (ii) *implies that Φ is a fundamental matrix.*

Proof Suppose $\det\{\Phi(t_0)\} = 0$ for some t_0. Then the columns of $\Phi(t_0)$ are linearly dependent: there exist constants $\alpha_1, \ldots, \alpha_n$, not all zero, such that $\alpha_1 \phi_1(t_0) + \ldots + \alpha_n \phi_n(t_0) = \mathbf{0}$. Define $\phi(t)$ by

$$\phi(t) = \alpha_1 \phi_1(t) + \ldots + \alpha_n \phi_n(t).$$

Then $\phi(t)$ is a solution satisfying the initial condition $\phi(t_0) = \mathbf{0}$. By the Uniqueness Theorem (Appendix A), $\phi(t) = \mathbf{0}$ for all t. Therefore

$$\alpha_1 \phi_1(t) + \alpha_2 \phi_2(t) + \ldots + \alpha_n \phi_n(t) = \mathbf{0}$$

for all t, and $\det\{\Phi(t)\}$ is therefore either zero everywhere, or non-zero everywhere.

Example 8.7

$$\begin{pmatrix} e^t \\ e^{2t} \end{pmatrix}, \quad \begin{pmatrix} e^{3t} \\ e^{4t} \end{pmatrix}$$

cannot be a pair of solutions of a second-order linear homogeneous system on $-\infty < t < \infty$.

For, writing

$$\Phi(t) = \begin{pmatrix} e^t & e^{2t} \\ e^{2t} & e^{4t} \end{pmatrix},$$

$\det\{\Phi(t)\} = e^{5t} - e^{4t}$, which is zero at $t = 0$ but nowhere else.

THEOREM 8.6 *The solution of* (8.14) *satisfying the initial conditions $\mathbf{x}(t_0) = \mathbf{x}_0$ is given by $\mathbf{x}(t) = \Phi(t)\Phi^{-1}(t_0)\mathbf{x}_0$, where Φ is any fundamental matrix for* (8.14).

Proof The solution must be of the form

$$\mathbf{x}(t) = \Phi(t)\mathbf{a} \tag{8.17}$$

where $\mathbf{a}$ is a constant vector, by Theorem 8.4. The initial conditions give $\mathbf{x}_0 = \Phi(t_0)\mathbf{a}$. The columns of $\Phi(t_0)$ (regarded as columns of constants) are linearly independent by Theorem 8.5, so $\Phi(t_0)$ has an inverse $\Phi^{-1}(t_0)$. Therefore $\mathbf{a} = \Phi^{-1}(t_0)\mathbf{x}_0$ and the result follows from (8.17).

Example 8.8 *Verify that*

$$\begin{pmatrix} 2 \\ e^t \end{pmatrix}, \quad \begin{pmatrix} e^{-t} \\ 1 \end{pmatrix}$$

are particular solutions of

$$\begin{pmatrix} \dot{x}_1 \\ \dot{x}_2 \end{pmatrix} = \begin{pmatrix} 1 & -2e^{-t} \\ e^t & -1 \end{pmatrix} \begin{pmatrix} x_1 \\ x_2 \end{pmatrix},$$

and find the solution $\mathbf{x}(t)$ *such that*

$$\mathbf{x}(0) = \begin{pmatrix} 3 \\ 1 \end{pmatrix}.$$

Direct substitution confirms that the given functions are solutions. They are linearly independent, so

$$\Phi(t) = \begin{pmatrix} 2 & e^{-t} \\ e^t & 1 \end{pmatrix}$$

is a fundamental matrix. We have

$$\Phi(0) = \begin{pmatrix} 2 & 1 \\ 1 & 1 \end{pmatrix}, \quad \Phi^{-1}(0) = \begin{pmatrix} 1 & -1 \\ -1 & 2 \end{pmatrix}.$$

Hence

$$\mathbf{x}(t) = \begin{pmatrix} 2 & e^{-t} \\ e^t & 1 \end{pmatrix} \begin{pmatrix} 1 & -1 \\ -1 & 2 \end{pmatrix} \begin{pmatrix} 3 \\ 1 \end{pmatrix} = \begin{pmatrix} 4 - e^{-t} \\ 2e^t - 1 \end{pmatrix}.$$

The general *inhomogeneous* linear system is

$$\dot{\mathbf{x}} = A(t)\mathbf{x} + \mathbf{f}(t), \tag{8.18}$$

where $\mathbf{f}(t)$ is a column vector. The *associated homogeneous system* is

$$\dot{\boldsymbol{\phi}} = A(t)\boldsymbol{\phi}. \tag{8.19}$$

The following properties are readily verified.

(I) Let $\mathbf{x} = \mathbf{x}_p(t)$ be any solution of (8.18) (called a *particular solution* of the given system) and $\boldsymbol{\phi} = \boldsymbol{\phi}_c(t)$ any solution of (8.19) (called a *complementary function* for the given system). Then $\mathbf{x}_p(t) + \boldsymbol{\phi}_c(t)$ is a solution of (8.18).

(II) Let $\mathbf{x}_{p_1}$ and $\mathbf{x}_{p_2}(t)$ be any solutions of (8.18). Then $\mathbf{x}_{p_1}(t) - \mathbf{x}_{p_2}(t)$ is a solution of (8.19); that is, it is a complementary function.

Theorem 8.6 follows immediately:

THEOREM 8.6   *Let* $\mathbf{x}_p(t)$ *be any one particular solution of* $\dot{\mathbf{x}} = A(t)\mathbf{x} + \mathbf{f}(t)$. *Then every solution of this equation is of the form* $\mathbf{x}(t) = \mathbf{x}_p(t) + \boldsymbol{\phi}_c(t)$, *where* $\boldsymbol{\phi}$ *is a complementary function, and conversely.*

The strategy for finding all solutions of (8.18) is therefore to obtain, somehow, any *one* solution of (8.18), then to find *all* the solutions of the associated homogeneous system (8.19).

Example 8.9 *Find all solutions of the system*

$$\dot{x}_1 = x_2, \quad \dot{x}_2 = -x_1 + t.$$

The corresponding homogeneous system is $\dot{\phi}_1 = \phi_2, \dot{\phi}_2 = -\phi_1$, which is equivalent to $\ddot{\phi}_1 + \phi_1 = 0$. The linearly independent solutions $\phi_1 = \cos t, \sin t$ correspond respectively to $\phi_2 = -\sin t, \cos t$. Therefore all solutions of the homogeneous system are given by the linear combinations of

$$\begin{pmatrix} \cos t \\ -\sin t \end{pmatrix}, \quad \begin{pmatrix} \sin t \\ \cos t \end{pmatrix};$$

these are given in matrix form by

$$\phi(t) = \begin{pmatrix} \cos t & \sin t \\ -\sin t & \cos t \end{pmatrix} \begin{pmatrix} a_1 \\ a_2 \end{pmatrix},$$

where a_1, a_2 are arbitrary.

It can be confirmed that $x_1 = t, x_2 = 1$ is a particular solution of the original system. Therefore all solutions are given by

$$\mathbf{x}(t) = \begin{pmatrix} t \\ 1 \end{pmatrix} + \begin{pmatrix} \cos t & \sin t \\ -\sin t & \cos t \end{pmatrix} \begin{pmatrix} a_1 \\ a_2 \end{pmatrix},$$

where a_1, a_2 are arbitrary.

THEOREM 8.7 *The solution of the system* $\dot{\mathbf{x}} = A(t)\mathbf{x} + \mathbf{f}(t)$ *with initial conditions* $\mathbf{x}(t_0) = \mathbf{x}_0$ *is given by*

$$\mathbf{x}(t) = \Phi(t)\Phi^{-1}(t_0)\mathbf{x}_0 + \Phi(t) \int_{t_0}^{t} \Phi^{-1}(s)\mathbf{f}(s)\,\mathrm{d}s,$$

where $\Phi(t)$ *is any fundamental solution matrix of* $\dot{\phi} = A(t)\phi$.

Proof Let $\mathbf{x}(t)$ be the required solution, for which the following form is postulated

$$\mathbf{x}(t) = \Phi(t)\Phi^{-1}(t_0)\{\mathbf{x}_0 + \phi(t)\}. \tag{8.20}$$

The inverses of $\Phi(t)$ and $\Phi^{-1}(t_0)$ exist since, by Theorem 8.4, they are non-singular. Then by the initial condition, $\mathbf{x}(t_0) = \mathbf{x}_0$, or $\mathbf{x}_0 + \phi(t_0)$ by (8.20) and so

$$\phi(t_0) = \mathbf{0}.$$

To find the equation satisfied by $\phi(t)$, substitute (8.20) into the equation, which becomes

$$\dot{\Phi}(t)\Phi^{-1}(t_0)\{\mathbf{x}_0 + \phi(t)\} + \Phi(t)\Phi^{-1}(t_0)\dot{\phi}(t)$$
$$= A(t)\Phi(t)\Phi^{-1}(t_0)\{\mathbf{x}_0 + \phi(t)\} + \mathbf{f}(t).$$

Since $\Phi(t)$ is a solution matrix of the homogeneous equation, $\dot{\Phi}(t)$ $= A(t)\Phi(t)$, and the previous equation then becomes

$$\Phi(t)\Phi^{-1}(t_0)\dot{\phi}(t) = \mathbf{f}(t).$$

Therefore

$$\dot{\phi}(t) = \Phi(t_0)\Phi^{-1}(t)\mathbf{f}(t),$$

whose solution satisfying the initial condition is

$$\phi(t) = \Phi(t_0) \int_{t_0}^t \Phi^{-1}(s)\mathbf{f}(s)\,\mathrm{d}s.$$

Therefore, by (8.20),

$$\mathbf{x}(t) = \Phi(t)\Phi^{-1}(t_0)\mathbf{x}_0 + \Phi(t) \int_{t_0}^t \Phi^{-1}(s)\mathbf{f}(s)\,\mathrm{d}s.$$

8.6. Stability and boundedness for linear systems

The following theorem requires a suitable norm for a matrix. For any matrix A, with elements a_{ij}, $i,j = 1,\ldots,n$, which are real or complex, we define

$$\|A\| = \sqrt{\left\{\sum_{i,j} |a_{ij}|^2\right\}}. \tag{8.21}$$

This has the property that

$$\|A\mathbf{a}\| \leqslant \|A\| \, \|\mathbf{a}\| \tag{8.22}$$

where $\mathbf{a}$ is a vector of dimension n.

THEOREM 8.8   *For the regular system* $\dot{\mathbf{x}} = A(t)\mathbf{x}$, *the zero solution (and hence all solutions) are stable on* $t \geqslant t_0$, t_0 *arbitrary, if and only if every solution is bounded. If A is constant and every solution is bounded, the solutions are uniformly stable.*

Proof   By Theorem 8.1 we need consider only the stability of the zero solution.

First, suppose that the zero solution, $\mathbf{x}^*(t) \equiv \mathbf{0}$ is stable. Choose any $\varepsilon > 0$. There must exist a corresponding δ for Definition 8.2. Let

$$\Psi(t) = (\boldsymbol{\psi}_1(t), \boldsymbol{\psi}_2(t),\ldots, \boldsymbol{\psi}_n(t))$$

be the fundamental matrix satisfying

$$\Psi(t_0) = \tfrac{1}{2}\delta I$$

where I is the unit matrix. Then each $\psi_i(t)$ is bounded, since by (8.3)

$$\|\boldsymbol{\psi}_i(t_0)\| = \tfrac{1}{2}\delta < \delta \Rightarrow \|\boldsymbol{\psi}_i(t)\| < \varepsilon, \quad t \geqslant t_0.$$

Therefore every solution is bounded since any other solution is a linear combination of the $\psi_i(t)$.

Suppose, conversely, that every solution is bounded. Let $\Phi(t)$ be any fundamental matrix; then there exists, by hypothesis, $M > 0$ such that $\|\Phi(t)\| < M$, $t \geqslant t_0$. Given any $\varepsilon > 0$ let $\delta = \varepsilon/M\|\Phi^{-1}(t_0)\|$. Let $\mathbf{x}(t)$ be any solution; we will test the stability of the zero solution. We have $\mathbf{x}(t) = \Phi(t)\Phi^{-1}(t_0)\mathbf{x}(t_0)$, (Theorem 8.6), and if

$$\|\mathbf{x}(t_0)\| < \delta,$$

then

$$\begin{aligned} \|\mathbf{x}(t)\| &\leqslant \|\Phi(t)\| \, \|\Phi^{-1}(t_0)\| \, \|\mathbf{x}(t_0)\| \\ &\leqslant M\frac{\varepsilon}{M\delta}\delta = \varepsilon. \end{aligned} \tag{8.23}$$

Thus Definition 8.2 of stability for the zero solution is satisfied.

When A is constant, the autonomous nature of the system ensures that stability is uniform.

Linearity is essential for this theorem: it predicts that the system $\dot{x}_1 = x_2, \dot{x}_2 = -x_1$ has only stable solutions; but the solutions of $\dot{x}_1 = x_2$, $\dot{x}_2 = -\sin x_1$ are not all stable. Note that the stability of *forced* solutions (Theorem 8.1) is not determined by whether they are bounded or not; it is the boundedness of the *unforced* solutions which determines this.

8.7. Stability of systems with constant coefficients

When the coefficients $a_{ij}(t)$ in (8.14) are functions of t, it will usually be impossible to construct an explicit fundamental matrix for the system. When the coefficients are all constants, however, the solutions are of comparatively elementary form, and we obtain a simple statement (Theorem 8.11) summarizing their stability properties.

Consider the system

$$\dot{\mathbf{x}} = A\mathbf{x} \tag{8.24}$$

where A is constant with real elements. We look for solutions of the form

$$\mathbf{x} = \mathbf{r}e^{\lambda t} \tag{8.25}$$

where λ is a constant. In order to satisfy (8.24) we must have

$$A\mathbf{r}e^{\lambda t} - \lambda\mathbf{r}e^{\lambda t} = (A - \lambda I)\mathbf{r}e^{\lambda t} = \mathbf{0}$$

for all t, or

$$(A - \lambda I)\mathbf{r} = \mathbf{0}. \tag{8.26}$$

Given a value for λ, this is equivalent to n linear equations for the components of **r**, and has non-trivial solutions if and only if

$$\det (A - \lambda I) = 0. \tag{8.27a}$$

In component form (8.27a) becomes

$$\begin{vmatrix} a_{11} - \lambda & a_{12} & \cdots & a_{1n} \\ a_{21} & a_{22} - \lambda & \cdots & a_{2n} \\ \cdots & \cdots & \cdots & \cdots \\ a_{n1} & a_{n2} & \cdots & a_{nn} - \lambda \end{vmatrix} = 0. \tag{8.27b}$$

Equation (8.27) is a polynomial equation of degree n for λ, called the *characteristic equation*. It therefore has n roots, real or complex, some of which may be repeated roots. If λ is a complex root, so is $\bar{\lambda}$ since A is a real matrix. The values of λ given by (8.27) are the *eigenvalues* of A. Equation (8.24) *therefore has a solution of the form* (8.25) *if and only if λ is an eigenvalue of A.*

Now suppose that *the eigenvalues are all different*, so that there are exactly n distinct eigenvalues, real, complex or zero, $\lambda_1, \lambda_2, \ldots, \lambda_n$. For each λ_i there exist non-zero solutions $\mathbf{r} = \mathbf{r}_i$ of equation (8.26). These are the *eigenvectors* corresponding to λ_i. It is known that in the present case of n distinct eigenvalues, all the eigenvectors of a particular eigenvalue are simply multiples of each other: therefore we have essentially n solutions of (8.24):

$$(\mathbf{r}_1 \, e^{\lambda_1 t}, \mathbf{r}_2 \, e^{\lambda_2 t}, \ldots, \mathbf{r}_n \, e^{\lambda_n t}),$$

where $\mathbf{r}_i$ is any one of the eigenvectors of λ_i. The solutions are linearly independent. Therefore we have

THEOREM 8.9   *For the system $\dot{\mathbf{x}} = A\mathbf{x}$, A a real, constant matrix, whose eigenvalues $\lambda_1, \lambda_2, \ldots, \lambda_n$ are all different,*

$$\Phi(t) = (\mathbf{r}_1 \, e^{\lambda_1 t}, \mathbf{r}_2 \, e^{\lambda_2 t}, \ldots, \mathbf{r}_n \, e^{\lambda_n t}) \tag{8.28}$$

is a fundamental matrix (complex in general), where $\mathbf{r}_i$ is any eigenvector corresponding to λ_i.

Example 8.10   *Find a fundamental matrix for the system*

$$\dot{x}_1 = x_2 - x_3, \quad \dot{x}_2 = x_3, \quad \dot{x}_3 = x_2.$$

We have $\dot{\mathbf{x}} = A\mathbf{x}$ with

$$A = \begin{pmatrix} 0 & 1 & -1 \\ 0 & 0 & 1 \\ 0 & 1 & 0 \end{pmatrix}.$$

To find the eigenvalues, we require

$$\det(A - \lambda I) = \begin{vmatrix} -\lambda & 1 & -1 \\ 0 & -\lambda & 1 \\ 0 & 1 & -\lambda \end{vmatrix} = -\lambda(\lambda-1)(\lambda+1),$$

so the eigenvalues are $\lambda = 0, 1, -1$. The equations for the eigenvectors, $\mathbf{r} = (r_1, r_2, r_3)^t$ are as follows.

For $\lambda = 0$, $r_2 - r_3 = 0$, $r_3 = 0$, $r_2 = 0$. One solution is $\mathbf{r} = (1, 0, 0)^t$.

For $\lambda = 1$, $-r_1 + r_2 - r_3 = 0$, $-r_2 + r_3 = 0$, $r_2 - r_3 = 0$. One solution is $\mathbf{r} = (0, 1, 1)^t$.

For $\lambda = -1$, $r_1 + r_2 - r_3 = 0$, $r_2 + r_3 = 0$, $r_2 + r_3 = 0$. One solution is $\mathbf{r} = (2, -1, 1)^t$.

A fundamental matrix is therefore given by

$$\Phi(t) = \begin{pmatrix} 1 & 0 & 2e^{-t} \\ 0 & e^t & -e^{-t} \\ 0 & e^t & e^{-t} \end{pmatrix}.$$

When eigenvalues are not all distinct the formal situation is more complicated, and for the theory the reader is referred, for example, to Wilson (1971). We illustrate some possibilities in the following examples.

Example 8.11 *Find a fundamental matrix for the system*

$$\dot{x}_1 = x_1, \quad \dot{x}_2 = x_2, \quad \dot{x}_3 = x_3.$$

The matrix for the system is

$$A = \begin{pmatrix} 1 & 0 & 0 \\ 0 & 1 & 0 \\ 0 & 0 & 1 \end{pmatrix}.$$

The characteristic equation is $\det(A - \lambda I) = 0$, which becomes $(1 - \lambda)^3 = 0$. Thus $\lambda = 1$ is a threefold repeated root. The equation for the eigenvectors, $(A - \lambda I)\mathbf{r} = \mathbf{0}$, gives no restriction on $\mathbf{r}$, so the linearly independent vectors $(1, 0, 0)^t$, $(0, 1, 0)^t$, $(0, 0, 1)^t$ can be chosen as eigenvectors. This leads to the fundamental matrix Φ given by

$$\Phi(t) = \begin{pmatrix} e^t & 0 & 0 \\ 0 & e^t & 0 \\ 0 & 0 & e^t \end{pmatrix}.$$

Example 8.12 *Find a fundamental matrix for the system*

$$\dot{x}_1 = x_1 + x_2, \quad \dot{x}_2 = x_2, \quad \dot{x}_3 = x_3.$$

We have

$$A = \begin{pmatrix} 1 & 1 & 0 \\ 0 & 1 & 0 \\ 0 & 0 & 1 \end{pmatrix}.$$

The characteristic equation becomes $(1 - \lambda)^3 = 0$, so $\lambda = 1$ is a threefold root. The equation for the eigenvectors gives simply $r_2 = 0$. We choose the two linearly independent eigenvectors, $(1, 0, 0)^t$ and $(0, 0, 1)^t$, satisfying this condition. The corresponding solutions are

$$(e^t, 0, 0)^t, \quad (0, 0, e^t)^t.$$

It is easy to confirm that a third solution, independent of the other two, is $(te^t, e^t, 0)^t$. A fundamental matrix Φ is therefore given by

$$\Phi(t) = \begin{pmatrix} e^t & 0 & te^t \\ 0 & 0 & e^t \\ 0 & e^t & 0 \end{pmatrix}.$$

Example 8.13   *Find a fundamental matrix for the system*

$$\dot{x}_1 = x_1 + x_2, \quad \dot{x}_2 = x_2 + x_3, \quad \dot{x}_3 = x_3.$$

We have

$$A = \begin{pmatrix} 1 & 1 & 0 \\ 0 & 1 & 1 \\ 0 & 0 & 1 \end{pmatrix}.$$

Once again, $\lambda = 1$ is a threefold root of the characteristic equation. The equation for the eigenvectors gives $r_2 = r_3 = 0$. One solution is $\mathbf{r} = (1, 0, 0)^t$ and there is no other which is linearly independent of this one. The corresponding solution is $\mathbf{x} = (e^t, 0, 0)^t$.

It can be confirmed that two more linearly independent solutions are

$$\mathbf{x} = (te^t, e^t, 0)^t, \quad \mathbf{x} = (\tfrac{1}{2}t^2 e^t, te^t, e^t)^t$$

leading to the fundamental matrix

$$\Phi(t) = \begin{pmatrix} e^t & te^t & \tfrac{1}{2}t^2 e^t \\ 0 & e^t & te^t \\ 0 & 0 & e^t \end{pmatrix}.$$

Inspection of these examples suggests the following theorem on the form of linearly independent solutions associated with multiple eigenvalues. We state the theorem without proof.

Theorem 8.10   *Corresponding to an eigenvalue of A, $\lambda = \lambda_i$, of multiplicity $m \leqslant n$ there are m linearly independent solutions of the system*

(8.24). *These are of the form*

$$\mathbf{p}_1(t)\,\mathrm{e}^{\lambda_i t},\ldots,\mathbf{p}_m(t)\,\mathrm{e}^{\lambda_i t}, \tag{8.29}$$

where the $\mathbf{p}_j(t)$ are vector polynomials of degree less than m.

Note that when an eigenvalue is complex, the eigenvectors and the polynomials in (8.29) will be complex, and the arrays consist of complex-valued solutions. Since the elements of A are real, the characteristic equation (8.27) has real coefficients. Therefore, if λ_i is one eigenvalue, $\bar{\lambda}_i$ is another. The corresponding polynomial coefficients in (8.29) are, similarly, complex conjugates. In place, therefore, of the pair of complex solutions $\phi_i(t)$, $\bar{\phi}_i(t)$ corresponding to λ_i and $\bar{\lambda}_i$ respectively, we could take the real solutions, $\mathrm{Re}\{\phi_i(t)\}$, $\mathrm{Im}\{\phi_i(t)\}$.

According to Theorems 8.9 and 8.10 it is possible to make a simple statement about the stability of the solutions of systems with constant coefficients, $\dot{\mathbf{x}} = A\mathbf{x}$.

THEOREM 8.11 *Let A be constant in the system $\dot{\mathbf{x}} = A\mathbf{x}$, with eigenvalues λ_i, $i = 1, 2, \ldots, n$.*
(i) *If the system is stable, then $\mathrm{Re}\{\lambda_i\} \leqslant 0$, $i = 1, 2, \ldots, n$.*
(ii) *If either $\mathrm{Re}\{\lambda_i\} < 0$, $i = 1, 2, \ldots, n$; or if $\mathrm{Re}\{\lambda_i\} \leqslant 0$, $i = 1, 2, \ldots, n$ and there is no zero repeated eigenvalue; then the system is uniformly stable.*
(iii) *The system is asymptotically stable if and only if $\mathrm{Re}\{\lambda_i\} < 0$, $i = 1, 2, \ldots, n$ (and then it is also uniformly stable, by (ii)).*

In connection with (ii), note that if there is a zero repeated eigenvalue the system may be stable or unstable. For example, the system

$$\dot{x}_1 = 0, \quad \dot{x}_2 = 0$$

has a fundamental matrix

$$\begin{pmatrix} 1 & 0 \\ 0 & 1 \end{pmatrix}$$

which implies boundedness and therefore stability, but

$$\dot{x}_1 = x_2, \quad \dot{x}_2 = 0$$

has a fundamental matrix

$$\begin{pmatrix} 1 & t \\ 0 & 1 \end{pmatrix}$$

showing that the system is unstable.

Exercises

1. Use the phase diagram for the pendulum equation, $\ddot{x} + \sin x = 0$, to say which

paths are not Poincaré stable. (See Fig. 1.2.)

2. Show that all the paths of $\dot{x} = x, \dot{y} = y$ are Poincaré unstable.

3. Find the limit cycles of the system

$$\dot{x} = -y + x \sin r, \quad \dot{y} = x + y \sin r, \quad r = \sqrt{(x^2 + y^2)}.$$

Which cycles are Poincaré stable?

4. Find the phase paths for $\dot{x} = x, \dot{y} = y \log y$, in the half-plane $y > 0$. Which paths are Poincaré stable?

5. Show that every non-trivial solution of $\dot{x} = x$ is unbounded and Liapunov unstable, but that every solution of $\dot{x} = 1$ is unbounded and stable.

6. Show that the solutions of the system $\dot{x} = 1, \dot{y} = 0$, are Poincaré and Liapunov stable, but that the system $\dot{x} = y, \dot{y} = 0$, is Poincaré but not Liapunov stable.

7. Solve the equations

$$\dot{x} = -y(x^2 + y^2), \quad \dot{y} = x(x^2 + y^2),$$

and show that the zero solution is Liapunov stable and that all other solutions are unstable.

Replace the coordinates x, y by r, ϕ where

$$x = r \cos(r^2 t + \phi), \quad y = r \sin(r^2 t + \phi)$$

and deduce that

$$\dot{r} = 0, \quad \dot{\phi} = 0.$$

Show that in this coordinate system the solutions are stable. (Change of coordinates can affect the stability of a system.)

8. Prove that Liapunov stability implies Poincaré stability of a solution (but not conversely: see Exercise 6).

9. Determine the stability of the solutions of
(i) $\dot{x}_1 = x_2 \sin t, \quad \dot{x}_2 = 0$.
(ii) $\dot{x}_1 = 0, \quad \dot{x}_2 = x_1 + x_2$.

10. Determine the stability of the solutions of
(i) $\begin{pmatrix} \dot{x}_1 \\ \dot{x}_2 \end{pmatrix} = \begin{pmatrix} -2 & 1 \\ 1 & -2 \end{pmatrix} \begin{pmatrix} x_1 \\ x_2 \end{pmatrix} + \begin{pmatrix} 1 \\ -2 \end{pmatrix} e^t;$
(ii) $\ddot{x} + e^{-t}\dot{x} + x = e^t.$

11. Show that every solution of the system

$$\dot{x} = -t^2 x, \quad \dot{y} = -ty$$

is asymptotically stable.

12. The motion of a heavy particle on a smooth surface of revolution with vertical axis z and shape $z = f(r)$ in cylindrical polar coordinates is

$$\frac{1}{r^4}\{1+f'^2(r)\}\frac{d^2r}{d\theta^2} + \left[\frac{1}{r^4}f'(r)f''(r) - \frac{2}{r^5}\{1+f'^2(r)\}\right]\left(\frac{dr}{d\theta}\right)^2 - \frac{1}{r^3} = -\frac{g}{h^2}f'(r),$$

where h is the angular momentum ($h = r^2\dot{\theta}$), a constant. Show that plane, horizontal motion $r = a, z = f(a)$, is stable for perturbations leaving h unaltered provided $3 + af''(a)/f'(a) > 0$.

13. Determine the linear dependence or independence of the following:
   (i) $(1, 1, -1)$, $(2, 1, 1)$, $(0, 1, -3)$.
   (ii) $(t, 2t)$, $(3t, 4t)$, $(5t, 6t)$.
   (iii) (e^t, e^{-t}), (e^{-t}, e^t). Could these both be solutions of a 2×2 homogeneous linear system?
   (iv) $(\sin(t-\alpha), \cos(t-\alpha))$, $(\sin(t-\beta), \cos(t-\beta))$, $(\sin(t-\gamma), \cos(t-\gamma))$.

14. Construct a complex and a real fundamental matrix Φ for the system $\dot{x} = y$, $\dot{y} = -x - 2y$. Deduce a fundamental matrix Ψ satisfying $\Psi(0) = I$.

15. Construct a fundamental matrix for the system $\dot{x}_1 = -x_1$, $\dot{x}_2 = x_1 + x_2 + x_3$, $\dot{x}_3 = -x_2$.

16. Construct a fundamental matrix for the system $\dot{x}_1 = x_2, \dot{x}_2 = x_1$, and deduce the solution satisfying $x_1 = 1$, $x_2 = 0$, at $t = 0$.

17. Construct a fundamental matrix for the system $\dot{x}_1 = x_2, \dot{x}_2 = x_3, \dot{x}_3 = -2x_1 + x_2 + 2x_3$, and deduce the solution of $\dot{x}_1 = x_2 + e^t$, $\dot{x}_2 = x_3$, $\dot{x}_3 = -2x_1 + x_2 + 2x_3$, with $\mathbf{x}(0) = (1, 0, 0)^t$.

18. Show that the differential equation $x^{(n)} + a_1 x^{(n-1)} + \ldots + a_n x = 0$ is equivalent to the system

$$\dot{x}_1 = x_2, \quad \dot{x}_2 = x_3, \ldots, \quad \dot{x}_{n-1} = x_n, \quad \dot{x}_n = -a_n x_1 - \ldots - a_1 x_n,$$

with $x = x_1$. Show that the equation for the eigenvalues is

$$\lambda^n + a_1 \lambda^{n-1} + \ldots + a_n = 0.$$

19. A bird population, $p(t)$, is governed by the differential equation $\dot{p} = \mu(t)p - kp$, where k is the death-rate and $\mu(t)$ represents a variable birth-rate with period 1 year. Derive a condition which ensures that the mean annual population remains constant. Assuming that this condition is fulfilled, does it seem likely that, *in practice*, the average population will remain constant? (This is asking a question about a particular kind of stability.)

20. Are the periodic solutions of

$$\ddot{x} + \operatorname{sgn}(x) = 0$$

(i) Poincaré stable?   (ii) Liapunov stable?

21. Give a descriptive argument to show that if the index of an equilibrium point is not unity, then the equilibrium point is not stable.

22. Show that the system

$$\dot{x} = x + y - x(x^2 + y^2), \quad \dot{y} = -x + y - y(x^2 + y^2), \quad \dot{z} = -z,$$

has a limit cycle $x^2 + y^2 = 1$, $z = 0$. Find the linear approximation at the origin and so confirm that the origin is unstable. Use cylindrical polar coordinates $r = \sqrt{(x^2 + y^2)}$, z to show that the limit cycle is stable. Sketch the phase diagram in x, y, z space.

23. Show that the nth-order non-autonomous system $\dot{x} = X(x, t)$ can be reduced to an $(n+1)$th-order autonomous system by introducing a new variable, $x_{n+1} = t$. (The $(n+1)$-dimensional phase diagram for the modified system is then of the type suggested by Fig. 8.9. The system has no equilibrium points.)

9 Determination of stability by solution perturbation

STABILITY or instability of nonlinear systems can often be tested by an approximate procedure which leads to a *linear* equation describing the growth of the difference between the test solution and its neighbours. By Theorem 8.8 the stability or instability of the original system resolves itself into the question of the boundedness or otherwise of the solutions of the linear equation. This 'variational equation' often turns out to have a periodic coefficient (Mathieu's equation) and the properties of such equations are derived in this chapter. The fact that the solutions to be tested are themselves usually known only approximately can also be assimilated into this theory.

9.1. The stability of forced oscillations by solution perturbation

Consider the general forced system

$$\dot{\mathbf{x}} = \mathbf{f}(\mathbf{x}, t). \tag{9.1}$$

The stability of a solution $\mathbf{x}^*(t)$ can be reduced to consideration of the zero solution of a related system. Let $\mathbf{x}(t)$ be any other solution, and write

$$\mathbf{x}(t) = \mathbf{x}^*(t) + \boldsymbol{\xi}(t). \tag{9.2}$$

Then $\boldsymbol{\xi}(t)$ represents a perturbation, or disturbance, of the original solution: it seems reasonable to see what happens to $\boldsymbol{\xi}(t)$, since the question of stability is whether such (small) disturbances grow or not. Equation (9.1) can be written in the form

$$\dot{\mathbf{x}}^* + \dot{\boldsymbol{\xi}} = \mathbf{f}(\mathbf{x}^*, t) + \{\mathbf{f}(\mathbf{x}^* + \boldsymbol{\xi}, t) - \mathbf{f}(\mathbf{x}^*, t)\}.$$

Since x^* satisfies (9.1), this becomes

$$\begin{aligned}\dot{\boldsymbol{\xi}} &= \mathbf{f}(\mathbf{x}^* + \boldsymbol{\xi}, t) - \mathbf{f}(\mathbf{x}^*, t) \\ &= \mathbf{h}(\boldsymbol{\xi}, t),\end{aligned} \tag{9.3}$$

say, since $\mathbf{x}^*(t)$ is assumed known. By (9.2), the stability properties of $\mathbf{x}^*(t)$ are the same as those of the zero solution of (9.3), $\boldsymbol{\xi}(t) \equiv \mathbf{0}$.

The right-hand side of (9.3) may have a linear approximation for small ξ, in which case

$$\dot{\boldsymbol{\xi}} = \mathbf{h}(\boldsymbol{\xi}, t) \approx A(t)\boldsymbol{\xi}, \tag{9.4}$$

and the properties of this linear system may correctly indicate that of the
zero solution of the exact system (9.3). The approximation (9.4) is
called the *first variational equation*. This process is not rigorous: it is
generally necessary to invoke an approximation not only at the stage
(9.4), but also in representing $\mathbf{x}^*(t)$, which, of course, will not generally
be known exactly.

We shall illustrate the procedure in the case of the forced, undamped
pendulum-type equation (a form of Duffing's equation)

$$\ddot{x} + x + \varepsilon x^3 = \Gamma \cos \omega t. \qquad (9.5)$$

In order to match the form of the theory of linear systems of Chapter 8
we will express it in the form

$$\dot{\mathbf{x}} = \begin{pmatrix} \dot{x} \\ \dot{y} \end{pmatrix} = \begin{pmatrix} y \\ -x - \varepsilon x^3 + \Gamma \cos \omega t \end{pmatrix}. \qquad (9.6)$$

To obtain the variational equation define $\boldsymbol{\xi} = (\xi, \eta)^t$ by

$$\boldsymbol{\xi} = \mathbf{x} - \mathbf{x}^* \qquad (9.7)$$

where

$$\mathbf{x}^* = (x^*, y^*)^t$$

and $\mathbf{x}^*$ is the solution to be tested. Substitution for x and y from (9.7)
into (9.6) gives

$$\dot{\xi} + \dot{x}^* = \eta + y^*,$$
$$\dot{\eta} + \dot{y}^* = -\xi - x^* - \varepsilon(\xi + x^*)^3 + \Gamma \cos \omega t.$$

By neglecting powers of ξ higher than the first, and using the fact that
x^*, y^* satisfy (9.6), the system simplifies to

$$\dot{\xi} = \eta, \qquad \dot{\eta} = -\xi - 3\varepsilon x^{*2}\xi, \qquad (9.8)$$

corresponding to (9.4).

From Section 7.2 we know that there are periodic solutions of (9.5)
which are approximately of the form

$$x = a \cos \omega t$$

where the amplitude a is given by the equation

$$\tfrac{3}{4}\varepsilon a^3 - (\omega^2 - 1)a - \Gamma = 0. \qquad (9.9)$$

We shall test the stability of one of these by treating it as being
sufficiently close to the corresponding exact form of x^* required by

(9.8). By eliminating η from (9.8) we obtain

$$\ddot{\xi} + (1 + 3\varepsilon x^{*2})\xi = 0.$$

When x^* is replaced by its appropriate estimate, $x^* = a \cos \omega t$, with a given, this equation becomes

$$\ddot{\xi} + (1 + \tfrac{3}{2}\varepsilon a^2 + \tfrac{3}{2}\varepsilon a^2 \cos 2\omega t)\xi = 0,$$

and we expect that the stability property of x^* and ξ will be the same. The last equation can be reduced to a standard form

$$\xi'' + (\alpha + \beta \cos \tau)\xi = 0 \qquad (9.10)$$

by the substitutions

$$\tau = 2\omega t, \ \xi' = \mathrm{d}\xi/\mathrm{d}\tau, \ \alpha = (2 + 3\varepsilon a^2)/8\omega^2, \ \beta = 3\varepsilon a^2/8\omega^2. \quad (9.11)$$

Equation (9.10) is known as Mathieu's equation. By Theorem 8.8 its solutions are stable for values of the parameters, α, β for which all its solutions are bounded. We shall return to the special case under discussion at the end of Section 9.4 after studying the stability of solutions of Mathieu's equation, to which problems of this kind may often be reduced.

9.2. Equations with periodic coefficients (Floquet theory)

Equation (9.11) is a particular example of an equation derived from the general first-order system

$$\dot{\mathbf{x}} = P(t)\mathbf{x}, \qquad (9.12)$$

where $P(t)$ is periodic with minimal period T; that is, T is the smallest positive number for which

$$P(t + T) = P(t), \quad -\infty < t < \infty. \qquad (9.13)$$

($P(t)$, of course, also has periods $2T, 3T, \ldots$.) The *solutions* are not necessarily periodic, as can be seen from the one-dimensional example

$$\dot{x} = P(t)x = (1 + \sin t)x;$$

the coefficient $P(t)$ has period 2π, but all solutions are given by

$$x = c\,e^{t - \cos t},$$

where c is any constant, so only the solution $x = 0$ is periodic.

In the following discussions remember that the displayed solution-vectors may consist of complex solutions.

THEOREM 9.1 (*Floquet's theorem.*) *The regular system* $\dot{\mathbf{x}} = P(t)\mathbf{x}$, *where P is an n × n matrix function with minimal period T, has at least one non-trivial solution* $\mathbf{x} = \chi(t)$ *such that*

$$\chi(t+T) = \mu\chi(t), \quad -\infty < t < \infty \tag{9.14}$$

where μ is a constant.

Proof Let $\Phi(t) = (\phi_{ij}(t))$ be a fundamental matrix for the system. Then $\dot{\Phi}(t) = P(t)\Phi(t)$. Since $P(t+T) = P(t)$, $\Phi(t+T)$ satisfies the same equation, and by Theorem 8.5, $\det \Phi(t+T) \neq 0$, so $\Phi(t+T)$ is another fundamental matrix. The columns (solutions) in $\Phi(t+T)$ are linear combinations of those in $\Phi(t)$ by Theorem 8.4:

$$\phi_{ij}(t+T) = \sum_{k=1}^{n} \phi_{ik}(t)e_{kj}$$

for some constants e_{kj}, so that

$$\Phi(t+T) = \Phi(t)E \tag{9.15}$$

where $E = (e_{kj})$. E is nonsingular, since $\det \Phi(t+T) = \det \Phi(t)\det(E)$, and therefore $\det(E) \neq 0$.

Let μ be an eigenvalue of E:

$$\det(E - \mu I) = 0, \tag{9.16}$$

and let $\mathbf{s}$ be an eigenvector corresponding to μ:

$$(E - \mu I)\mathbf{s} = \mathbf{0}. \tag{9.17}$$

Consider the solution $\mathbf{x} = \Phi(t)\mathbf{s} = \chi(t)$ (being a linear combination of the columns of Φ, χ is a solution of (9.12)). Then

$$\chi(t+T) = \Phi(t+T)\mathbf{s}$$
$$= \Phi(t)E\mathbf{s} = \Phi(t)\mu\mathbf{s}, \qquad \text{(by (9.17))}$$
$$= \mu\chi(t). \qquad \text{(by (9.14))}$$

The eigenvalues of E are called *characteristic numbers* of equation (9.12). The importance to us of this theorem is the possibility of a characteristic number with a special value implying the existence of a periodic solution (though not necessarily of period T).

Example 9.1 *Find a fundamental matrix for the periodic differential equation*

$$\begin{pmatrix} \dot{x}_1 \\ \dot{x}_2 \end{pmatrix} = \begin{pmatrix} 1 & 1 \\ 0 & h(t) \end{pmatrix} \begin{pmatrix} x_1 \\ x_2 \end{pmatrix} \tag{9.18}$$

where $h(t) = (\cos t + \sin t)/(2 + \sin t - \cos t)$, and determine the characteristic numbers.

From (9.18),

$$(2+\sin t-\cos t)\dot{x}_2 = (\cos t+\sin t)x_2,$$

which has solutions

$$x_2 = b(2+\sin t-\cos t)$$

where b is any constant. Then x_1 satisfies

$$\dot{x}_1-x_1 = x_2 = b(2+\sin t-\cos t)$$

and therefore

$$x_1 = a e^t - b(2+\sin t).$$

A fundamental matrix $\Phi(t)$ can be obtained by putting, say, $a = 0$, $b = 1$, and $a = 1$, $b = 0$:

$$\Phi(t) = \begin{pmatrix} -2-\sin t & e^t \\ 2+\sin t-\cos t & 0 \end{pmatrix}.$$

$P(t)$ has minimal period $T = 2\pi$, and E in (9.15) must satisfy $\Phi(t+2\pi) = \Phi(t)E$ for all t. Therefore $\Phi(2\pi) = \Phi(0)E$ and

$$E = \Phi^{-1}(0)\Phi(2\pi) = \begin{pmatrix} 1 & 0 \\ 0 & e^{2\pi} \end{pmatrix}.$$

The eigenvalues μ of E satisfy

$$\begin{vmatrix} 1-\mu & 0 \\ 0 & e^{2\pi}-\mu \end{vmatrix} = 0,$$

so $\mu = 1$ or $e^{2\pi}$.

From (9.14), since one eigenvalue is unity there exist solutions such that $\chi(t+2\pi) = \chi(t)$: that is, solutions with period 2π. We have already found these: they correspond to $a = 0$.

THEOREM 9.2 *The constants μ in Theorem 9.1 are independent of the choice of Φ.*

Proof Let $\Phi(t), \Phi^*(t)$ be two fundamental matrices; then

$$\Phi^*(t) = \Phi(t)C, \tag{9.19}$$

where C is some constant, nonsingular matrix (nonsingular since $\Phi(t)$ and $\Phi^*(t)$ are nonsingular by Theorem 8.5). Let T be the minimal period of $P(t)$. Then

$$\begin{aligned} \Phi^*(t+T) &= \Phi(t+T)C, & \text{(by (9.19))} \\ &= \Phi(t)EC, & \text{(by (9.15))} \\ &= \Phi^*(t)C^{-1}EC, & \text{(by (9.19))} \\ &= \Phi^*(t)D, \end{aligned}$$

say. But E and D are *similar* matrices and so have the same eigenvalues.

We can therefore properly refer to '*the* characteristic numbers of the system'. Note that when Φ is chosen as real, E is real, and the characteristic equation for the numbers μ has real coefficients. Therefore if μ (complex) is a characteristic number, then so is $\bar{\mu}$.

DEFINITION 9.1   *A solution of* (9.12) *satisfying* (9.14) *is called a **normal solution**.*

DEFINITION 9.2 (*Characteristic exponent*)   *Let μ be a characteristic number of* (9.12), *corresponding to the minimal period T of $P(t)$. Then ρ, defined by*

$$e^{\rho T} = \mu, \tag{9.20}$$

is called a characteristic exponent of the system. Note that ρ is defined only to an additive multiple of $2\pi i/T$. It will be fixed by requiring $-\pi < \text{Im}\,(\rho T) \leqslant \pi$, or by $\rho = 1/T\,\text{Log}\,(\mu)$, where the principal value of the logarithm is taken.

THEOREM 9.3   *Suppose that E of Theorem 9.1 has n distinct eigenvalues, μ_i, $i = 1, 2, \ldots, n$. Then* (9.12) *has n linearly independent normal solutions of the form*

$$\mathbf{x}_i = \mathbf{p}_i(t)\,e^{\rho_i t} \tag{9.21}$$

(ρ_i *are the characteristic exponents corresponding to μ_i), where the $\mathbf{p}_i(t)$ are functions with period T.*

Proof   To each μ_i corresponds a solution $\mathbf{x}_i(t)$ satisfying (9.14): $\mathbf{x}_i(t+T) = \mu_i \mathbf{x}_i(t) = e^{\rho_i T} x_i(t)$. Therefore, for every t,

$$\mathbf{x}_i(t+T)\,e^{-\rho_i(t+T)} = \mathbf{x}_i(t)\,e^{-\rho_i t}. \tag{9.22}$$

Writing

$$\mathbf{p}_i(t) = e^{-\rho_i t}\mathbf{x}_i(t),$$

(9.22) implies that $\mathbf{p}_i(t)$ has period T.

The linear independence of the $\mathbf{x}_i(t)$ is implied by their method of construction in Theorem 9.1: from (9.17), they are given by $\mathbf{x}_i(t) = \Phi(t)\mathbf{s}_i$; $\mathbf{s}_i$ are the eigenvectors corresponding to the *different* eigenvalues μ_i, and are therefore linearly independent. Since $\Phi(t)$ is nonsingular it follows that the $\mathbf{x}_i(t)$ are also linearly independent.

When the eigenvalues of E are not all distinct, the coefficients corresponding to the $\mathbf{p}_i(t)$ are more complicated. Under the conditions of Theorem 9.3, periodic solutions of period T exist when E has an eigenvalue

$$\mu = 1.$$

The corresponding normal solutions have period T, the minimal period of $P(t)$. This can be seen from (9.14) or from the fact that the corresponding ρ is zero.

There are periodic solutions whenever E has an eigenvalue μ which is one of the mth roots of unity:

$$\mu = 1^{1/m}, \quad m \text{ a positive integer.} \tag{9.23a}$$

In this case, from (9.14),

$$\chi(t+mT) = \mu\chi\{t+(m-1)T\} = \ldots = \mu^m\chi(t) = \chi(t), \tag{9.23b}$$

so that $\chi(t)$ has period mT.

In the preceding theory, $\det \Phi(t)$ appeared repeatedly, where Φ is a fundamental matrix of the regular system $\dot{\mathbf{x}} = A(t)\mathbf{x}$. This has a simple representation, as follows.

DEFINITION 9.3   *Let* $(\boldsymbol{\phi}_1(t), \boldsymbol{\phi}_2(t), \ldots, \boldsymbol{\phi}_n(t))$ *be a matrix whose columns are any solutions of the system* $\dot{\mathbf{x}} = A(t)\mathbf{x}$. *Then*

$$W(t) = \det(\boldsymbol{\phi}_1(t), \boldsymbol{\phi}_2(t), \ldots, \boldsymbol{\phi}_n(t)) \tag{9.24}$$

is called the Wronskian of this set of solutions, taken in order.

THEOREM 9.4   *Let* $\Phi(t)$ *be any matrix of solutions for* $\dot{\mathbf{x}} = A(t)\mathbf{x}$; *then, for any* t_0,

$$W(t) = W(t_0)\exp\left(\int_{t_0}^{t} \operatorname{tr}\{A(s)\}\,\mathrm{d}s\right), \tag{9.25}$$

where $\operatorname{tr}\{A(s)\}$ *is the trace of* $A(t)$: *the sum of the elements of its principal diagonal.*

Proof   If the solutions are linearly dependent, $W(t) \equiv 0$ by Theorem 8.5, and the result is true trivially.

If not, let $\Phi(t)$ be any fundamental matrix of solutions, with $\Phi(t) = (\phi_{ij}(t))$. Then $\mathrm{d}W/\mathrm{d}t$ is equal to the sum of n determinants $\Delta_k, k = 1, 2, \ldots, n$, where Δ_k is the same as $\det(\phi_{ij}(t))$, except for having $\dot{\phi}_{kj}(t), j = 1, 2, \ldots, n$ in place of $\phi_{kj}(t)$ in its kth row. Consider one of the Δ_k, say Δ_1:

$$\Delta_1 = \begin{vmatrix} \dot\phi_{11} & \dot\phi_{12} & \cdots & \dot\phi_{1n} \\ \phi_{21} & \phi_{22} & \cdots & \phi_{2n} \\ \cdots & \cdots & \cdots & \cdots \\ \phi_{n1} & \phi_{n2} & & \phi_{nn} \end{vmatrix} =$$

$$\begin{vmatrix} \sum_{m=1}^{n} a_{1m}\phi_{m1} & \sum_{m=1}^{n} a_{1m}\phi_{m2} & \cdots & \sum_{m=1}^{n} a_{1n}\phi_{mn} \\ \phi_{21} & \phi_{22} & \cdots & \phi_{2n} \\ \cdots & \cdots & \cdots & \cdots \\ \phi_{n1} & \phi_{n2} & \cdots & \phi_{nn} \end{vmatrix}$$

(from eqn (8.14))

$$= \sum_{m=1}^{n} a_{1m} \begin{vmatrix} \phi_{m1} & \phi_{m2} & \cdots & \phi_{mn} \\ \phi_{21} & \phi_{22} & \cdots & \phi_{2n} \\ \cdots & \cdots & \cdots & \cdots \\ \phi_{n1} & \phi_{n2} & & \phi_{nn} \end{vmatrix}$$

$$= a_{11} \begin{vmatrix} \phi_{11} & \phi_{12} & \cdots & \phi_{1n} \\ \phi_{21} & \phi_{22} & \cdots & \phi_{2n} \\ \cdots & \cdots & \cdots & \cdots \\ \phi_{n1} & \phi_{n2} & & \phi_{nn} \end{vmatrix} = a_{11} W(t),$$

since all the other determinants have repeated rows, and therefore vanish. In general $\Delta_k = a_{kk} W(t)$. Therefore

$$\frac{dW(t)}{dt} = \text{tr}\{A(t)\}W(t),$$

which is a differential equation for W having solution (9.25).

For periodic systems we have the following result.

THEOREM 9.5   *For the system* $\dot{\mathbf{x}} = P(t)\mathbf{x}$, *where* $P(t)$ *has principal period* T, *let the characteristic numbers of the system be* $\mu_1, \mu_2, \ldots, \mu_n$. *Then*

$$\mu_1 \mu_2 \ldots \mu_n = \exp\left(\int_0^T \text{tr}\{P(s)\}\, ds \right),$$

a repeated characteristic number being counted according to its multiplicity.

Proof Let $\Psi(t)$ be the fundamental matrix of the system for which

$$\Psi(0) = I. \tag{9.26}$$

Then, (eqn (9.15)),

$$\Psi(T) = \Psi(0)E = E, \tag{9.27}$$

in the notation of Theorem 9.1. The characteristic numbers μ are the eigenvalues of E

$$\det(E - \mu I) = 0.$$

This is an nth-degree polynomial in μ, and the product of the roots is equal to the constant term: that is, equal to the value taken when $\mu = 0$. Thus, by (9.27),

$$\mu_1\mu_2\ldots\mu_n = \det(E) = \det\Psi(T) = W(T),$$

but by Theorem 9.4 with $t_0 = 0$ and $t = T$,

$$W(t) = W(0)\int_0^T \text{tr}\{P(s)\}\,\mathrm{d}s$$

and $W(0) = 1$ by (9.26).

9.3. Mathieu's equation arising from a Duffing equation

We now return to look in more detail at Mathieu's equation (9.11)

$$\ddot{x} + (\alpha + \beta\cos t)x = 0. \tag{9.28}$$

As a first-order system it can be expressed as

$$\begin{pmatrix} \dot{x} \\ \dot{y} \end{pmatrix} = \begin{pmatrix} 0 & 1 \\ -\alpha - \beta\cos t & 0 \end{pmatrix}\begin{pmatrix} x \\ y \end{pmatrix}, \tag{9.29}$$

where, in the notation of the previous section,

$$P(t) = \begin{pmatrix} 0 & 1 \\ -\alpha - \beta\cos t & 0 \end{pmatrix}. \tag{9.30}$$

Clearly $P(t)$ is periodic with minimal period 2π. The general structure of the solution is determined by Theorem 9.3, whilst the question of the stability of the solution can be decided, through Theorem 8.8, by the boundedness or otherwise of the solution for given values of the parameters α and β. We are not particularly interested in periodic solutions as such, though we shall need them to settle the latter point.

From eqn (9.30),

$$\operatorname{tr}\{P(t)\} = 0. \tag{9.31}$$

Therefore, by Theorem 9.5,

$$\mu_1\mu_2 - e^0 = 1, \tag{9.32}$$

where μ_1, μ_2 are the characteristic numbers of $P(t)$. Therefore they are solutions of a quadratic characteristic equation (9.16), with real coefficients, which by (9.32) has the form

$$\mu^2 - \phi(\alpha, \beta)\mu + 1 = 0.$$

The solutions μ are given by

$$\mu_1, \mu_2 = \tfrac{1}{2}[\phi \pm \sqrt{(\phi^2 - 4)}]. \tag{9.33}$$

Although $\phi(\alpha, \beta)$ is not known explicitly (see Exercise 7), we can make the following deductions by reference to Theorem 9.3.

(i) $\phi > 2$. The characteristic numbers are real, different, and positive, and by (9.32), one of them, say μ_1, exceeds unity. The corresponding characteristic exponents (9.20) are real and have the form $\rho_1 = \sigma > 0$, $\rho_2 = -\sigma < 0$. The general solution is therefore of the form (Theorem 9.3)

$$x(t) = c_1 e^{\sigma t} p_1(t) + c_2 e^{-\sigma t} p_2(t),$$

where c_1, c_2 are constants and p_1, p_2 have minimal period 2π. The parameter region $\phi(\alpha, \beta) > 2$ therefore contains unbounded solutions, and is called an *unstable parameter region*.

(ii) $\phi = 2$. Then $\mu_1 = \mu_2 = 1$, $\rho_1 = \rho_2 = 0$. By (9.21), there is *one solution of period 2π on the curves* $\phi(\alpha, \beta) = 2$. (The other is unbounded.)

(iii) $-2 < \phi < 2$. The characteristic numbers are complex, and $\mu_2 = \bar{\mu}_1$. Since also $|\mu_1| = |\mu_2| = 1$, we must have $\rho_1 = iv, \rho_2 = -iv, v$ real. The general solution is of the form

$$x(t) = c_1 e^{ivt} p_1(t) + c_2 e^{-ivt} p_2(t)$$

and *all solutions in the parameter region* $-2 < \phi(\alpha, \beta) < 2$ *are bounded*. This is called the *stable parameter region*. The solutions are oscillatory, but not in general periodic, since the two frequencies v and 2π are present.

(iv) $\phi = -2$. Then $\mu_1 = \mu_2 = -1$ ($\rho_1 = \rho_2 = \tfrac{1}{2}i$), so by Theorem 9.1, eqn (9.14), there *is one solution with period* 4π *at every point on* $\phi(\alpha, \beta) =$ -2. (The other solution is in fact unbounded.)

(v) $\phi < -2$. Then μ_1 and μ_2 are real and negative. Since, also, $\mu_1\mu_2 = 1$, the general solution is of the form

$$x(t) = c_1 e^{(\sigma + \frac{1}{2}i)t} p_1(t) + c_2 e^{(-\sigma + \frac{1}{2}i)t} p_2(t)$$

where $\sigma > 0$ and p_1, p_2 have period 2π. For later purposes it is important to notice that *the solutions have the alternative form*

$$c_1 e^{\sigma t} q_1(t) + c_2 e^{-\sigma t} q_2(t), \tag{9.34}$$

where q_1, q_2 *have period* 4π.

From (i) to (v) it can be seen that certain curves, of the form

$$\phi(\alpha, \beta) = \pm 2,$$

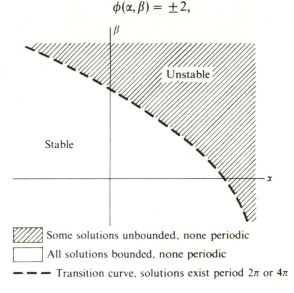

FIG. 9.1.

separate parameter regions where unbounded solutions exist ($|\phi(\alpha, \beta)|$ > 2) from regions where all solutions are bounded ($|\phi(\alpha, \beta)| < 2$) (Fig. 9.1). We do not know the function $\phi(\alpha, \beta)$ explicitly, but we do know that these are also the curves on which periodic solutions, period 2π or 4π, occur. Therefore, *if we can establish, by any method, the parameter values for which such periodic solutions can occur, then we have also found the* boundaries between the stable and unstable region by Theorem 8.8. These boundaries are called *transition curves*.

First, we find what parameter values α, β give periodic solutions of period 2π. Represent such a solution by the complex Fourier series

$$x(t) = \sum_{n=-\infty}^{\infty} c_n e^{int}.$$

After substituting into Mathieu's equation, (9.10) (with $\cos t = \frac{1}{2}(e^{it} + e^{-it})$), we obtain after rearrangement

$$\sum_{n=-\infty}^{\infty} \{\tfrac{1}{2}\beta c_{n+1} + (\alpha - n^2)c_n + \tfrac{1}{2}\beta c_{n-1}\} e^{int} = 0$$

for all t. This can be satisfied only if the coefficients are all zero:

$$\tfrac{1}{2}\beta c_{n+1} + (\alpha - n^2)c_n + \tfrac{1}{2}\beta c_{n-1} = 0, \quad n = 0, \pm 1, \pm 2, \dots. \tag{9.35}$$

This infinite set of homogeneous equations for $\{c_n\}$ has non-zero solutions if the infinite determinant (Whittaker and Watson 1962) formed by the coefficients is zero; when $\alpha \neq n^2$ for any n the following form is convenient:

$$\begin{vmatrix} \cdot & \cdot & \cdot & \cdot & \cdot & \cdot & \cdot \\ \cdot & \gamma_1 & 1 & \gamma_1 & 0 & 0 & \cdot \\ \cdot & 0 & \gamma_0 & 1 & \gamma_0 & 0 & \cdot \\ \cdot & 0 & 0 & \gamma_1 & 1 & \gamma_1 & \cdot \\ \cdot & \cdot & \cdot & \cdot & \cdot & \cdot & \cdot \end{vmatrix} = 0, \tag{9.36}$$

where

$$\gamma_n = \beta/2(\alpha - n^2), \quad n = 0, 1, 2, \dots.$$

This is not explicitly the relation $\phi(\alpha, \beta) = 2$, but it must be equivalent to it, since it specifies transition lines on the α, β-plane.

To find the equations for transition lines corresponding to solutions of (9.28) with period 4π, start with the series

$$x(t) = \sum_{n=-\infty}^{\infty} d_n e^{\frac{1}{2}int}.$$

The equations for the coefficients are

$$\tfrac{1}{2}\beta d_{n+2} + (\alpha - \tfrac{1}{4}n^2)d_n + \tfrac{1}{2}\beta d_{n-2} = 0, \tag{9.37}$$

$n = 0, \pm 1, \pm 2, \dots$. This set of equations splits into two independent sets for $\{d_n\}$ with n even and odd respectively. The set for n even gives a Hill determinant (as these are called) of the form in (9.36). We may therefore specify $d_{2m+1} = 0$ for all m and solve (9.36); this produces the functions of period 2π already obtained. Otherwise, specify $d_{2m} = 0$, all m; for the remaining equations in (9.37) we require

$$
\begin{vmatrix}
\cdot & \cdot & \cdot & \cdot & \cdot & \cdot & \cdot \\
\cdot & \gamma_2 & 1 & \gamma_2 & 0 & 0 & 0 & \cdot \\
\cdot & 0 & \gamma_1 & 1 & \gamma_1 & 0 & 0 & \cdot \\
\cdot & 0 & 0 & \gamma_1 & 1 & \gamma_1 & 0 & \cdot \\
\cdot & 0 & 0 & 0 & \gamma_2 & 1 & \gamma_2 & \cdot \\
\cdot & \cdot & \cdot & \cdot & \cdot & \cdot & \cdot
\end{vmatrix} = 0 \qquad (9.38)
$$

where

$$
\gamma_1 = \beta/2\{\alpha - \tfrac{1}{4}1^2\}, \quad \gamma_2 = \beta/2\{\alpha - \tfrac{1}{4}3^2\}, \ldots,
$$

(provided $\alpha \neq \tfrac{1}{4}(2m+1)^2$ for any m), which specifies the transition curves

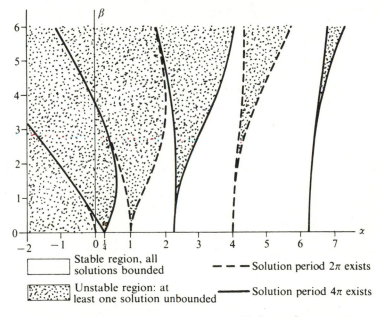

FIG. 9.2. Stability diagram for Mathieu's equation

corresponding to solutions of period 4π.

Given α, β, the determinants in (9.36), (9.38) can rapidly be computed from recurrence relations between the determinants of orders differing by 2 at each step, starting with the mid-determinant (see Exercises 8 and 9). Zeros can be found by specifying α (or β) and searching for the corresponding β (or α) which gives a result sufficiently close to zero. Convergence is rapid enough to produce the data for Fig. 9.2 to two or three decimals in a very short time. The transition curves corresponding to eqn (9.36) pass through the points $\beta = 0$, $\alpha = n^2$, $n = 1, 2, \ldots$, and on these curves solutions of period 2π exist. The curves associated with eqn (9.38) pass through $\beta = 0$, $\alpha = \frac{1}{4}m^2$, $m = 1, 3, 5, \ldots$, and on these solutions of period 4π exist. For $\beta < 0$ the diagram corresponding to Fig. 9.2 is the mirror image in the α axis of that shown.

9.4. Transition curves for Mathieu's equation by perturbation

For small values of $|\beta|$ the perturbation method can be used to establish the transition curves. In the equation

$$\ddot{x} + (\alpha + \beta \cos t)x = 0, \tag{9.39}$$

suppose that the transition curves are given by

$$\alpha = \alpha(\beta) = \alpha_0 + \beta\alpha_1 + \ldots, \tag{9.40}$$

and that the corresponding solutions have the form

$$x(t) = x_0(t) + \beta x_1(t) + \ldots, \tag{9.41}$$

where $x_0, x_1, \ldots$ have period 2π or 4π.

When (9.40) and (9.41) are substituted into (9.39) and the coefficients of powers of β equated to zero in the usual way, we have

$$\ddot{x}_0 + \alpha_0 x_0 = 0, \tag{9.42a}$$

$$\ddot{x}_1 + \alpha_0 x_1 = -(\alpha_1 + \cos t)x_0, \tag{9.42b}$$

$$\ddot{x}_2 + \alpha_0 x_2 = -x_0\alpha_2 - (\alpha_1 + \cos t)x_1, \tag{9.42c}$$

and so on.

The solutions of (9.42a) have minimum period 2π if $\alpha_0 = n^2$, $n = 0, 1, \ldots$; and minimum period 4π if $\alpha_0 = (n + \frac{1}{2})^2$, $n = 0, 1, \ldots$. Both classes of solution can be obtained together by considering

$$\alpha_0 = \tfrac{1}{4}n^2, \quad n = 0, 1, \ldots. \tag{9.43}$$

We consider only $n = 0$ and 1 (see also Exercise 10).

(i) $n = 0$. Equation (9.42a) has the periodic solution $x_0(t) = a_0$. Then, from (9.42b),

$$\ddot{x}_1 = -(\alpha_1 + \cos t)a_0,$$

which has periodic solutions only if $\alpha_1 = 0$, the solutions being of the form

$$x_1(t) = a_0 \cos t + a_1.$$

Equation (9.42c) becomes

$$\ddot{x}_2 = -a_0\alpha_2 - \tfrac{1}{2}a_0 - \tfrac{1}{2}a_0 \cos 2t - a_1 \cos t,$$

which has a periodic solution (of period 2π) only if $a_0\alpha_2 + \tfrac{1}{2}a_0 = 0$ or if $\alpha_2 = -\tfrac{1}{2}$. Therefore, for small $|\beta|$

$$\alpha = -\tfrac{1}{2}\beta^2 + O(\beta^3). \tag{9.44}$$

(ii) $n = 1$. Here $\alpha_0 = \tfrac{1}{4}$, and $x_0(t) = a_0 \cos \tfrac{1}{2}t + b_0 \sin \tfrac{1}{2}t$. (9.42b) becomes

$$\ddot{x}_1 + \tfrac{1}{4}x_1 = -a_0(\alpha_1 + \tfrac{1}{2})\cos \tfrac{1}{2}t - b_0(\alpha_1 - \tfrac{1}{2})\sin \tfrac{1}{2}t - \tfrac{1}{2}a_0 \cos \tfrac{3}{2}t - \tfrac{1}{2}b_0 \sin \tfrac{3}{2}t.$$

There are periodic solutions (of period 4π) only if $b_0 = 0, \alpha_1 = -\tfrac{1}{2}$, or if $a_0 = 0, \alpha_1 = \tfrac{1}{2}$. Thus, to a first approximation, for small β

$$\alpha = \tfrac{1}{4} \pm \tfrac{1}{2}\beta \tag{9.45}$$

describes the transition curves near $\alpha = \tfrac{1}{4}$, $\beta = 0$. Figure 9.2 may be compared with these results.

We are now in a position to settle the question of the stability of the forced, undamped Duffing equation in Section 9.1, at least for ε small and ω not too far from 1. In the variational equation (9.10)

$$\alpha = (2 + 3\varepsilon a^2)/8\omega^2, \qquad \beta = 3\varepsilon a^2/8\omega^2.$$

Since $\varepsilon \ll 1$, the stability boundaries near $\alpha = \tfrac{1}{4}$ are the relevant ones. Suppose that ε is positive: then for stability, since β is also positive, α must satisfy $\alpha < \tfrac{1}{4} - \tfrac{1}{2}\beta$ or $\alpha > \tfrac{1}{4} + \tfrac{1}{2}\beta$ (see (9.45) and Fig. 9.2). In terms of the original variables these conditions become

$$\omega^2 < 1 + \tfrac{3}{4}\varepsilon a^2 \qquad \omega^2 > 1 + \tfrac{9}{4}\varepsilon a^2.$$

These equations must be taken in conjunction with (9.9) which determines the values of a and ω to be associated.

9.5. Mathieu's damped equation arising from a Duffing equation

As we saw in Section 9.1, the variational equation for the undamped forced pendulum is Mathieu's equation. With dissipation included, the Duffing equation in standardized form is

$$\ddot{x} + k\dot{x} + x + \varepsilon x^3 = \Gamma \cos \omega t. \tag{9.46}$$

In Chapter 7 we also showed that this equation has periodic solutions which are approximately of the form $a \cos \omega t + b \sin \omega t$ where $r = \sqrt{(a^2 + b^2)}$ satisfies

$$\{(\omega^2 - 1 - \tfrac{3}{4}\varepsilon r^2)^2 + \omega^2 k^2\} r^2 = \Gamma^2, \tag{9.47}$$

which reduces to eqn (7.23) since $\varepsilon = \beta$. Following the notation and procedure of Section 9.1, write (9.46) as the first-order system

$$\dot{\mathbf{x}} = \begin{pmatrix} \dot{x} \\ \dot{y} \end{pmatrix} = \begin{pmatrix} y \\ -ky - x - \varepsilon x^3 + \Gamma \cos \omega t \end{pmatrix} \tag{9.48}$$

and put (approximately)

$$x^* = a \cos \omega t + b \sin \omega t, \quad y^* = -a\omega \sin \omega t + b\omega \cos \omega t.$$

The variations $\xi = x - x^*$ and $\eta = y - y^*$ satisfy

$$\dot{\xi} + \dot{x}^* = \eta + y^*,$$

$$\dot{\eta} + \dot{y}^* = -k(\eta + y^*) - (\xi + x^*) - \varepsilon(\xi + x^*)^3 + \Gamma \cos \omega t.$$

By using (9.48) and retaining only the first powers of ξ and η we obtain corresponding linearized equations

$$\dot{\xi} = \eta, \qquad \dot{\eta} = -k\eta - \xi - 3\varepsilon x^{*2}\xi.$$

Elimination of η leads to the second-order equation

$$\ddot{\xi} + k\dot{\xi} + (1 + 3\varepsilon x^{*2})\xi = 0.$$

By substituting for x^* its approximate form $a \cos \omega t + b \sin \omega t$ we obtain

$$\ddot{\xi} + k\dot{\xi} + \{1 + \tfrac{3}{2}\varepsilon r^2 + \tfrac{3}{2}\varepsilon r^2 \cos(2\omega t + 2\gamma)\}\xi = 0, \tag{9.49}$$

where r, γ are defined by

$$a \cos \omega t + b \sin \omega t = r \cos(\omega t + \gamma).$$

We can reduce the equation to a 'standard' form by putting

$$\tau = 2\omega t + 2\gamma, \qquad \xi' \equiv d\xi/d\tau,$$

$$\kappa = k/2\omega, \qquad \nu = (2 + 3\varepsilon r^2)/8\omega^2, \qquad \beta = 3\varepsilon r^2/8\omega^2,$$

so that

$$\xi'' + \kappa \xi' + (v + \beta \cos \tau)\xi = 0. \qquad (9.50)$$

This equation is known as Mathieu's equation with damping.

A simple transformation carries (9.50) into the ordinary Mathieu equation. If we write

$$\xi(\tau) = e^{-\frac{1}{2}\kappa \tau} \rho(\tau), \qquad (9.51)$$

then (9.50) becomes

$$\rho'' + (v - \tfrac{1}{4}\kappa^2 + \beta \cos \tau)\rho = 0, \qquad (9.52)$$

which is Mathieu's equation with $\alpha = v - \frac{1}{4}\kappa^2$. Whenever α and β take values such that only bounded solutions of (9.52) exist, then by (9.51) only bounded solutions of (9.50) exist, since $\kappa > 0$. Therefore the regions of stability in the plane for (9.50) will be extensions of those for (9.52).

We shall assume that $\varepsilon \ll 1$, $\omega \approx 1$ (near-resonance), and $k = O(\varepsilon)$. This places us near $v = \frac{1}{4}$, $\beta = 0$: for eqn (9.52) this is near to the first unstable region of Fig. 9.2 with $\alpha = 0$. On the boundaries of this unstable region we know from the previous section that there exist periodic solutions $\rho(\tau)$ having period 4π, corresponding to case (iv), p. 252, for which $\phi(\alpha, \beta) = -2$. Inside the unstable region we must therefore have case (v), for which $\phi(\alpha, \beta) < -2$, and the solutions have the form (9.34). Hence, by (9.51), the corresponding $\xi(\tau)$ take the form

$$\xi(\tau) = c_1 e^{(\sigma - \frac{1}{2}\kappa)\tau} q_1(\tau) + c_2 e^{(-\sigma - \frac{1}{2}\kappa)\tau} q_2(\tau), \qquad (9.53)$$

where $\sigma > 0$, and $q_1(\tau)$, $q_2(\tau)$ both have period 4π. The boundary of the unstable region for (9.50) in the (v, β)-plane consists of the points for which periodic solutions exist. From (9.53) they occur when $\sigma - \frac{1}{2}\kappa = 0$, $c_2 = 0$, and such solutions have period 4π. Therefore we can identify the parameter values defining the boundary by searching by an independent method for solutions of period 4π.

Since $\varepsilon \ll 1$ and $k = O(\varepsilon)$, then $|\beta| \ll 1$, and κ is of the same order of magnitude: write $\kappa = \beta \kappa_2$. As in Section 9.3 let the unknown stability boundary be given by

$$v = v(\beta) = v_0 + \beta v_1 + \ldots, \qquad (9.54)$$

and the solutions of period 4π by

$$\xi(\tau) = \xi_0(\tau) + \beta \xi_1(\tau) + \ldots. \qquad (9.55)$$

The perturbation method requires

$$\xi_0'' + v_0\xi_0 = 0, \tag{9.56a}$$

$$\xi_1'' + v_0\xi_1 = -\kappa_1\xi_0' - (v_1 + \cos\tau)\xi_0, \tag{9.56b}$$

and so on. Equation (9.56a), with the periodicity condition, requires $v_0 = \frac{1}{4}$; then $\xi_0(\tau) = a_0\cos\frac{1}{2}\tau + b_0\sin\frac{1}{2}\tau$. By substituting these results into (9.56b) and equating the coefficients of $\cos\frac{1}{2}\tau$, $\sin\frac{1}{2}\tau$ on the right to zero to ensure only periodic solutions we find that

$$(v_1 + \tfrac{1}{2})a_0 + \tfrac{1}{2}\kappa_1 b_0 = 0, \qquad \tfrac{1}{2}\kappa_1 a_0 - (v_1 - \tfrac{1}{2})b_0 = 0.$$

The existence of a non-zero solution requires $(v_1 - \frac{1}{2})(v_1 + \frac{1}{2}) + (\frac{1}{2}\kappa_1)^2 = 0$, or $v_1 = \pm\frac{1}{2}\sqrt{(1 - \kappa_1^2)}$. Thus the stability boundaries near $v = \frac{1}{4}$ are given by

$$v = \tfrac{1}{4} \pm \tfrac{1}{2}\beta\sqrt{(1 - \kappa_1^2)} \tag{9.57}$$

(so long as $\kappa_1^2 < 1$; otherwise the whole region becomes stable for $|\beta| \ll 1$). Solutions are bounded, and the system stable, if

$$v < \tfrac{1}{4} - \tfrac{1}{2}\beta\sqrt{(1 - \kappa_1^2)} \quad \text{or} \quad v > \tfrac{1}{4} + \tfrac{1}{2}\beta\sqrt{(1 - \kappa_1^2)},$$

or if

$$(v - \tfrac{1}{4})^2 - \tfrac{1}{4}\beta^2(1 - \kappa_1^2) > 0. \tag{9.58}$$

Note that, by virtue of (9.51), there are regions in which the solutions of the damped Mathieu equation tend to zero: and from the analysis of the various regions in Section 9.3 and the form of the corresponding solutions these regions are exactly the stable regions of (9.49).

Returning now to eqn (9.49), we have shown that if $\varepsilon \ll 1$, $\omega \approx 1$ (near-resonance), and $k = O(\varepsilon)$, then the solutions are stable when

$$\left(\frac{2 + 3\varepsilon r^2}{8\omega^2} - \frac{1}{4}\right)^2 - \frac{1}{4}\left(\frac{3\varepsilon r^2}{8\omega^2}\right)^2\left[1 - \left(\frac{4k\omega}{3\varepsilon r^2}\right)^2\right] > 0,$$

or when

$$(1 - \omega^2)^2 + 3\varepsilon(1 - \omega^2)r^2 + \tfrac{27}{16}\varepsilon^2 r^4 + k^2\omega^2 > 0. \tag{9.59}$$

Since the solutions of the damped Mathieu equation tend to zero in the stable region, asymptotic stability is predicted, confirming the analysis of Chapter 7.

We can also confirm the remark made in Section 5.4(vi) that stability is to be expected when

$$\frac{d(\Gamma^2)}{d(r^2)} > 0 \tag{9.60}$$

or when an increase or decrease in magnitude of Γ results in an increase or decrease respectively in the amplitude. From (9.48) it is easy to verify that $d(\Gamma^2)/d(r^2)$ is equal to the expression on the left of (9.59), and the speculation is therefore confirmed.

In general, when periodic solutions of the original equation are expected the reduced equation (9.4) is the more complicated Hill type (see Exercise 11). The stability regions for this equation and examples of the corresponding stability estimates may be found in Hayashi (1964).

Exercises

1. The system
$$\dot{x}_1 = (-\sin 2t)x_1 + (\cos 2t - 1)x_2, \quad \dot{x}_2 = (\cos 2t + 1)x_1 + (\sin 2t)x_2$$
has a fundamental matrix of normal solutions:
$$\begin{pmatrix} e^t(\cos t - \sin t) & e^{-t}(\cos t + \sin t) \\ e^t(\cos t + \sin t) & e^{-t}(-\cos t + \sin t) \end{pmatrix}.$$
Obtain the corresponding E-matrix (Theorem 9.1), the characteristic numbers, and the characteristic exponents.

2. Let the system $\dot{x} = P(t)x$ have a matrix of coefficients P with minimal period T (and therefore also with periods $2T, 3T, \ldots$). Follow the argument of Theorem 8.11, using period mT, $m > 1$, to show that $\Phi(t + mT) = \Phi(t)E^m$. Assuming that if the eigenvalues of E are μ_i, then those of E^m are μ_i^m, deduce that no new periodic solutions are disclosed by this process.

3. Obtain Wronskians for the following linear systems:
(i) $\dot{x}_1 = x_1 \sin t + x_2 \cos t, \ \dot{x}_2 = -x_1 \cos t + x_2 \sin t,$
(ii) $\dot{x}_1 = f(t)x_2, \ \dot{x}_2 = g(t)x_1.$

4. By substituting $x = c + a\cos t + b\sin t$ into Mathieu's equation
$$\ddot{x} + (\alpha + \beta \cos t)x = 0,$$
obtain by harmonic balance an approximation to the transition curve near $\alpha = 0$, $\beta = 0$ (compare with Section 9.1).

By substituting $x = c + a\cos \frac{1}{2}t + b\sin \frac{1}{2}t$, find the transition curves near $\alpha = \frac{1}{4}$, $\beta = 0$.

5. The figure represents a mass m attached to two identical linear elastic strings of stiffness λ and natural length l. The ends of the strings pass through frictionless guides A and B at a distance $2L$, $l < L$, apart. The particle is set into lateral

FIG. 9.3.

motion at the mid-point, and symmetrical displacements $a + b \cos \omega t$, $a > b$, are imposed on the ends of the string. Show that, for $x \ll L$,

$$\ddot{x} + \left(\frac{2\lambda(L - l + a)}{mL} + \frac{2\lambda b}{mL} \cos \omega t \right) x = 0.$$

Analyse the motion in terms of suitable parameters, using the information of Sections 9.3 and 9.4 on the growth or decay, periodicity and near-periodicity of the solutions of Mathieu's equation in the regions of its parameter plane.

6. A pendulum with a light, rigid suspension is placed upside-down on end, and the point of suspension is caused to oscillate vertically with displacement y given by $y = \varepsilon \cos \omega t$, $\varepsilon \ll 1$. Show that the equation of motion is

$$\ddot{\theta} + \left(-\frac{g}{a} + \frac{1}{a} \ddot{y} \right) \sin \theta = 0,$$

where a is the length of the suspension, g is gravitational acceleration, and θ the inclination to the vertical. Linearize the equation for small amplitudes and show that the vertical position is stable (that is, the motion of the pendulum restricts itself to the neighbourhood of the vertical: it does not topple over) provided $\varepsilon^2 C^2 / 2ag > 1$.

7. Let $\Phi(t) = (\phi_{ij}(t))$, $i, j = 1, 2$, be the fundamental matrix for the system $\dot{x}_1 = x_2$, $\dot{x}_2 = -(\alpha + \beta \cos t)x_1$, satisfying $\Phi(0) = I$ (Mathieu's equation). Show that the characteristic numbers μ satisfy the equation

$$\mu^2 - \mu\{\phi_{11}(2\pi) + \phi_{22}(2\pi)\} + 1 = 0.$$

8. In eqn (9.36), for the transition curves of Mathieu's equation for solutions period 2π, let

$$D_{m,n} = \begin{vmatrix} 1 & \gamma_m & & & & & \\ \gamma_{m-1} & 1 & \gamma_{m-1} & & & & \\ & & \cdots & & & & \\ & & & \gamma_0 & 1 & \gamma_0 & \\ & & & & \cdots & & \\ & & & & \gamma_{n-1} & 1 & \gamma_{n-1} \\ & & & & & \gamma_n & 1 \end{vmatrix}$$

for $m \geqslant 0$, $n \geqslant 0$. Show that

$$D_{m,n} = D_{m-1,n} - \gamma_m \gamma_{m-1} D_{m-2,n}.$$

Let $E_n = D_{n,n}$ and verify that

$$E_0 = 1, \quad E_1 = 1 - 2\gamma_0\gamma_1, \quad .E_2 = (1 - \gamma_1\gamma_2)^2 - 2\gamma_0\gamma_1(1 - \gamma_1\gamma_2).$$

Prove that, for $n \geqslant 1$,

$$E_{n+2} = (1 - \gamma_{n+1}\gamma_{n+2})E_{n+1} + \gamma_{n+1}\gamma_{n+2}(1 - \gamma_{n+1}\gamma_{n+2})E_n + \gamma_n^2\gamma_{n+1}^3\gamma_{n+2}E_{n-1}.$$

This recurrence relation can be used to evaluate the determinant (9.36).

9. In equation (9.38), for the transition curves for Mathieu's equation for solutions of period 4π, let

$$F_{m,n} = \begin{vmatrix} 1 & \gamma_m & & & & \\ \gamma_{m-1} & 1 & \gamma_{m-1} & & & \\ & & \cdots & & & \\ & & & \gamma_{n-1} & 1 & \gamma_{n-1} \\ & & & & \gamma_n & 1 \end{vmatrix}.$$

Show as in the last exercise that $G_n = F_{n,n}$ satisfies the same recurrence relation as E_n for $n \geqslant 2$. Verify that

$$G_1 = 1 - \gamma_1^2,$$
$$G_2 = (1 - \gamma_1\gamma_2)^2 - \gamma_1^2,$$
$$G_3 = (1 - \gamma_1\gamma_2 - \gamma_2\gamma_3)^2 - \gamma_1^2\gamma_2\gamma_3(1 - \gamma_2\gamma_3).$$

10. Show, by the perturbation method, that the transition curves for Mathieu's equation

$$\ddot{x} + (\alpha + \beta \cos t)x = 0,$$

$\alpha \approx 1, \beta \approx 0$, are given by $\alpha = 1 + \frac{1}{12}\beta^2, \alpha = 1 - \frac{5}{12}\beta^2$.

11. Consider Hill's equation $\ddot{x} + f(t)x = 0$, where f has period 2π, and

$$f(t) = \alpha + \sum_{r=1}^{\infty} \beta_r \cos rt$$

is its Fourier expansion, with $\alpha \approx \frac{1}{4}$ and $|\beta_r| \ll 1$, $r = 1, 2, \dots$. Assume an approximate solution $e^{\sigma t}q(t)$, where σ is real and q has period 4π as in (9.34). Show that

$$\ddot{q} + 2\sigma\dot{q} + \left(\sigma^2 + \alpha + \sum_{r=1}^{\infty} \beta_r \cos rt\right)q = 0.$$

Take $q \approx \sin(\frac{1}{2}t + \gamma)$ as the approximate form for q and match terms in $\sin\frac{1}{2}t$, $\cos\frac{1}{2}t$, on the assumption that these terms dominate. Deduce that

$$\sigma^2 = -(\alpha + \frac{1}{4}) + \frac{1}{2}\sqrt{(4\alpha + \beta_1^2)}$$

and that the transition curves near $\alpha = \frac{1}{4}$ are given by $\alpha = \frac{1}{4} \pm \frac{1}{2}\beta_1$. ($\beta_n$ is similarly the dominant coefficient for transition curves near $\alpha = \frac{1}{4}n^2$, $n \geqslant 1$.)

12. Obtain, as in Section 9.4, the boundary of the stable region in the neighbourhood of $v = 1, \beta = 0$ for Mathieu's equation with damping,

$$\ddot{x} + \kappa\dot{x} + (v + \beta \cos t)x = 0$$

where $\kappa = O(\beta^2)$.

13. Solve Meissner's equation

$$\ddot{x} + (\alpha + \beta f(t))x = 0$$

where $f(t) = 1, 0 \leqslant t < \pi; f(t) = -1, \pi \leqslant t < 2\pi;$ and $f(t + 2\pi) = f(t)$ for all t. Find the conditions on α, β, for periodic solutions by putting $x(0) = x(2\pi), \dot{x}(0) = \dot{x}(2\pi)$ and by making x and $\dot{x}$ continuous at $t = \pi$. Find a determinant equation for α and β.

14. By using the harmonic balance method of Chapter 4, show that the van der Pol equation with parametric excitation,

$$\ddot{x} + \varepsilon(x^2 - 1)\dot{x} + (1 + \beta \cos t)x = 0$$

has a 2π-periodic solution with approximately the same amplitude as the unforced van der Pol equation.

15. The male population M and female population F or a bird community have a constant death-rate k and a variable birth-rate $\mu(t)$ which has period T, so that

$$\dot{M} = -kM + \mu(t)F, \quad \dot{F} = -kF + \mu(t)F.$$

If the births are seasonal, so that

$$\mu(t) = \begin{cases} \delta, & 0 < t \leqslant \varepsilon; \\ 0, & \varepsilon < t \leqslant T, \end{cases}$$

show that periodic solutions of period T exist for M and F if $kT = \delta\varepsilon$.

16. A pendulum bob is suspended by a string of length a, and the support is constrained to move vertically with displacement $\zeta(t)$. Show (by using the Lagrange's equation method or otherwise) that the equation of motion is

$$a\ddot{\theta} + (g - \ddot{\zeta}(t))\sin \theta = 0$$

where θ is the angle of inclination. Examine the stability of the motion for the case when $\zeta(t) = c \sin \omega t$, on the assumption that it is permissible to put $\sin \theta \approx \theta$.

17. In a predator–prey model, the prey x and the predator y satisfy

$$\dot{x} = -\alpha x + \beta x^2 - \gamma xy, \quad \dot{y} = -\delta y + \varepsilon xy$$

where $\alpha, \beta, \gamma, \delta,$ and ε are positive constants. How does the system behave when there is exactly one equilibrium point, at $(0,0)$? Prove that the solution $x = 0$, $y = 0$ is asymptotically stable. Discuss the implications of this result for the prey and the predator.

18. A pendulum, with bob of mass m and suspension of length a, hangs from a support which is constrained to move with vertical and horizontal displacements $\zeta(t)$ and $\eta(t)$ respectively. Show that the inclination θ of the pendulum satisfies the equation

$$a\ddot{\theta} + (g - \ddot{\zeta})\sin\theta + \ddot{\eta}\cos\theta = 0.$$

Let $\zeta = \alpha\sin\omega t$ and $\eta = \beta\sin 2\omega t$, where $\omega = \sqrt{(g/a)}$. Show that after linearizing this equation for small amplitudes, the resulting equation has a solution

$$\theta = (8\beta/\alpha)\cos\omega t.$$

Determine the stability of this solution.

19. The equation

$$\ddot{x} + (\tfrac{1}{4} - 2\varepsilon b\cos^2 \tfrac{1}{2}t)x + \varepsilon x^3 = 0$$

has the exact solution $x^*(t) = \sqrt{(2b)}\cos\tfrac{1}{2}t$. Show that the solution is stable by constructing the variational equation.

20. Consider the equation $\ddot{x} + (\alpha + \beta\cos t)x = 0$, where $|\beta| \ll 1$ and $\alpha = \tfrac{1}{4} + \beta c$. In the unstable region near $\alpha = \tfrac{1}{4}$ (Section 9.4) this equation has solutions of the form $c_1 e^{\sigma t}q_1(t) + c_2 e^{-\sigma t}q_2(t)$, where σ is real, $\sigma > 0$ and q_1, q_2 have period 4π. Construct the equation for q_1, q_2, and show that $\sigma = \pm\beta\sqrt{(\tfrac{1}{4} - c^2)}$.

21. By using the method of Section 9.5 show that a solution of the equation

$$\ddot{x} + \varepsilon(x^2 - 1)\dot{x} + x = \Gamma\cos\omega t$$

where $|\varepsilon| \ll 1$, $\omega = 1 + \varepsilon\omega_1$, of the form $x^* = r_0\cos(\omega t + \alpha)$ (α constant) is asymptotically stable when

$$4\omega_1^2 + \tfrac{3}{16}r_0^4 - r_0^2 + 1 < 0.$$

(Use the result of Exercise 20.)

22. The equation

$$\ddot{x} + \alpha x + \varepsilon x^3 = \varepsilon\gamma\cos\omega t$$

has the exact subharmonic solution

$$x = (4\gamma)^{1/3}\cos\tfrac{1}{3}\omega t$$

when

$$\omega^2 = 9\left(\alpha + \frac{3}{4^{1/3}}\varepsilon\gamma^{2/3}\right).$$

Show that the solution is stable.

23. Analyse the stability of the equation

$$\ddot{x} + \varepsilon x\dot{x}^2 + x = \Gamma\cos\omega t$$

for small ε. (First find approximate solutions of the form $\alpha \cos \omega t$ by the harmonic balance method of Chapter 4, then perturb the solution by the method of Section 9.4.)

24. The equation $\ddot{x} + x + \varepsilon x^3 = \Gamma \cos \omega t$ ($\varepsilon \ll 1$) has an approximate solution $x^*(t) = a \cos \omega t$ where (eqn (7.10)) $\frac{3}{4}\varepsilon a^3 - (\omega^2 - 1)a - \Gamma = 0$. Show that the first variational equation (Section 9.4) is $\ddot{\xi} + \{1 + 3\varepsilon x^{*2}(t)\}\xi = 0$. Reduce this to Mathieu's equation and find conditions for stability of $x^*(t)$.

25. The equation $\ddot{x} + x - \frac{1}{6}x^3 = 0$ has an approximate solution $a \cos \omega t$ where $\omega^2 = 1 - \frac{1}{8}a^2$, $a \ll 1$ (Example 4.12). Use the method of Section 9.4 to show that the solution is unstable.

26. Van der Pol's equation $\ddot{x} + \varepsilon(x^2 - 1)\dot{x} + x = 0$, $\varepsilon \ll 1$, has a limit cycle given approximately by $x^*(t) = 2 \cos t$. Use the method of Section 9.4 to show that this solution is stable.

10 Liapunov methods for determining stability

HERE WE describe methods for deciding the stability of equilibrium points, which at the simplest level amount to tracing the general tendency of the paths in the locality by referring them to another family of curves strategically matched to the individual problem so as to deliver the right kind of information. We shall discuss two-dimensional autonomous systems only, but the method can be extended to higher-dimensional and non-autonomous ones (see Cesari 1971 and Willems 1970). The method also allows domains of initial conditions which generate stable solutions to be found.

10.1. Liapunov's direct method

We show how this works by means of an example.

Example 10.1 *Investigate the stability of the zero solution of the system*

$$\dot{x} = -y - x^3, \qquad \dot{y} = x - y^3.$$

We try to find a family of closed, nested curves surrounding the origin which have the property that, if we choose any point on any one of them, then the phase path through that point is directed towards the interior of the closed curve (Fig. 10.1). If such a family can be found then each half-path will cross in turn all the members of the family, and so enter the origin. This implies stability, and indeed asymptotic stability, of the zero solution.

To test the family of curves

$$V(x, y) = x^2 + y^2 = c, \qquad c > 0$$

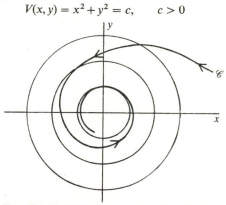

FIG. 10.1. Path $\mathscr{C}$ crossing the family of circles

for this property, consider any particular path $\mathscr{P}$ corresponding to the solution $x(t)$, $y(t)$. Then on this path

$$V(x, y) = V\{x(t), y(t)\},$$

and

$$\left(\frac{dV}{dt}\right)_{\mathscr{P}} = \left(\frac{\partial V}{\partial x} \dot{x} + \frac{\partial V}{\partial y} \dot{y}\right)_{\mathscr{P}} = 2x(-y-x^3) + 2y(x-y^3)$$

$$= -2(x^4 + y^4).$$

With the exception of the origin this is negative everywhere: V therefore decreases along every path so that the representative point moves constantly across circles with smaller values of c, and so enters the origin. Asymptotic stability follows from Definition 8.4. (The linear approximation predicts a centre—it was noted in Chapter 2 (see, for example, Exercise 2.7) that the nonlinear terms may be decisive in breaking up the centre into a spiral.)

Such a family of curves, designed to give an indication of the path directions, is called a *topographic system* for the problem. $V(x, y)$ is called a *Liapunov function* for the zero solution of the given system. To produce a form of V which will decide the stability in a particular case is the difficult practical feature of the method.

10.2. Liapunov functions

In the following definitions $V(\mathbf{x})$ is a scalar function of the n components $x_1, x_2, \ldots, x_n$ of $\mathbf{x}$.

DEFINITION 10.1 $V(\mathbf{x})$ *is positive (negative) definite in a neighbourhood* $\mathscr{S}$ *of the origin if* $V(\mathbf{x}) > 0$ ($V(\mathbf{x}) < 0$) *for all* $\mathbf{x} \neq \mathbf{0}$ *in* $\mathscr{S}$, *and* $V(\mathbf{0}) = 0$. $V(\mathbf{x})$ *is positive (negative) semidefinite in a neighbourhood* $\mathscr{S}$ *of the origin if* $V(\mathbf{x}) \geqslant 0$ ($V(\mathbf{x}) \leqslant 0$) *for all* $\mathbf{x} \neq \mathbf{0}$ *in* $\mathscr{S}$, *and* $V(\mathbf{0}) = 0$.

Thus $V(x, y) = x^2 + y^2$ is positive definite on the whole plane, and $V(x, y) = y^2(x^2 - 1)$ is negative semidefinite on the strip $|x| < 1$.

The *total derivative* of a function $V(\mathbf{x})$ along a path $\mathscr{P}$ of the autonomous system $\dot{\mathbf{x}} = \mathbf{X}(\mathbf{x})$ corresponding to a solution $\mathbf{x}(t)$ is

$$\left(\frac{dV}{dt}\right)_{\mathscr{P}} = \frac{d}{dt} V\{\mathbf{x}(t)\} = \sum_{i=1}^{n} \frac{\partial V}{\partial x_i} X_i(\mathbf{x}).$$

When we wish to treat (10.6) as a function of position we write $\dot{V}(\mathbf{x})$, where

$$\dot{V}(\mathbf{x}) = \sum_{i=1}^{n} \frac{\partial V}{\partial x_i} X_i(\mathbf{x}). \tag{10.1}$$

Thus we may say 'V is continuous', or '$\dot{V}$ is positive definite', meaning with respect to $\mathbf{x}$.

We adopt a geometrical approach to certain aspects of the two theorems which follow. Suppose that $V(\mathbf{x})$ is positive definite. A simple model for such a function in 2 dimensions is Fig. 10.2(a). The level curves, $V(x, y) = c$, form a topographic system consisting of simple closed curves surrounding the origin, those for smaller c being nearer the origin, as is notionally indicated; and as $c \to 0$ the origin is approached.

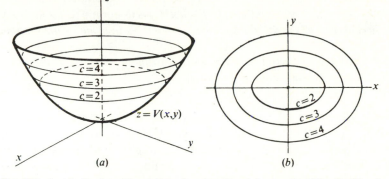

Fig. 10.2. An idealized positive definite function V and the associated family $V(x, y) = c$

In general a positive definite function will not be so simple, but may have such a form as in Fig. 10.3(a), with several minima. Even in such a case, there is a neighbourhood of the origin in which V has the simple structure of Fig. 10.2. In the theorems we confine the domain of operations to such a neighbourhood of the origin. In higher dimensions the situation cannot be visualized, but when $V(\mathbf{x})$ is positive definite and has continuous partial derivatives, then for all small enough positive c and for $\mathbf{x}$ in some sufficiently small neighbourhood of the origin, an analogous result holds.

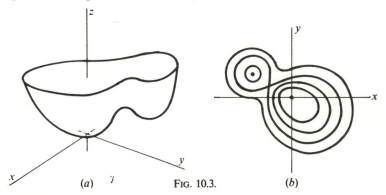

Fig. 10.3.

In Theorem 10.1 we assume that all constructions are carried out in a neighbourhood $\mathscr{V}^*$ of the origin defined by $V(\mathbf{x}) < c^*$ for which the above is true. The family of neighbourhoods $\mathscr{V}_c$ defined by $V(\mathbf{x}) < c$ for c small enough provides a suitable means for recognizing stability. In place of the neighbourhoods $\|\mathbf{x}\| < \varepsilon$, etc., we use differently-shaped neighbourhoods defined by $V(\mathbf{x}) < \varepsilon \leqslant c^*$.

THEOREM 10.1   *Let* $\mathbf{x}^*(t) = \mathbf{0}, t \geqslant t_0$, *be the zero solution of the regular system* $\dot{\mathbf{x}} = \mathbf{X}(\mathbf{x})$, *where* $\mathbf{X}(\mathbf{0}) = \mathbf{0}$. *Then* $\mathbf{x}^*(t)$ *is uniformly stable for* $t \geqslant t_0$ *if there exists* $V(\mathbf{x})$ *with the following properties in some neighbourhood* $\mathscr{V}^*$ *of* $\mathbf{x} = \mathbf{0}$:
 (i) $V(\mathbf{x})$ *and its partial derivatives are continuous*;
 (ii) $V(\mathbf{x})$ *is positive definite*;
 (iii) $\dot{V}(\mathbf{x})$ *is negative semidefinite*.

Proof   Choose ε arbitrarily, subject to the condition $\mathscr{V}_\varepsilon \subset \mathscr{V}^*$ above. In $\mathscr{V}_\varepsilon$, $\dot{V}(x) \leqslant 0$ by (iii); therefore any half-path starting in $\mathscr{V}_\varepsilon$ remains in it. The same holds for every small enough ε and hence for every sufficiently small neighbourhood of the origin. The zero solution is therefore stable (and uniformly stable, since the system is autonomous).

Such a function V is called a *weak Liapunov function* for the system.

THEOREM 10.2   *Let all the conditions of Theorem 10.1 apply, except that* (iii) *is replaced by*

(iii)' $\dot{V}(\mathbf{x})$ *is negative definite*.

Then the zero solution is asymptotically stable. (*Such a function* V *is called a* strong Liapunov function *for the system.*)

Proof   Theorem 10.1 shows that the solution is stable (uniformly). Since $\dot{V}$ is negative definite, $V\{\mathbf{x}(t)\}$ must decrease monotonically on every half-path $\mathbf{x}(t)$. Therefore either $V\{\mathbf{x}(t)\} \to 0$ as $t \to \infty$ and the half-paths all approach the origin, or there is (at least) one half-path $\mathbf{x}(t)$ such that

$$V\{\mathbf{x}(t)\} \geqslant m > 0 \quad \text{for all } t \geqslant t_0, \tag{10.2}$$

for some value of $m > 0$. Since $\mathbf{x}(t)$ is stable, there exists an annular region D, defined by $m \leqslant V(\mathbf{x}) \leqslant c$, containing this half-path. The region is closed, and $\dot{V}$ is continuous on it by (i) and the regularity of the system. $\dot{V}$ therefore attains its upper bound in D, and this is a negative constant, $-M$, say

$$\dot{V}(\mathbf{x}) \leqslant -M < 0, \quad \mathbf{x} \in D, \quad M > 0. \tag{10.3}$$

Integration of (10.3) from t_0 to t gives

$$V(\mathbf{x}(t)) - V(\mathbf{x}(t_0)) \leqslant -M(t-t_0) \quad \text{for all } t > t_0$$

and this contradicts (10.2). There is therefore no path which fails to reach the origin.

Example 10.2 *Find a strong Liapunov function for the system*

$$\dot{x} = -x - 2y^2, \quad \dot{y} = xy - y^3.$$

Try the function $V(x, y) = x^2 + ay^2$, with a to be specified. Then by (10.1)

$$\dot{V}(x, y) = -2x^2 + 2(a-2)xy^2 - 2ay^4.$$

When $a = 2$ this becomes

$$\dot{V}(x, y) = -2(x^2 + 2y^4).$$

Then $V(x, y) = x^2 + 2y^2$ is positive definite and $\dot{V}(x, y)$ is negative definite. Theorem 10.2 tells that the zero solution is asymptotically stable.

Example 10.3 *Find a weak Liapunov function for the system*

$$\dot{x} = y, \quad \dot{y} = -\{f(x, y)y + \lambda x\}, \quad \lambda > 0,$$

where $f(x, y) \geqslant 0$ in a neighbourhood of the origin.

The system is equivalent to $\ddot{x} + f(x, \dot{x})\dot{x} + \lambda x = 0$, representing a linear mass-spring system modified by a nonlinear damping term, giving positive damping for small amplitudes. Try the positive definite function

$$V(x, y) = \tfrac{1}{2}(\lambda x^2 + y^2).$$

Then

$$\dot{V}(x, y) = -y^2 f(x, y), \quad \leqslant 0$$

and V is a weak Liapunov function. By Theorem 10.1, the zero solution is stable.

In the last example V fails to be a strong Liapunov function because $\dot{V} = 0$ at the points of the topographic system where $y = 0$. Although the paths are tangential at these isolated points it can be seen that they can only continue inwards, crossing curves corresponding to smaller values of V. The argument of Theorem 10.2 therefore still applies and the origin represents an asymptotically stable solution. The choice for V is not arbitrary, but recognizable as the 'total energy' of the system; that is, the sum of the kinetic and potential energy; which we expect will be dissipated to zero by the damping. The level curves of the Liapunov function are curves of constant energy of the system. The paths pierce them inwards except along the y-axis. This confirms that the motion is damped (energy is lost), and we expect the system to come to rest.

It can be inferred from the proof of Theorem 10.2 that not only is the zero solution asymptotically stable, but that all solutions starting in certain neighbourhoods of the origin are asymptotically stable. Such a neighbourhood is called a *domain of asymptotic stability*, or a *domain of attraction*. A domain of asymptotic stability can be identified by determining the region in which V has the neighbourhood property on p. 269; though this will not in general determine the largest domain, since the region found will depend on the particular Liapunov function chosen. For the equation $\dot{x} = -x(1-x^2-y^2)$, $\dot{y} = -y(1-x^2-y^2)$, with $V = x^2 + y^2$ we find that $\dot{V} = -2(x^2+y^2)(1-x^2-y^2)$. V and $\dot{V}$ have the required property in the disc $x^2 + y^2 < 1$ and this is therefore a domain of asymptotic stability, which happens to be the largest domain in this case. When the domain of attraction is the whole x, y-plane, the system is said to be *globally asymptotically stable*.

10.3. A test for instability

The general approach of the Liapunov direct method can be adapted to test instability of zero solutions.

THEOREM 10.3   Let $\mathbf{x}(t) = \mathbf{0}, t \geqslant t_0$, be the zero solution of the regular system $\dot{\mathbf{x}} = \mathbf{X}(\mathbf{x})$, where $\mathbf{X}(\mathbf{0}) = \mathbf{0}$. If there exists a function $U(\mathbf{x})$ such that in some neighbourhood $\|\mathbf{x}\| \leqslant k$

  (i) $U(\mathbf{x})$ and its partial derivatives are continuous;

  (ii) $U(\mathbf{0}) = 0$;

  (iii) $\dot{U}(\mathbf{x})$ is positive definite for the given system;

  (iv) in every neighbourhood of the origin there exists at least one point $\mathbf{x}$ at which $\mathrm{U}(\mathbf{x}) > 0$;

then the zero solution is unstable.

Proof   By (iv), given any $\delta, 0 < \delta < k$, there exists $\mathbf{x}_\delta$ such that $0 < \|\mathbf{x}_\delta\| < \delta$ and $U(\mathbf{x}_\delta) > 0$. Suppose that $\mathbf{x}(\delta, t), t \geqslant t_0$, is the solution satisfying $\mathbf{x}(\delta, t_0) = \mathbf{x}_\delta$. Its path cannot enter the origin as $t \to \infty$, because $U(\mathbf{0}) = 0$ (by (ii)) but $\dot{U}(\mathbf{x}) > 0$, $\mathbf{x} \neq \mathbf{0}$. Since $\dot{U}$ is positive definite by (iii), and continuous (by (i) and the regularity condition), and since the path is bounded away from the origin there exists a number $m > 0$ such that

$$\dot{U}\{\mathbf{x}(\delta, t)\} \geqslant m > 0, \quad t \geqslant t_0 \tag{10.4}$$

so long as

$$\|\mathbf{x}(\delta, t)\| \leqslant k. \tag{10.5}$$

Therefore

$$U\{x(\delta,t)\} - U\{x(\delta,t_0)\} = \int_{t_0}^{t} \dot{U}\{x(\delta,t)\}\,dt \geqslant m(t-t_0). \quad (10.6)$$

$U(x)$ is continuous by (i) and therefore bounded on $\|x\| \leqslant k$; but the right-hand side of (10.6) is unbounded. Therefore $x(\delta,t)$ cannot remain in $\|x\| \leqslant k$ and so the path reaches the boundary $\|x\| = k$.

Therefore, given any ε, $0 < \varepsilon < k$, then for any δ, however small, there is at least one solution $x(\delta,t)$ with $\|x(\delta,t_0)\| < \delta$ but $\|x(\delta,t)\| > \varepsilon$ for some t. Therefore the zero solution is unstable.

Typically (by (iv)), $U(x)$ will take both positive and negative values close to $x = 0$. In the plane, simple functions of the type xy, $x^2 - y^2$, may be successful, as in Example 10.4.

Example 10.4 *Show that* $x(t) = 0, t \geqslant t_0$, *is an unstable solution of the equation* $\ddot{x} - x + \dot{x}\sin x = 0$.

The equivalent system is $\dot{x} = y$, $\dot{y} = x - y\sin x$. Consider

$$U(x,y) = xy$$

satisfying (i), (ii), and (iv) for every neighbourhood. Then

$$\dot{U}(x,y) = x^2 + y^2 - xy\sin x.$$

It is easy to show that this has a minimum at the origin and since $\dot{U}(0,0) = 0$, $\dot{U}$ is positive definite. Therefore the zero solution is unstable. (To show by calculus methods that there is a minimum at the origin is sometimes the simplest way to identify such a function.)

10.4. Stability and the linear approximation in two dimensions

The specific question discussed is the following. For the two-dimensional system

$$\dot{x} = X(x) = Ax + h(x) \quad (10.7)$$

where A is constant and $h(x)$ is of smaller order of magnitude than Ax, does the instability or asymptotic stability of the zero solution of $\dot{x} = Ax$ imply the same property for the zero solutions of the system $\dot{x} = X(x)$? To answer this question we shall construct explicit Liapunov functions for the linearized system $\dot{x} = Ax$ and show that they also work for the original system.

Suppose that the linear approximate system

$$\dot{\mathbf{x}} = A\mathbf{x}, \quad \text{or} \quad \begin{pmatrix} \dot{x} \\ \dot{y} \end{pmatrix} = \begin{pmatrix} a & b \\ c & d \end{pmatrix} \begin{pmatrix} x \\ y \end{pmatrix}, \tag{10.8}$$

is asymptotically stable. We show that a strong Liapunov function $V(\mathbf{x})$ for this system can be constructed which is a quadratic form:

$$V(\mathbf{x}) = \mathbf{x}^t K \mathbf{x}. \tag{10.9}$$

It follows that

$$\dot{V}(\mathbf{x}) = \dot{\mathbf{x}}^t K \mathbf{x} + \mathbf{x}^t K \dot{\mathbf{x}} = \mathbf{x}^t A^t K \mathbf{x} + \mathbf{x}^t K A \mathbf{x}$$

$$= \mathbf{x}^t (A^t K + K A) \mathbf{x}. \tag{10.10}$$

We now arrange for $V(\mathbf{x})$ to be positive definite and $\dot{V}(\mathbf{x})$ negative definite: the second requirement is satisfied if, for example, we can choose K so that

$$A^t K + K A = -I, \tag{10.11}$$

for then $\dot{V}(\mathbf{x}) = -x^2 - y^2$.

In the notation of Chapter 2, let

$$p = a + d, \quad q = ad - bc. \tag{10.12}$$

We shall show that there is a solution of (10.11) of the form

$$K = m(A^t)^{-1} A^{-1} + nI, \tag{10.13}$$

where m and n are constants. The left of (10.11) becomes

$$A^t K + K A = m\{A^{-1} + (A^t)^{-1}\} + n(A + A^t).$$

Since

$$A = \begin{pmatrix} a & b \\ c & d \end{pmatrix}, \quad A^{-1} = \begin{pmatrix} d & -b \\ -c & a \end{pmatrix} \Big/ q,$$

then by (10.11) we require

$$A^t K + K A = \frac{1}{q} \begin{pmatrix} 2md + 2naq & (nq - m)(b + c) \\ (nq - m)(b + c) & 2ma + 2ndq \end{pmatrix} = \begin{pmatrix} -1 & 0 \\ 0 & -1 \end{pmatrix}.$$

This equation is satisfied if

$$m = -q/2p, \quad n = -1/2p.$$

Then from (10.13)

$$K = -\frac{1}{2pq} \begin{pmatrix} c^2 + d^2 + q & -ac - bd \\ -ac - bd & a^2 + b^2 + q \end{pmatrix}. \tag{10.14}$$

The associated Liapunov function is given by (10.9):

$$V(\mathbf{x}) = -\{(c^2 + d^2 + q)x^2 - 2(ac + bd)xy + (a^2 + b^2 + q)y^2\}/2pq$$
$$= -\{(dx - by)^2 + (cx - ay)^2 + q(x^2 + y^2)\}/2pq, \qquad (10.15)$$

which is positive definite at least when

$$p < 0, \quad q > 0. \qquad (10.16)$$

Thus V is a strong Liapunov function for the *linearized* system (10.8) when (10.16) is satisfied. Now, the stability regions for (10.8) are already known, and displayed in Fig. 2.8, which is partly reproduced below as Fig. 10.4. We see that (10.16) in fact exhausts the regions I of asymptotic

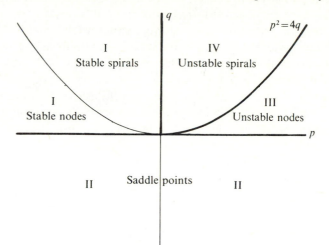

FIG. 10.4. Stability regions in the p, q plane for $\dot{x} = ax + by$, $\dot{y} = cx + dy$, $p = a + d$, $q = ad - bc$ (see Fig. 2.8)

stability of eqn (10.8). Therefore Theorem 10.4 may be given as follows.

THEOREM 10.4 *Let* $(0, 0)$ *be an equilibrium point of the regular system*

$$\begin{pmatrix} \dot{x} \\ \dot{y} \end{pmatrix} = \begin{pmatrix} a & b \\ c & d \end{pmatrix} \begin{pmatrix} x \\ y \end{pmatrix} + \begin{pmatrix} h_1(x, y) \\ h_2(x, y) \end{pmatrix}, \qquad (10.17)$$

where

$$h_1(x, y) = O(x^2 + y^2), \quad h_2(x, y) = O(x^2 + y^2) \qquad (10.18)$$

as $x^2 + y^2 \to 0$. *Then the zero solution of* (10.17) *is asymptotically stable when its linear approximation is asymptotically stable.*

Proof   We shall show that V given by (10.15) is also a strong Liapunov function for the system (10.17). It is positive definite when $p < 0, q > 0$; that is, whenever the linearized system is asymptotically stable.

Also, for the given system (10.17),

$$\dot{V} = \mathbf{x}^t(A^tK + KA)\mathbf{x} + \mathbf{h}^t(\mathbf{x})K\mathbf{x} + \mathbf{x}^tK\mathbf{h}(\mathbf{x})$$
$$= -\mathbf{x}^tI\mathbf{x} + 2\mathbf{h}^t(\mathbf{x})K\mathbf{x}$$
$$= -(x^2 + y^2) + \sum_{i,j=1}^{2} K_{ij}h_i(\mathbf{x})x_i, \qquad (10.19)$$

where the symmetry of K has been used.

By (10.18), for any p, q, there is clearly a neighbourhood of the origin in which the first term of (10.19) predominates, that is to say, where $\dot{V}$ is negative definite. Therefore, V is a strong Liapunov function for (10.17) in the parameter region $p < 0, q > 0$; that is, when the linearized system is asymptotically stable. The result follows by Theorem 10.2.

The open regions II, III, IV in Fig. 10.4 are, we know, where the solutions of $\dot{\mathbf{x}} = A\mathbf{x}$ are *unstable*. To prove a statement in similar terms to the last theorem, only relating to the *instability* of the zero solution of $\dot{\mathbf{x}} = A\mathbf{x} + \mathbf{h}(\mathbf{x})$, we require a function $U(\mathbf{x})$ with the properties stated in Theorem 10.3. There are two principal cases according to the eigenvalues of A, given as (i) and (ii) below:

(i) *The eigenvalues of A are real and different* (II *and* III *in Fig.* 10.4)

Since we hope that the stability of $\dot{\mathbf{x}} = A\mathbf{x}$ will determine that of $\dot{\mathbf{x}} = A\mathbf{x} + \mathbf{h}(\mathbf{x})$, we are interested in the case where the solutions of $\dot{\mathbf{x}} = A\mathbf{x}$ are *unstable*: then (Theorem 8.11) *at least one eigenvalue of A is positive*. This covers the regions II and III of Fig. 10.4. (We will not consider the case when the eigenvalues are equal: the line $p^2 = 4q$ in Fig. 10.4; nor when one of the eigenvalues is zero: $q = 0$.) In this case we can reduce the problem to a simpler one by a transformation

$$\mathbf{x} = C\mathbf{u}, \qquad (10.20)$$

where C is nonsingular, and $\mathbf{u} = (u_1, u_2)^t$; whence the equation satisfied by $\mathbf{u}$ is

$$\dot{\mathbf{u}} = C^{-1}AC\mathbf{u}. \qquad (10.21)$$

It follows directly from Definition 8.1 that when $\mathbf{x}(t)$ is a stable solution so is $\mathbf{u}(t)$, and conversely (since $\mathbf{u} = C^{-1}\mathbf{x}$). Therefore the same holds for instability. In the cases being considered the eigenvalues of A are real and different, and it is known that C can then be chosen so that

$$C^{-1}AC = \begin{pmatrix} \lambda_1 & 0 \\ 0 & \lambda_2 \end{pmatrix} = D, \text{ say,} \qquad (10.22)$$

where the columns of C are eigenvectors corresponding to λ_1, λ_2. Suitable C are given by

$$C = \begin{pmatrix} -b & -b \\ a-\lambda_1 & a-\lambda_2 \end{pmatrix}, \quad b \neq 0;$$

$$C = \begin{pmatrix} -c & -c \\ d-\lambda_1 & d-\lambda_2 \end{pmatrix}, \quad b = 0, \quad c \neq 0,$$

and if $b = c = 0$, the equations are already in a suitable form. Then

$$\dot{\mathbf{u}} = D\mathbf{u}, \qquad (10.23)$$

and we wish to establish the instability of the zero solution of this system.

Consider U defined by

$$U(\mathbf{u}) = \mathbf{u}^t D^{-1} \mathbf{u} = u_1^2/\lambda_1 + u_2^2/\lambda_2. \qquad (10.24)$$

Then $U(\mathbf{u}) > 0$ at some point in every neighbourhood of $\mathbf{u} = \mathbf{0}$ (since instability requires λ_1 or λ_2 positive). $U(\mathbf{u})$ therefore satisfies conditions (i), (ii), and (iv) of Theorem 10.3 in the open regions II and III of Fig. 10.4. Also

$$\dot{U}(\mathbf{u}) = \dot{\mathbf{u}}^t D^{-1} \mathbf{u} + \mathbf{u}^t D^{-1} \dot{\mathbf{u}} = \mathbf{u}^t D D^{-1} \mathbf{u} + \mathbf{u}^t D^{-1} D \mathbf{u} = 2(u_1^2 + u_2^2),$$
$$(10.25)$$

which is positive definite, so that Theorem 10.3 is satisfied.

(ii) *The eigenvalues of A are conjugate complex, with positive real part (region IV of Fig. 10.4)*

Write

$$\lambda_1 = \alpha + i\beta, \quad \lambda_2 = \alpha - i\beta.$$

The diagonalization process above gives $\dot{\mathbf{u}} = D\mathbf{u}$ where D and $\mathbf{u}$ are complex. Since the theorems refer to real functions, these cannot immediately be used. However, instead of diagonalizing, we can reduce A by a matrix G so that

$$G^{-1}AG = \begin{pmatrix} \alpha & -\beta \\ \beta & \alpha \end{pmatrix}, \qquad (10.26)$$

say: an appropriate G is given by

$$G = C\begin{pmatrix} 1 & i \\ 1 & -i \end{pmatrix}$$

where C is as before. The system becomes

$$\dot{\mathbf{u}} = \begin{pmatrix} \alpha & -\beta \\ \beta & \alpha \end{pmatrix}\mathbf{u}, \tag{10.27}$$

with

$$\mathbf{x} = G\mathbf{u}. \tag{10.28}$$

Suppose we define $U(\mathbf{u})$ by

$$U(\mathbf{u}) = \mathbf{u}^t\mathbf{u}. \tag{10.29}$$

This satisfies (i), (ii), and (iv) of Theorem 10.3 in region (IV) of Fig. 10.4. Also

$$\dot{U}(\mathbf{u}) = 2\alpha(u_1^2 + u_2^2) \tag{10.30}$$

as in case (i), so that $\dot{U}$ is positive definite.

By using the prior knowledge of Chapter 2 as to the regions in which the system is unstable, we have therefore determined that wherever $\dot{\mathbf{x}} = A\mathbf{x}$ is unstable, U given by either (10.24) or (10.29) as appropriate, is a function of the type required for (i), (ii), and (iv) of Theorem 10.2. Theorem 10.5 shows that $\dot{U}$ has the final property (iii) for the nonlinear system $\dot{\mathbf{x}} = A\mathbf{x} + \mathbf{h}(\mathbf{x})$.

THEOREM 10.5   *Let $(0,0)$ be an equilibrium point of the regular two-dimensional system*

$$\dot{\mathbf{x}} = A\mathbf{x} + \mathbf{h}(\mathbf{x}) \tag{10.31}$$

where

$$\mathbf{h}(\mathbf{x}) = \begin{pmatrix} h_1(\mathbf{x}) \\ h_2(\mathbf{x}) \end{pmatrix},$$

and $h_1(\mathbf{x}), h_2(\mathbf{x}) = \mathrm{O}(|\mathbf{x}|^2)$ as $|\mathbf{x}| \to 0$. Then when the eigenvalues of A are different, non-zero, and at least one has positive real part, the zero solution of (10.31) is unstable.

Proof   We shall assume that the eigenvalues of A are real and that one of them is positive, the other case (λ complex with positive real part) being closely similar. Assume that equation (10.20) is used to reduce (10.31) to the form

$$\dot{\mathbf{u}} = D\mathbf{u} + \mathbf{g}(\mathbf{u}); \tag{10.32}$$

then if $\mathbf{u}(t) = \mathbf{0}$ is unstable, so is $\mathbf{x}(t) = \mathbf{0}$, and conversely. Also

$$g_1(\mathbf{u}), g_2(\mathbf{u}) = O(x^2 + y^2), \quad x^2 + y^2 \to 0. \tag{10.33}$$

It is sufficient to display a function $U(\mathbf{u})$ for the reduced system (10.32). U given by (10.24) satisfies conditions (i), (ii), and (iv) of Theorem 10.2. Also

$$\begin{aligned}
\dot{U}(\mathbf{u}) &= \dot{\mathbf{u}}^t D^{-1} \mathbf{u} + \mathbf{u}^t D^{-1} \mathbf{u} \\
&= \mathbf{u}^t D D^{-1} \mathbf{u} + \mathbf{u}^t D^{-1} D \mathbf{u} + \mathbf{g}^t D^{-1} \mathbf{u} + \mathbf{u}^t D^{-1} \mathbf{g} \\
&= 2\{u_1^2 + u_2^2 + \lambda_1 g_1(u) + \lambda_2 g_2(u)\}.
\end{aligned}$$

From (10.33), it is clear that $\dot{U}$ is positive definite in a small enough neighbourhood of the origin, satisfying (iii) of Theorem 10.3. Therefore, by Theorem 10.3, the zero solution $\mathbf{u}(t) = \mathbf{0}$ is unstable in regions II and III; thus $\mathbf{x}(t) = \mathbf{0}$ is unstable.

The calculation for region IV of Fig. 10.4 is similar, using U given by (10.29).

Example 10.5 *Show that the zero solution of van der Pol's equation*

$$\ddot{x} + e(x^2 - 1)\dot{x} + x = 0$$

is uniformly and asymptotically stable when $e < 0$ and unstable when $e > 0$. Construct a Liapunov function for the stable case.

The equivalent system is $\dot{x} = y, \dot{y} = -x + ey - ex^2 y$. The linearized system is $\dot{x} = y, \dot{y} = -x + ey$. The eigenvalues of

$$A = \begin{pmatrix} 0 & 1 \\ -1 & e \end{pmatrix}$$

are given by $\lambda_1, \lambda_2 = \frac{1}{2}(e \pm \sqrt{(e^2 - 4)})$.

Suppose $e < 0$. Then the linearized system is asymptotically stable. Also $\mathbf{h}^t(\mathbf{x}) = (0, -ex^2)$. Therefore, by Theorem 10.4, the zero solution of the given equation is asymptotically stable.

By (10.9), (10.14), $V = (x, y)K(x, y)^t$ where

$$K = -\frac{1}{2e} \begin{pmatrix} 2 + e^2 & -e \\ -e & 2 \end{pmatrix}.$$

Hence

$$V(x, y) = -\{(2 + e^2)x^2 - 2exy + 2y^2\}/2e.$$

It can be confirmed that $\dot{V} = -x^2 - y^2 - ex^3y + 2x^2y^2$, which is negative definite near the origin. The topographic family

$$(1 + \tfrac{1}{2}e^2)x^2 - exy + y^2 = \text{constant}$$

would be preferable to the simpler family of circles $x^2 + y^2 = constant$, which lead to a $\dot{V}$ which is only semi-definite.

Suppose $e > 0$. The linearized system is unstable and Theorem 10.5 applies.

Example 10.6 *Investigate the stability of the equilibrium points of*

$$\dot{x} = y(x+1), \quad \dot{y} = x(1 + y^3).$$

There are two equilibrium points, at $(0,0)$ and $(-1, -1)$. Near the origin the linear approximation is $\dot{x} = y, \dot{y} = x$, with eigenvalues $\lambda = \pm 1$. Hence, the linear approximation has a saddle point at $(0,0)$. By Theorem 10.5, the zero solution of the given equation is also unstable.

Near the point $(-1, -1)$ put $x = -1 + \xi, y = -1 + \eta$. Then

$$\dot{\xi} = \xi(\eta - 1), \quad \dot{\eta} = (\xi - 1)\{1 + (\eta - 1)^3\}.$$

The linear approximation is $\dot{\xi} = -\xi, \dot{\eta} = -3\eta$. The eigenvalues are $\lambda = -1$, -3, both negative. By Theorem 10.4 the solution $\xi(t) = \eta(t) = 0$ is therefore asymptotically stable; so is the solution $x(t) = y(t) = -1$ of the original solution.

Figure 10.5 shows the computed phase diagram. The shaded region shows the *domain of attraction* of the point $(-1, -1)$.

The foregoing theorems refer to cases where the linearized system is asymptotically stable, or is unstable, and make no prediction as to the effect of modifying, by nonlinear terms, a case of mere stability. In this case the effect may go either way, as shown by Example 10.7.

Example 10.7 *Investigate the stability of the following systems*

(i) $\dot{x} = y - x(x^2 + y^2), \quad \dot{y} = -x - y(x^2 + y^2)$;

(ii) $\dot{x} = y + x(x^2 + y^2), \quad \dot{y} = -x + y(x^2 + y^2)$.

The linear approximation shows a centre in both cases.

(i) Let $V = x^2 + y^2$, positive definite. Then

$$\dot{V} = -2(x^2 + y^2)^2$$

which is negative definite and the zero solution is asymptotically stable, by Theorem 10.2.

(ii) Let $U = x^2 + y^2$ (positive in every neighbourhood of the origin). Then

$$\dot{U} = 2(x^2 + y^2)^2$$

which is positive definite. Theorem 10.3 predicts instability.

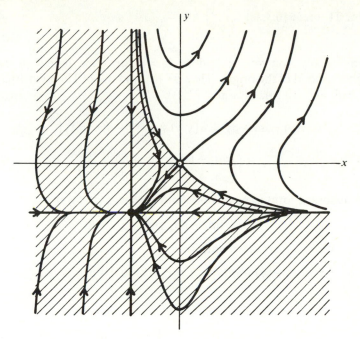

FIG. 10.5. Example 10.6: domain of attraction of $(-1, -1)$ shaded.

The extension of the theory to n dimensions $(n > 2)$ requires the solution of eqn (10.11) in n dimensions. For this, other techniques are needed: details are given by Barnett (1975).

10.5. Special systems

Quadratic systems. Consider the strongly nonlinear system whose lowest-order terms are of the second degree:

$$\dot{x} = X(x, y) \approx a_1 x^2 + 2b_1 xy + c_1 y^2, \tag{10.34}$$

$$\dot{y} = Y(x, y) \approx a_2 x^2 + 2b_2 xy + c_2 y^2. \tag{10.35}$$

Neglecting any higher-order terms, the equations can be written

$$\dot{x} = (x, y)A_1 \begin{pmatrix} x \\ y \end{pmatrix}, \quad \dot{y} = (x, y)A_2 \begin{pmatrix} x \\ y \end{pmatrix}, \tag{10.36}$$

where

$$A_1 = \begin{pmatrix} a_1 & b_1 \\ b_1 & c_1 \end{pmatrix}, \quad A_2 = \begin{pmatrix} a_2 & b_2 \\ b_2 & c_2 \end{pmatrix}.$$

For Theorem 10.3, let

$$U(x, y) = \alpha x + \beta y. \tag{10.37}$$

Then for any non-zero α and β, $U > 0$ for some points in every neighbourhood of the origin. Thus (i), (ii), and (iv) of Theorem 10.3 are satisfied. For (10.36), $\dot{U}$ is given by

$$\dot{U}(x, y) = \alpha \dot{x} + \beta \dot{y} = \alpha(x, y) A_1 \begin{pmatrix} x \\ y \end{pmatrix} + \beta(x, y) A_2 \begin{pmatrix} x \\ y \end{pmatrix}$$

$$= (x, y) \begin{pmatrix} \alpha a_1 + \beta a_2 & \alpha b_1 + \beta b_2 \\ \alpha b_1 + \beta b_2 & \alpha c_1 + \beta c_2 \end{pmatrix} \begin{pmatrix} x \\ y \end{pmatrix}.$$

This quadratic form is positive definite (see eqn (10.5)) if

$$\alpha a_1 + \beta a_2 > 0; \tag{10.38}$$

and, with some rearrangement, if

$$\Delta(\alpha, \beta) = \alpha^2 (a_1 c_1 - b_1^2) + \alpha \beta (a_1 c_2 + a_2 c_1 - 2b_1 b_2) + \beta^2 (a_2 c_2 - b_2^2) > 0,$$

or

$$\Delta(\alpha, \beta) = (\alpha, \beta) \begin{pmatrix} a_1 c_1 - b_1^2 & \frac{1}{2}(a_1 c_2 + a_2 c_1 - 2b_1 b_2) \\ \frac{1}{2}(a_1 c_2 + a_2 c_1 - 2b_1 b_2) & a_2 c_2 - b_2^2 \end{pmatrix} \begin{pmatrix} \alpha \\ \beta \end{pmatrix} > 0.$$

$$\tag{10.39}$$

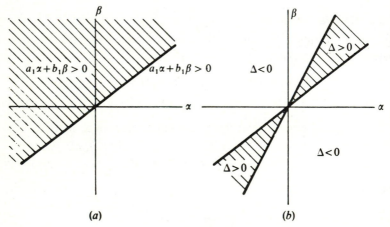

FIG. 10.6.

Consider whether α, β exist satisfying (10.38) and (10.39). On a plane for α, β, $a_1\alpha + b_1\beta = 0$ is a line, and (10.38) is represented by a half-plane, like that shown in Fig. 10.6(*a*). Now $\Delta(\alpha, \beta)$ is a quadratic form and is either positive definite, positive on a ray or in a sector as in Fig. 10.6(*b*), or negative definite. Unless $\Delta(\alpha, \beta)$ is negative definite, or unless it is negative except on the line $a_1\alpha + b_1\beta = 0$ where it is zero, there exist *some* values of α, β making $U(\mathbf{x})$ positive definite and therefore satisfying the final conditions of Theorem 10.3. The zero solution of (10.34) and (10.35) is therefore unstable, except possibly in the particular cases, depending on exact relations between the coefficients, mentioned above. We shall not investigate these cases.

Hamiltonian problems. Conservative problems, particularly in dynamics, can be expressed in the form

$$\dot{p}_i = -\frac{\partial \mathcal{H}}{\partial q_i}, \quad \dot{q}_i = \frac{\partial \mathcal{H}}{\partial p_i}, \quad i = 1, 2, \ldots n, \tag{10.40}$$

where $\mathcal{H}$ is a given function called the Hamiltonian of the system, q_i is a generalized coordinate, and p_i a generalized momentum. The Hamiltonian is defined by

$$\mathcal{H}(\mathbf{p}, \mathbf{q}) = \mathcal{T}(\mathbf{p}, \mathbf{q}) + \mathcal{V}(\mathbf{q}),$$

where $\mathcal{T}$ is the kinetic energy and $\mathcal{V}$ the potential energy: it is assumed that $\mathcal{V}(\mathbf{0}) = 0$. $\mathcal{T}$ is a positive definite quadratic form in p_i, so $\mathcal{H}(\mathbf{0}, \mathbf{0}) = 0$.

Suppose that $\mathbf{q} = \mathbf{0}$ is a *minimum* of $\mathcal{V}(\mathbf{q})$ so that $\mathcal{V}$, and hence $\mathcal{H}$, is positive definite in a neighbourhood of the origin. Then

$$\dot{\mathcal{H}}(\mathbf{p}, \mathbf{q}) = \sum_{i=1}^{n} \frac{\partial \mathcal{H}}{\partial p_i} \dot{p}_i + \sum_{i=1}^{n} \frac{\partial \mathcal{H}}{\partial q_i} \dot{q}_i = 0$$

by (10.40). Thus, $\mathcal{H}$ is a weak Liapunov function for (10.40). Therefore, by Theorem 10.2, *the zero solution* $\mathbf{p} = \mathbf{0}, \mathbf{q} = \mathbf{0}$, *a position of equilibrium, is stable when it is at a minimum of* $\mathcal{V}$.

Now, suppose the origin is a *maximum* of $\mathcal{V}$, and that $\mathcal{V}$ has the expansion

$$\mathcal{V}(\mathbf{q}) = P_N(\mathbf{q}) + P_{N+1}(\mathbf{q}) + \ldots, \quad N \geqslant 2$$

where $P_N(q)$ is a homogeneous polynomial of degree $N \geqslant 1$ in the q_i. Consider a function $U(\mathbf{p}, \mathbf{q})$, of the type in Theorem 10.3, defined by

$$U(\mathbf{p}, \mathbf{q}) = \sum_{i=1}^{n} p_i q_i.$$

Then U satisfies conditions (i), (ii), and (iv) of Theorem 10.3. Also

$$
\begin{aligned}
\dot{U}(\mathbf{p}, \mathbf{q}) &= \sum_{i=1}^{n} p_i \dot{q}_i + \sum_{i=1}^{n} \dot{p}_i q_i \\
&= \sum_{i=1}^{n} p_i \frac{\partial \mathscr{H}}{\partial p_i} - \sum_{i=1}^{n} q_i \frac{\partial \mathscr{H}}{\partial q_i} \\
&= \sum_{i=1}^{n} p_i \frac{\partial \mathscr{T}}{\partial p_i} - \sum_{i=1}^{n} q_i \frac{\partial \mathscr{T}}{\partial q_i} - \sum_{i=1}^{n} q_i \frac{\partial \mathscr{V}}{\partial q_i} \\
&= 2\mathscr{T} - \sum_{i=1}^{n} \frac{\partial \mathscr{T}}{\partial q_i} - N P_N(\mathbf{q}) - (N+1) P_{N+1}(\mathbf{q}) + \dots,
\end{aligned}
$$

where Euler's theorem on homogeneous functions has been applied to $\mathscr{T}, P_N, P_{N+1}, \dots$. The dominating terms in the series near $\mathbf{p} = \mathbf{q} = \mathbf{0}$ are $2\mathscr{T} - N P_N(\mathbf{q})$. Since, by hypothesis, P_N has a maximum at $\mathbf{q} = \mathbf{0}$ and $\mathscr{T}$ is positive definite, $\dot{U}$ is negative definite in a neighbourhood of the origin. Therefore, by Theorem 10.3, *the equilibrium is unstable when it is at a maximum of $\mathscr{V}$.*

The Liénard equation. Consider the equation

$$\ddot{x} + f(x)\dot{x} + g(x) = 0, \tag{10.41}$$

or the equivalent system

$$\dot{x} = y - F(x), \quad \dot{y} = -g(x); \tag{10.42}$$

where

$$F(x) = \int_{0}^{x} f(u)\,\mathrm{d}u.$$

Suppose that f and g are continuous and that
 (i) $f(x)$ is positive in a deleted neighbourhood of the origin;
(ii) $g(x)$ is positive/negative when x is positive/negative (implying $g(0) = 0$).
   Now let

$$G(x) = \int_{0}^{x} g(u)\,\mathrm{d}u, \tag{10.43}$$

and consider the function

$$V(x, y) = G(x) + \tfrac{1}{2}y^2,$$

as a possible strong Liapunov function. It is clearly positive definite. Also by (10.42)

$$\dot{V}(x, y) = g(x)\dot{x} + y\dot{y} = -g(x)F(x).$$

This is negative definite. The zero solution is therefore uniformly and asymptotically stable.

Exercises

1. Find a simple U or V function (Theorems 10.1 and 10.3) to establish the stability or instability respectively of the zero solution of the following equations:

(i) $\dot{x} = -x + y - xy^2$, $\dot{y} = -2x - y - x^2 y$;

(ii) $\dot{x} = y^3 + x^2 y$, $\dot{y} = x^3 - xy^2$;

(iii) $\dot{x} = 2x + y + xy$, $\dot{y} = x - 2y + x^2 + y^2$;

(iv) $\dot{x} = -x^3 + y^4$, $\dot{y} = -y^3 + y^4$;

(v) $\dot{x} = \sin y$, $\dot{y} = -2x - 3y$;

(vi) $\dot{x} = x + e^{-y} - 1$, $\dot{y} = x$;

(vii) $\dot{x} = e^x - \cos y$, $\dot{y} = x$;

(viii) $\dot{x} = \sin(y + x)$, $\dot{y} = -\sin(y - x)$;

(ix) $\ddot{x} = x^3$;

(x) $\dot{x} = x + 4y$, $\dot{y} = -2x - 5y$;

(xi) $\dot{x} = -x + 6y$, $\dot{y} = 4x + y$.

2. Show that α may be chosen so that $V = x^2 + \alpha y^2$ is a strong Liapunov function for the system

$$\dot{x} = y - \sin^3 x, \quad \dot{y} = -4x - \sin^3 y.$$

3. Find domains of asymptotic stability for the following systems, using $V = x^2 + y^2$:

(i) $\dot{x} = -\frac{1}{2}x(1 - y^2)$, $\dot{y} = -\frac{1}{2}y(1 - x^2)$;

(ii) $\dot{x} = y - x(1 - x)$, $\dot{y} = -x$.

4. Find a strong Liapunov function at $(0, 0)$ for the system

$$\dot{x} = x(y - b), \quad \dot{y} = y(x - a)$$

and confirm that all solutions starting in the domain $(x/a)^2 + (y/b)^2 < 1$ approach the origin.

5. Show that the origin of the system

$$\dot{x} = xP(x, y), \quad \dot{y} = yQ(x, y)$$

is asymptotically stable when $P(x, y) < 0$, $Q(x, y) < 0$ in a neighbourhood of the origin.

6. Prove that the equations
   (i) $\ddot{x} - \dot{x}^2 \operatorname{sgn}(\dot{x}) + x = 0$;
   (ii) $\dot{x} = y + xy^2$, $\dot{y} = -x + x^2 y$;
   have an unstable zero solution.

7. Investigate the stability of the zero solution of

$$\dot{x} = x^2 - y^2, \quad \dot{y} = -2xy$$

by using the function $U(x, y) = \alpha x y^2 + \beta x^3$ for suitable constants α and β (Section 10.3).

8. Show that the origin of the system

$$\dot{x} = -y - x\sqrt{(x^2 + y^2)}, \quad \dot{y} = x - y\sqrt{(x^2 + y^2)}$$

is a centre in the linear approximation, but in fact is a stable spiral. Find a Liapunov function for the zero solution.

9. Euler's equations for a body spinning freely about a fixed point under no forces are

$$A\dot{\omega}_1 - (B - C)\omega_2\omega_3 = 0,$$
$$B\dot{\omega}_2 - (C - A)\omega_3\omega_1 = 0,$$
$$C\dot{\omega}_3 - (A - B)\omega_1\omega_2 = 0,$$

where A, B, and C are the principal moments of inertia, and $(\omega_1, \omega_2, \omega_3)$ is the spin of the body in principal axes fixed in the body. Find all the states of steady spin of the body.

Consider perturbations about the steady state $(\omega_0, 0, 0)$ by putting $\omega_1 = \omega_0 + x_1, \omega_2 = x_2, \omega_3 = x_3$, and show that the linear approximation is

$$\dot{x}_1 = 0, \quad \dot{x}_2 = \{(C - A)/B\}\omega_0 x_3, \quad \dot{x}_3 = \{(A - B)/C\}\omega_0 x_2.$$

Deduce that this state is unstable if $C < A < B$ or $B < A < C$.
  Show that

$$V = \{B(A - B)x_2^2 + C(A - C)x_3^2\} + \{Bx_2^2 + Cx_3^2 + A(x_1^2 + 2\omega_0 x_1)\}^2$$

is a Liapunov function for the case when A is the largest moment of inertia, so that this state is stable. Suggest a Liapunov function which will establish the stability of the case in which A is the smallest moment of inertia. Are these states asymptotically stable?
  Why would you expect V as given above to be a first integral of the Euler equations? Show that each of the terms in braces is such an integral.

10. Show that the zero solution of the equation

$$\ddot{x} + h(x, \dot{x})\dot{x} + x = 0$$

is stable if $h(x, y) \geqslant 0$ in a neighbourhood of the origin.

11. The n-dimensional system

$$\dot{\mathbf{x}} = \mathbf{grad}\ W(\mathbf{x})$$

has an isolated equilibrium point at $\mathbf{x} = \mathbf{0}$. Show that the zero solution is asymptotically stable if W has a maximum at $\mathbf{x} = \mathbf{0}$. Give a condition for instability of the zero solution.

12. A particle of mass m and position vector $\mathbf{r} = (x, y, z)$ moves in a potential field $W(x, y, z)$, so that its equation of motion is

$$m\ddot{\mathbf{r}} = -\mathbf{grad}\ W.$$

By putting $\dot{x} = u$, $\dot{y} = v$, $\dot{z} = w$, express this in terms of first-order derivatives. Suppose that W has a minimum at $\mathbf{r} = \mathbf{0}$. Show that the origin of the system is stable, by using the Liapunov function

$$V = W + \tfrac{1}{2}m(u^2 + v^2 + w^2).$$

What do the level curves of V represent physically? Is the origin asymptotically stable?

An additional non-conservative force $\mathbf{f}(u, v, w)$ is introduced, so that

$$m\ddot{\mathbf{r}} = -\mathbf{grad}\ W + \mathbf{f}.$$

Use the same Liapunov function to give a sufficient condition for $\mathbf{f}$ to be of frictional type.

13. Use the test for instability to show that if

$$\dot{x} = X(x, y), \quad \dot{y} = Y(x, y)$$

has an equilibrium point at the origin, then the zero solution is unstable if there exist constants α and β such that

$$\alpha X(x, y) + \beta Y(x, y) > 0$$

in a deleted neighbourhood of the origin.

14. Use the result of Exercise 13 to show that the origin is unstable for each of the following:
   (i) $\dot{x} = x^2 + y^2$, $\dot{y} = x + y$;
   (ii) $\dot{x} = y \sin y$, $\dot{y} = xy + x^2$;
   (iii) $\dot{x} = y^{2m}$, $\dot{y} = x^{2n}$ (m, n positive integers).

15. For the system

$$\dot{x} = y, \quad \dot{y} = f(x, y),$$

where $f(0,0) = 0$, show that V given by

$$V(x, y) = \tfrac{1}{2}y^2 - \int_0^x f(u, 0)\, du$$

is a weak Liapunov function for the zero solution when

$$\{f(x, y) - f(x, 0)\}y \leqslant 0, \quad \int_0^x f(u, 0)\, du < 0,$$

in a neighbourhood of the origin.

16. Use the result of Exercise 15 to show the stability of the zero solutions of the following:
(i) $\ddot{x} = -x^3 - x^2\dot{x}$;
(ii) $\ddot{x} = -x^3/(1 - x\dot{x})$;
(iii) $\ddot{x} = -x + x^3 - x^2\dot{x}$.

17. Let

$$\dot{x} = -\alpha x + \beta f(y), \quad \dot{y} = \gamma x - \delta f(y),$$

where $f(0) = 0$, $yf(y) > 0$ ($y \neq 0$), and $\beta\gamma - \alpha\delta > 0$, where $\alpha, \beta, \gamma, \delta$ are positive. Show that, for suitable values of A and B,

$$V = \tfrac{1}{2}Ax^2 + B\int_0^y f(u)\, du$$

is a strong Liapunov function for the zero solutions (and hence that these are asymptotically stable).

18. A particle moving under a central attractive force $f(r)$ per unit mass has the equations of motion

$$\ddot{r} - r\dot{\theta}^2 = f(r), \quad \frac{d}{dt}(r^2\dot{\theta}) = 0.$$

For a circular orbit, $r = a$, show that $r^2\dot{\theta} = h$, a constant, and $h^2 + a^3 f(a) = 0$. The orbit is subjected to a small radial perturbation $r = a + \rho$, in which h is kept constant. Show that the variational equation for ρ is

$$\ddot{\rho} - \frac{h^2}{(a+\rho)^3} - f(a+\rho) = 0.$$

Show that

$$V(\rho, \dot{\rho}) = \tfrac{1}{2}\dot{\rho}^2 + \frac{h^2}{2(a+\rho)^2} - \int_0^\rho f(a+u)\, du - \frac{h^2}{2a^2}$$

is a Liapunov function for the zero solution of this equation provided that $3h^2 > a^4 f'(a)$, and that the gravitational orbit is stable in this sense.

19. Show that
$$V(x, y) = \max(|X|, |Y|)$$
is a Liapunov function for the zero solution of the system $\dot{x} = X(x, y)$, $\dot{y} = Y(x, y)$, provided that $X_x < 0$, $Y_y < 0$, $|X_x| > |X_y|$, and $|Y_y| > |Y_x|$ in a neighbourhood of the origin.

20. Show that the following Liénard-type equations have zero solutions which are asymptotically stable:
   (i) $\ddot{x} + |x|(\dot{x} + x) = 0$;
   (ii) $\ddot{x} + \dfrac{\sin x}{x}\dot{x} + x^3 = 0$;
   (iii) $\dot{x} = y - x^3$, $\dot{y} = -x^3$.

21. Give a geometrical account of Theorem 10.3.

22. For the system
$$\dot{x} = f(x) + \beta y, \quad \dot{y} = \gamma x + \delta y, \quad (f(0) = 0),$$
establish that V given by
$$V(x, y) = (\delta x - \beta y)^2 + 2\delta \int_0^x f(u)\,du - \beta\gamma x^2$$
is a strong Liapunov function for the zero solution when, in some neighbourhood of the origin,
$$\delta\frac{f(x)}{x} - \beta\gamma > 0, \quad \frac{f(x)}{x} + \delta < 0$$
for $x \neq 0$. (Barbashin 1970.)
   Deduce that for initial conditions in the circle $x^2 + y^2 < 1$, the solutions of the system
$$\dot{x} = -x^3 + x^4 + y, \quad \dot{y} = -x,$$
tend to zero.

23. For the system
$$\dot{x} = f(x) + \beta y, \quad \dot{y} = g(x) + \delta y, \quad f(0) = g(0) = 0,$$
show that V given by
$$V(x, y) = (\delta x - \beta y)^2 + 2\int_0^x \{\delta f(u) - \beta g(u)\}\,du$$

is a strong Liapunov function for the zero solution when, in some neighbourhood of the origin,

$$\{\delta f(x) - \beta g(x)\}x > 0, \quad xf(x) + \delta x^2 < 0$$

for $x \neq 0$. (Barbashin 1970.)

Deduce that the zero solution of the system

$$\dot{x} = -x^3 + 2x^4 + y, \quad \dot{y} = -x^4 - y$$

is asymptotically stable. Show how to find a domain of initial conditions from which the solutions tend to the origin.

24. Consider van der Pol's equation, $\ddot{x} + \varepsilon(x^2 - 1)\dot{x} + x = 0$, for $\varepsilon < 0$, in the Liénard phase plane, eqn (10.42):

$$\dot{x} = y - \varepsilon(\tfrac{1}{3}x^3 - x), \quad \dot{y} = -x.$$

Show that, in this plane, $V = \tfrac{1}{2}(x^2 + y^2)$ is a strong Liapunov function for the zero solution, which is therefore asymptotically stable. Show that all solutions starting from initial conditions inside the circle $x^2 + y^2 = 3$ tend to the origin (and hence the limit cycle lies outside this region for every $\varepsilon < 0$). Sketch this 'domain of asymptotic stability' in the ordinary phase plane with $\dot{x} = y$.

25. Give an intuitive argument, based on the general reasoning of Theorem 10.2, to show that the system is globally asymptotically stable if, additionally, $V \to \infty$ as $\|x\| \to \infty$.

Show that the system

$$\dot{x} = -xy^2, \quad \dot{y} = -x^2 y$$

is globally asymptotically stable, by guessing a suitable Liapunov function.

26. Assuming that the conditions of Exercise 22 are satisfied, obtain further conditions which ensure that the system is globally asymptotically stable.

Show that the system

$$\dot{x} = y - x^3, \quad \dot{y} = -x - y$$

is globally asymptotically stable.

27. Assuming that the conditions of Exercise 23 are satisfied, obtain further conditions which ensure that the system is globally asymptotically stable.

Show that the system

$$\dot{x} = -x^3 - x + y, \quad \dot{y} = -x^3 - y$$

is globally asymptotically stable.

28. Give conditions on the functions f and g of the Liénard equation, $\ddot{x} + f(x)\dot{x} + g(x) = 0$ which ensure that the corresponding system $\dot{x} = y - F(x), \dot{y} = -g(x)$ (Section 10.5) is globally asymptotically stable.

Show that all solutions of the equation $\ddot{x} + x^2\dot{x} + x^3 = 0$ tend to zero.

29. Consider the system
$$\dot{x} = \gamma y - \delta x - \beta y(x^2 + y^2),$$
$$\dot{y} = -\gamma x - \delta y + \beta x(x^2 + y^2) - 1.$$

Let (x_0, y_0) be an equilibrium point of the system and set $x = x_0 + \xi, y = y_0 + \eta$, where
$$x_0 = r\cos\theta, \quad y_0 = r\sin\theta, \quad \xi = \rho\cos(\theta + \psi), \quad \eta = \rho\sin(\theta + \psi).$$

Let
$$V = \tfrac{1}{2}\beta r^2\rho^2(1 + 2\cos^2\psi - \gamma/\beta r^2) + \beta r\rho^3\cos\psi + \tfrac{1}{4}\beta\rho^4.$$

Show that
$$\dot{V} = -\delta\beta r^2\rho^2(1 + 2\cos^2\psi - \gamma/\beta r^2) - 3\delta\beta r\rho^3\cos\psi - \delta\beta\rho^4.$$

Deduce that V and $\dot{V}$ are definite with opposite sign near the equilibrium point if
(i) $\delta > 0, \gamma < \beta r^2$;
(ii) $\delta > 0, \gamma > 3\beta r^2$.

30. (Zubov's method.) Suppose that a function $W(x, y)$, negative definite in the whole plane, is chosen as the time derivative $\dot{V}$ of a possible Liapunov function for a system $\dot{x} = X(x, y), \dot{y} = Y(x, y)$, for which the origin is an asymptotically stable equilibrium point. Show that $V(x, y)$ satisfies the linear partial differential equation
$$X\frac{\partial V}{\partial x} + Y\frac{\partial V}{\partial y} = W$$
with $V(0, 0) = 0$.

Show also that for the path $x(t), y(t)$ starting at (x_0, y_0) at time t_0
$$V\{x(t), y(t)\} - V(x_0, y_0) = \int_{t_0}^t W\{x(t), y(t)\}\, dt.$$

Deduce that the boundary of the domain of asymptotic stability (the domain of initial conditions from which the solutions go into the origin) is the set of points (x, y) for which $V(x, y)$ is infinite, by considering the behaviour of the integral as $t \to \infty$, firstly when (x_0, y_0) is inside this domain and then when it is outside. (Therefore the solution $V(x, y)$ of the partial differential equation above could be used to give the boundary of the domain directly. However, solving this equation is equivalent in difficulty to finding the paths: the characteristics are in fact the paths themselves.)

31. For the system

$$\dot{x} = X(x, y) = -\tfrac{1}{2}x(1-x^2)(1-y^2), \quad \dot{y} = Y(x, y) = -\tfrac{1}{2}y(1-x^2)(1-y^2)$$

show that the Liapunov function $V = x^2 + y^2$ leads to $\dot{V} = -(x^2+y^2)(1-x^2)(1-y^2)$ and explain why the domain of asymptotic stability (see Exercise 30) contains at least the unit circle $x^2 + y^2 = 1$.

Alternatively, start with $\dot{V} = -(x^2+y^2)$ and obtain V from the equation

$$X\frac{\partial V}{\partial x} + Y\frac{\partial V}{\partial y} = \dot{V}, \quad V(0,0) = 0$$

(see Exercise 30). It is sufficient to verify that $V = \log\{(1-x^2)(1-y^2)\}$. Explain why the square $|x| < 1, |y| < 1$ is the complete domain of asymptotic stability for the zero solution.

11 The existence of periodic solutions

SUPPOSE THAT the phase diagram for a differential equation contains a single, unstable equilibrium point and a limit cycle surrounding it, as in the case of the van der Pol equation. Then in practice all initial states lead to the periodic oscillation represented by the limit cycle. In such cases the limit cycle is the principal feature of the system from the practical point of view, and it is desirable to be able to decide with certainty whether it is there or not. Hitherto our attitude to this question has been intuitive; we assemble qualitative evidence that there is a limit cycle, from energy considerations or geometrical arguments, then attempt to estimate the radius by the methods of Chapters 4, 5, and 7, a definite result being taken as further confirmation that the limit cycle is really there. The present chapter contains theorems and methods for positively proving the existence of limit cycles and centres for certain types of equation. The cases chosen can be interpreted physically as involving a balance between energy gain and loss on various regions in the phase plane; they include the type discussed in Sections 1.5 and 4.1; also others with different symmetry. The latter serve as a guide to intuition for related types which do not quite fit the conditions of the theorems. Further theorems can be found, for example, in Andronov *et al.* (1973) and Cesari (1971).

11.1. The Poincaré–Bendixson Theorem

We shall be concerned with second-order autonomous differential equations. A limit cycle is an isolated periodic motion which appears as an isolated closed curve in the phase plane. We take Fig. 3.17 as showing the general nature of a limit cycle; neighbouring paths resemble spirals which approach, or recede from, the limit cycle, though the appearance of actual limit cycles may be somewhat different (see, for example, Fig. 3.23).

In Section 3.4 we gave criteria whereby it can sometimes be shown that *no* limit cycle can exist in a certain region, and also a necessary condition for a limit cycle (the index requirement), but we have no tests giving *sufficient* conditions for existence. We first give without proof a plausible theorem on which several of the results of this chapter are based. For the proof see Cesari (1971) and Andronov *et al.* (1973).

THEOREM 11.1   (*The Poincaré–Bendixson Theorem.*) *Let $\mathscr{R}$ be a closed, bounded region consisting of nonsingular points of a 2×2 system $\dot{\mathbf{x}} = \mathbf{X}(\mathbf{x})$ such that some positive half-path $\mathscr{H}$ of the system lies entirely within $\mathscr{R}$. Then either $\mathscr{H}$ is itself a closed path, or it approaches a closed path, or it terminates at an equilibrium point.*

If we can isolate a region from which some path cannot escape, the theorem describes what may happen to it: roughly speaking, given some control of its movement through regularity conditions on the differential

FIG. 11.1. Possible behaviour of half-paths restricted to a bounded region

equations, it cannot wander about at random for ever. The possibilities are illustrated in Fig. 11.1.

The theorem implies, in particular, that if $\mathscr{R}$ contains no equilibrium points, and some $\mathscr{H}$ remains in $\mathscr{R}$, then $\mathscr{R}$ must contain a periodic solution. The theorem can be used in the following way. Suppose (Fig. 11.2) that we can find two closed curves, $\mathscr{C}_1$ and $\mathscr{C}_2$, with $\mathscr{C}_2$ inside $\mathscr{C}_1$, such that *all* paths crossing $\mathscr{C}_1$ point towards its interior, and *all* paths crossing $\mathscr{C}_2$ point outwards from it. Then no path which enters the annular region between them can ever escape again. The annulus is therefore a region $\mathscr{R}$ for the theorem. If, further, we can arrange that $\mathscr{R}$ has no equilibrium points in it, then the theorem predicts at least one closed path $\mathscr{L}$ somewhere in $\mathscr{R}$. Evidently $\mathscr{L}$ must wrap round the inner curve as shown, for its index is 1 and it must therefore have an equilibrium point interior to it, and $\mathscr{R}$ contains no equilibrium points. For the same reason, there must exist suitable equilibrium points interior to $\mathscr{C}_2$ for all this to be possible.

The same result is true if paths are all outward across $\mathscr{C}_1$ and inward across $\mathscr{C}_2$ (by reversing the time variable).

The practical difficulty is in finding, for a given system, a suitable $\mathscr{C}_1$ and $\mathscr{C}_2$ to substantiate a belief in the existence of a limit cycle. There is a

Fig. 11.2.

similarity between this process and that of finding Liapunov functions (Chapter 10): we normally search for a topographic system (*q.v.*) containing $\mathscr{C}_1$ and $\mathscr{C}_2$.

Example 11.1 *Show that the system*

$$\dot{x} = x - y - (x^2 + \tfrac{3}{2}y^2)x, \quad \dot{y} = x + y - (x^2 + \tfrac{1}{2}y^2)y$$

has a periodic solution.

We shall try to find two circles, centred on the origin, with the required properties. In Fig. 11.3, $\mathbf{n} = (x, y)$ is a normal, pointing outward at P from the circle radius r, and $\mathbf{X} = (X, Y)$ is in the direction of the path through P. Also $\cos \phi = \mathbf{n} \cdot \mathbf{X}/|\mathbf{n}|\,|\mathbf{X}|$ and therefore $\mathbf{n} \cdot \mathbf{X}$ is positive or negative according to whether $\mathbf{X}$ is pointing outwards or inwards. We have

$$\begin{aligned}
\mathbf{n} \cdot \mathbf{X} &= xX + yY = x^2 + y^2 - x^4 - \tfrac{1}{2}y^4 - \tfrac{5}{2}x^2y^2 \\
&= r^2 - r^4 + \tfrac{1}{2}y^2(y^2 - x^2)\text{,} \\
&= r^2 - r^4(1 + \tfrac{1}{4}\cos 2\theta - \tfrac{1}{4}\cos^2 2\theta).
\end{aligned} \tag{11.1}$$

When, for example, $r = \tfrac{1}{2}$, this is positive for all θ and so all paths are directed outwards on this circle, and when $r = 2$ it is negative, with all paths directed inwards. Therefore, somewhere between $r = \tfrac{1}{2}$ and $r = 2$ there is at least one closed path, or a periodic solution.

We can look for the pair of circles which pin down *the narrowest annular region in which we can predict that a closed path exists.* The interval of r in which a closed path might lie is that for which, given a value of r, **X** points inward on some points

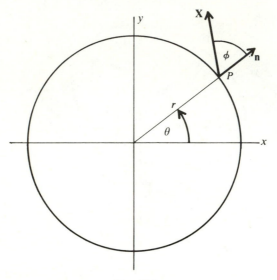

FIG. 11.3.

of the circle, outward on others; that is to say, values of r in this region give circles to which **X** is tangential at some point ('cycles with contact': Andronov *et al.* 1973). For such values of r the equation

$$r^2 - r^4(1 + \tfrac{1}{4}\cos 2\theta - \tfrac{1}{4}\cos^2 2\theta) = 0$$

has a solution (r, θ). Write this in the form

$$4\left(\frac{1}{r^2} - 1\right) = \cos 2\theta - \cos^2 2\theta.$$

The range of the right-hand side is $(-2, \tfrac{1}{4})$, and solutions will exist for values of the left-hand side which lie in this range, that is, when

$$\frac{4}{\sqrt{17}} \leqslant r \leqslant \sqrt{2}, \quad \text{or} \quad 0.97 \leqslant r \leqslant 1.41.$$

Another point of view, closely resembling that of the Liapunov functions of Chapter 10, is the following. Let

$$v(x, y) = c > 0, \quad c_1 < c < c_2,$$

be a 'band' of closed curves for some range of c including $c = c_0$, the outer curves corresponding to larger c (Fig. 11.4). If

$$\left[\frac{dv}{dt}\right]_P = \dot{v}(x, y) = X\frac{\partial v}{\partial x} + Y\frac{\partial v}{\partial y} > 0, \quad P \in \mathscr{C}_0,$$

then all paths point outwards on $\mathscr{C}_0$ (see Fig. 11.4).

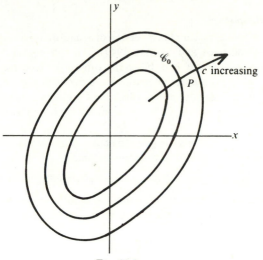

FIG. 11.4.

Example 11.2 *Show that the system*

$$\dot{x} = y - x^3 + x, \quad \dot{y} = -x - y^3 + y$$

has at least one closed path in the phase plane.

Put $v(x, y) = x^2 + y^2$. Then

$$\frac{dv}{dt} = 2x(y - x^3 + x) + 2y(-x - y^3 + y) = 2(x^2 + y^2 - x^4 - y^4).$$

If $x^2 + y^2 < 1$ then

$$x^2 + y^2 > (x^2 + y^2)^2 = x^4 + 2x^2y^2 + y^4 \geq x^4 + y^4.$$

Hence dv/dt is strictly positive on the circle $x^2 + y^2 = r$ for any r such that $0 < r < 1$. On the other hand, for $x^2 + y^2 > 2$,

$$2(x^4 + y^4) \geq (x^2 + y^2)^2 > 2(x^2 + y^2).$$

As a result dv/dt is strictly negative on the circle $x^2 + y^2 = r$ for any $r < 2$. Therefore the system has at least one closed path lying between the two concentric circles.

It is generally very difficult to find curves with the required properties. It may be possible, however, to construct supporting arguments which are not rigorous, as in the following example.

Example 11.3 *Show that the system*

$$\dot{x} = y, \quad \dot{y} = -4x - 5y + \frac{6y}{1+x^2}$$

has a periodic solution.

These equations are a simplified version of equations for a tuned-grid vacuum-tube circuit (Andronov and Chaikin 1949). The only equilibrium point on the finite plane is at $x = 0$, $y = 0$. Near the origin, the linear approximation is $\dot{x} = y$, $\dot{y} = -4x + y$, an unstable spiral. As in the last example, the existence of a Liapunov function for the stable spiral (see Chapter 10) guarantees that there is in the present case a closed curve surrounding the origin over which the paths point outward.

To find what happens at a great distance use, for example, the mapping $z = 1/x, v = y/x$ (eqn (3.15)); the system becomes

$$\dot{z} = -vz, \quad \dot{v} = -4 - 5v - v^2 + 6vz^2/(z^2 + 1).$$

The equilibrium points on the horizon $z = 0$ are at $v = -1$ and $v = -4$ and these are unstable. Viewed in the spherical projection of Section 3.3 the picture is

Fig. 11.5.

as in Fig. 11.5, giving a reasonable assurance of a periodic solution, shown by the heavy line in the figure.

The Bendixson principle can be employed to obtain theorems covering broad types of differential equation, of which the following is an example.

THEOREM 11.2   *The differential equation*

$$\ddot{x} + f(x, \dot{x})\dot{x} + g(x) = 0 \tag{11.2}$$

(the Liénard equation), or the equivalent system

$$\dot{x} = y, \quad \dot{y} = -f(x, y)y - g(x),$$

where f and g are continuous, has at least one periodic solution under the following conditions:

(i) *there exists a > 0 such that $f(x, y) > 0$ when $\sqrt{(x^2 + y^2)} > a$;*

(ii) *$f(0, 0) < 0$ (hence $f(x, y) < 0$ in a neighbourhood of the origin);*

(iii) *$g(0) = 0$, $g(x) > 0$ when $x > 0$, and $g(x) < 0$ when $x < 0$;*

(iv) *$G(x) = \displaystyle\int_0^x g(u)\,du \to \infty$ as $x \to \infty$.*

Proof   (iii) implies that there is a single equilibrium point, at the origin.

Consider the function

$$\mathscr{E}(x, y) = \tfrac{1}{2}y^2 + G(x). \tag{11.3}$$

This represents the energy of the system (potential plus kinetic) when it is regarded, say, as representing a spring–particle system with dissipative external forces. Clearly, $G(0) = 0$, $G(x) > 0$ when $x \neq 0$, and G is monotonic increasing to infinity (by (iv)); and is continuous. Therefore $\mathscr{E}(0, 0) = 0$, and $\mathscr{E}(x, y) > 0$ for $x \neq 0$ or $y \neq 0$ ($\mathscr{E}$ is positive definite). Also $\mathscr{E}$ is continuous and increases monotonically in every radial direction from the origin. Therefore (Fig. 11.6) the family of contours of $\mathscr{E}$ with parameter $c > 0$:

$$\mathscr{E}(x, y) = c \tag{11.4}$$

consists of simple closed curves encircling the origin. As c tends to zero they approach the origin and as $c \to \infty$ they become infinitely remote (the principal consequence of (iv)).

We can choose c, $c = c_1$, small enough for the corresponding contour, $\mathscr{C}_1$, to be entirely within the neighbourhood of the origin where, by (ii), $f(x, y) < 0$. We will examine a half-path $\mathscr{H}$ starting at a point on $\mathscr{C}_1$.

FIG. 11.6.

Consider $\dot{\mathscr{E}}(x, y)$ on $\mathscr{H}$:

$$\dot{\mathscr{E}}(x, y) = g(x)\dot{x} + y\dot{y} = g(x)y + y\{-f(x, y)y - g(x)\}$$
$$= -y^2 f(x, y). \quad (11.5)$$

This is positive, except at $y = 0$, on $\mathscr{C}_1$. Choose $\mathscr{H}$ to start at a point other than $y = 0$ on $\mathscr{C}_1$. Then it leaves $\mathscr{C}_1$ in the outward direction. It is clear that it can never reappear inside $\mathscr{C}_1$, since to do so it must cross some interior contours in the inward direction, which is impossible since, by (11.5), $\dot{\mathscr{E}} \geqslant 0$ on all contours near to $\mathscr{C}_1$, as well as on $\mathscr{C}_1$.

Now consider a contour $\mathscr{C}_2$ for large c, $c = c_2$ say. $\mathscr{C}_2$ can be chosen, by (iv), to lie entirely outside the circle $x^2 + y^2 = a^2$, so that, by (i), $f(x, y) > 0$ on $\mathscr{C}_2$. By (11.5), with $f(x, y) > 0$, all paths crossing $\mathscr{C}_2$ cross inwardly, or are tangential (at $y = 0$), and by a similar argument to the above, no positive half-path, once inside $\mathscr{C}_2$, can escape.

Therefore $\mathscr{H}$ remains in the region bounded by $\mathscr{C}_1$ and $\mathscr{C}_2$, and by Theorem 11.1, there is a periodic solution in this region.

The theorem can be interpreted as in Section 1.5: near to the origin the 'damping' coefficient, f, is negative, and we expect paths to spiral outwards due to intake of energy. Further out, $f(x, y) > 0$, so there is loss of energy and paths spiral inwards. Between the families we expect a closed path.

Example 11.4 *Show that the equation* $\ddot{x} + e(x^2 + \dot{x}^2 - 1)\dot{x} + x^3 = 0$ *has a limit cycle, and locate it between two curves* $\mathscr{E}(x, y) = constant$.

For the theorem, $f(x, y) = x^2 + y^2 - 1$, $g(x) = x^3$, $G(x) = \frac{1}{4}x^4$. Therefore (11.4) gives the contours of $\mathscr{E}$: $\frac{1}{4}x^4 + \frac{1}{2}y^2 = c$. The periodic solution located by the theorem lies between two such contours, one inside, the other outside, of the curve $f(x, y) = 0$, or $x^2 + y^2 = 1$, and is most closely fixed by finding the smallest

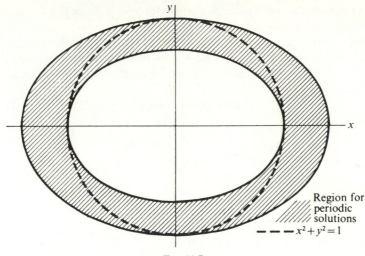

Region for periodic solutions
$--- x^2+y^2=1$

FIG. 11.7.

contour lying outside this circle and the largest lying inside. We require respectively min/max of $x^2 + y^2$ subject to $\frac{1}{4}x^4 + \frac{1}{2}y^2 = c$, c being then chosen so that the min/max is equal to 1. The calculation by means of a Lagrange multiplier gives $c = \frac{1}{2}$ and $c = \frac{1}{4}$ respectively. (See Fig. 11.7.)

11.2. A theorem on the existence of a centre

The following theorem involves ingredients different from those previously considered. (For a theorem of similar form but with less restrictive conditions see Minorsky 1962, p. 113.)

THEOREM 11.3 *The origin is a centre for the equation*

$$\ddot{x} + f(x)\dot{x} + g(x) = 0,$$

or for the equivalent system

$$\dot{x} = y, \quad \dot{y} = -f(x)y - g(x), \tag{11.6}$$

when, in some neighbourhood of the origin, f and g are continuous, and
(i) $f(x)$ *is odd, and of one sign in the half-plane* $x > 0$;
(ii) $g(x) > 0$, $x > 0$, *and* $g(x)$ *is odd* (*implying* $g(0) = 0$);
(iii) $g(x) > \alpha f(x)F(x)$ *for* $x > 0$, *where* $F(x) = \displaystyle\int_0^x f(u)\,\mathrm{d}u$, *and* $\alpha > 1$.

Proof We may assume that $f(x) > 0$ when $x > 0$, since the other case, $f(x) < 0$, reduces to this, putting $-t$ for t. The equation may describe a particle-spring system, with positive damping (loss of energy) for $x > 0$, and negative damping (gain in energy) for $x < 0$. Since f and g are odd, the paths are symmetrical about the y axis (put $x = -z$, $t = -\tau$ into (11.6)). As in Theorem 11.2, (ii) ensures that there is a single equilibrium point, at the origin. Figure 11.8 shows possible types of closed path satisfying these conditions: we have to exclude possibilities such as (b).

Let

$$\mathscr{E}(x, y) = \tfrac{1}{2}y^2 + G(x), \tag{11.7}$$

where

$$G(x) = \int_0^x g(u)\,\mathrm{d}u.$$

Fig. 11.8.

(a) (b)

Fig. 11.9.

As in Theorem 11.2, the family of contours

$$\mathscr{E}(x, y) = c > 0, \tag{11.8}$$

where c is a parameter, define a family of closed curves about the origin (but this time only in some neighbourhood of the origin where the conditions hold, or for small enough c), and as $c \to 0$ the closed curves approach the origin.

Let $\mathscr{C}_0$ be an arbitrary member of the family (11.8) in the prescribed neighbourhood of the origin, and consider the half-path $\mathscr{H}$ starting at A on the intersection of $\mathscr{C}_0$ with the y axis (Fig. 11.9). On $\mathscr{H}$,

$$\dot{\mathscr{E}}(x, y) = -y^2 f(x). \tag{11.9}$$

While $y > 0$, $\dot{x} > 0$, so $\mathscr{H}$ constantly advances to the right in the first quadrant. Also, by (11.9), $\dot{\mathscr{E}}(x, y) < 0$, so y moves continuously to curves with smaller $\mathscr{E}$. Since $\mathscr{H}$ remains within $\mathscr{C}_0$ it leaves the quadrant at some point B on the positive x axis, inside $\mathscr{C}_0$.

On the axis $y = 0$, $x > 0$ we have, by (ii),

$$\dot{x} = 0, \quad \dot{y} = -g(x) < 0.$$

Therefore all paths cut the positive x axis vertically downward, and so cannot re-enter the first quadrant from the fourth. Moreover, since $\dot{\mathscr{E}}(x, y) < 0$ in the fourth quadrant, $\mathscr{H}$ continues to lie inside $\mathscr{C}_0$; therefore it must either enter the origin (the only equilibrium point), or cross the y axis at some point C for which $y < 0$.

We shall show that it does not enter the origin. In the fourth quadrant,

$$\dot{x} < 0 \tag{11.10}$$

but

$$\dot{y} = -f(x)y - g(x), \tag{11.11}$$

which is sometimes negative and sometimes positive. We can deal with (11.11) as follows (Fig. 11.9). OQ is the curve on which $\dot{y} = 0$ (and $dy/dx = 0$), dividing the region where $\dot{y} > 0$ from that where $\dot{y} < 0$ (eqn (11.11)). Assume initially that this curve approaches the origin, that is, that

$$\lim_{x \to 0} \{-g(x)/f(x)\} = 0. \tag{11.12}$$

For this case, $\mathscr{H}$ turns over in the manner indicated; is horizontal at R, and has negative slope up to C, where $dy/dx = 0$ again. We have to show

that $y_0 < 0$. On RC

$$y_1 - y_0 = \int_0^{x_1} \frac{dy}{dx} dx.$$

Therefore,

$$y_0 = y_1 + \int_0^{x_1} \{f(x) + y^{-1}g(x)\} dx \qquad \text{(by (11.6))},$$

$$= -\frac{g(x_1)}{f(x_1)} + F(x_1) + \int_0^{x_1} y^{-1}g(x) dx \quad \text{(since } F(0) = 0\text{)},$$

$$< -\frac{g(x_1)}{f(x_1)} + F(x_1) \quad \text{(since } g(x) > 0, y \leqslant 0 \text{ on } 0 \leqslant x \leqslant x_1\text{)},$$

$$< -\alpha F(x_1) + F(x_1) < 0 \qquad \text{(by (iii))}.$$

Therefore $y_0 < 0$, and $ABRC$ is half of a closed path which is its own reflection in the y axis.

Now return to the possibility that the curve $y = -g(x)/f(x)$ does not pass through the origin. Since $g(x) > 0$ and $f(x) > 0$ when $x > 0$, it lies entirely below the x axis, and Fig. 11.10 shows how $\mathcal{H}$ must in this case also remain below the x axis, and therefore cannot enter the origin.

Finally, since the argument is independent of A near enough to the origin, the origin is a centre.

FIG. 11.10.

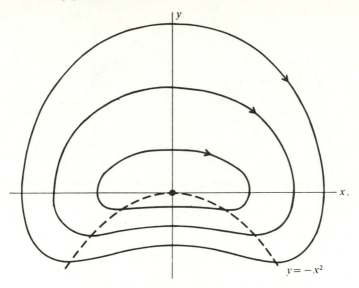

FIG. 11.11. Paths for $\ddot{x} + x\dot{x} + x^3 = 0$

Example 11.5 *The equation $\ddot{x} + x\dot{x} + x^3 = 0$ has a centre at the origin.*

Conditions (i) and (ii) are satisfied. Also

$$f(x)F(x) = \tfrac{1}{2}x^3.$$

Therefore

$$g(x) > 2f(x)F(x),$$

so (iv) is satisfied with $\alpha = 2$ (see Fig. 11.11).

11.3. A theorem on the existence of a limit cycle

We consider the equation

$$\ddot{x} + f(x)\dot{x} + g(x) = 0, \tag{11.13}$$

where, broadly speaking, $f(x)$ is positive when $|x|$ is large, and negative when $|x|$ is small, and where g is such that, in the absence of the damping term $f(x)\dot{x}$ we expect periodic solutions for small x. Van der Pol's equation, $\ddot{x} + e(x^2 - 1)\dot{x} + x = 0$, $e > 0$, is of this type. Effectively the theorem demonstrates a pattern of expanding and contracting spirals about a limit cycle. Paths far from the origin spend part of their time in regions of energy input and part in regions of energy loss, so the physical argument for a limit cycle is less compelling than in the case of Theorem 11.2.

FIG. 11.12.

The proof is carried out on a different phase plane from the usual one:

$$\dot{x} = y - F(x), \quad \dot{y} = -g(x), \tag{11.14}$$

(the Liénard plane) where

$$F(x) = \int_0^x f(u)\,\mathrm{d}u.$$

The use of this plane enables the shape of the paths to be simplified (under the conditions of Theorem 11.4, $\dot{y} = 0$ only on $x = 0$) without losing symmetry, and, additionally, allows the burden of the conditions to rest on F rather than f, f being thereby less restricted.

THEOREM 11.4   *The equation* $\ddot{x} + f(x)\dot{x} + g(x) = 0$ *has a unique periodic solution if f and g are continuous, and*
(i) $F(x)$ *is an odd function;*
(ii) $F(x)$ *is zero only at* $x = 0$, $x = a$, $x = -a$, *for some* $a > 0$;
(iii) $F(x) \to \infty$ *as* $x \to \infty$ *monotonically for* $x > a$;
(iv) $g(x)$ *is an odd function, and* $g(x) > 0$ *for* $x > 0$.
(These conditions imply that $f(x)$ is even, $f(0) < 0$ and $f(x) > 0$ for $x > a$.)

Proof   The general permitted shape of $f(x)$ and $F(x)$ is shown in Fig. 11.12.

The general pattern of the paths can be obtained from the following considerations.
(a) If $x(t)$, $y(t)$ is a solution, so is $-x(t), -y(t)$ (since F and g are odd); therefore the phase diagram is symmetrical about the origin (but not necessarily the individual phase paths).
(b) The slope of a path is given by

$$\frac{\mathrm{d}y}{\mathrm{d}x} = \frac{-g(x)}{y - F(x)}$$

FIG. 11.13.

so the paths are horizontal only on $x = 0$ (from (iv)), and are vertical only on the curve $y = F(x)$. Above $y = F(x)$, $\dot{x} > 0$, and below, $\dot{x} < 0$.

(c) $\dot{y} < 0$ for $x > 0$, and $\dot{y} > 0$ for $x < 0$, by (iv).

Some typical paths are shown in Fig. 11.13. A path $YY'Y''$ is closed if and only if Y and Y'' coincide. The symmetry condition (a) implies that this is the case if and only if

$$OY = OY'. \tag{11.15}$$

We prove the Theorem by showing that for the path $Y_0 Y_0' Y_0''$ through A, $(x = a)$,

$$OY_0 - OY_0' < 0,$$

and that as Y recedes to infinity along the positive y axis

$$OY - OY' \to \infty$$

monotonically. There is therefore exactly one point Y for which $OY - OY'$ is zero, and this identifies the (single) closed path. Thus we are really confirming a pattern of spirals converging on a limit cycle.

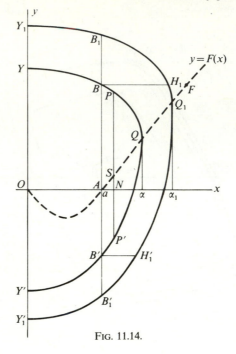

FIG. 11.14.

Now let

$$v(x, y) = \int_0^x g(u)\,\mathrm{d}u + \tfrac{1}{2}y^2. \tag{11.16}$$

Then

$$OY - OY' = 0 \;\Leftrightarrow\; V_{YY'} = 0, \tag{11.17}$$

where we write for a typical path in Fig. 11.14

$$V_{YY'} = v_{Y'} - v_Y = \oint_{YQY'} \mathrm{d}v.$$

Along an element of the path we have

$$\mathrm{d}v = y\,\mathrm{d}y + g\,\mathrm{d}x = y\,\mathrm{d}y + g\frac{\mathrm{d}x}{\mathrm{d}y}\mathrm{d}y$$

$$= y\,\mathrm{d}y + g\frac{y - F}{-g}\mathrm{d}y \quad \text{(from (11.14))}$$

$$= F\,\mathrm{d}y. \tag{11.18}$$

Let BB' (through $x = a$) separate the parts of the path where $F(x)$ is positive and where it is negative. We consider these parts separately in estimating $V_{YY'}$:

$$V_{YY'} = V_{YB} + V_{BB'} + V_{B'Y'}. \tag{11.19}$$

The proof is carried out through the steps (A) and (D) below.

(A) *As Q moves out from A along AF, $V_{YB} + V_{B'Y'}$ is positive and monotonic decreasing.*

Choose any fixed path YQY', with Q at the point $(\alpha, F(\alpha))$ as shown, and another, $Y_1 Q_1 Y_1'$, with Q at the point $(\alpha_1, F(\alpha_1))$, where $\alpha_1 > \alpha$.

By (11.14), since $y > 0$ and $F(x) < 0$ on YB, $Y_1 B_1$,

$$0 < \left(\frac{\mathrm{d}y(x)}{\mathrm{d}x} \right)_{YB} < \left(\frac{\mathrm{d}y(x)}{\mathrm{d}x} \right)_{Y_1 B_1} \tag{11.20}$$

for every x. Similarly, since $y < F(x)$ (from (b)) and $F(x) < 0$ on $B'Y'$, $B_1' Y_1'$,

$$0 < \left(\frac{\mathrm{d}y(x)}{\mathrm{d}x} \right)_{B_1' Y_1'} < \left(\frac{\mathrm{d}y(x)}{\mathrm{d}x} \right)_{B'Y'}. \tag{11.21}$$

(That is to say, $Y_1 B_1$ and $B_1' Y_1'$ are shallower than YB and $B'Y'$ respectively.) Therefore from (11.18), with $F(x) < 0$,

$$V_{YB} = \oint_{YB} F \, \mathrm{d}y = \oint_{YB} (-F) \left(-\frac{\mathrm{d}y}{\mathrm{d}x} \right) \mathrm{d}x > \oint_{Y_1 B_1} (-F) \left(-\frac{\mathrm{d}y}{\mathrm{d}x} \right) \mathrm{d}x$$

$$= V_{Y_1 B_1} > 0 \tag{11.22}$$

using (11.20). Similarly,

$$V_{B'Y'} > V_{B_1' Y_1'} > 0. \tag{11.23}$$

The result (A) follows from (11.21) and (11.23).

(B) *As Q moves out from A along AF, $V_{BB'}$ is monotonic decreasing.*

Choose BQB' arbitrarily to the right of A and let $B_1 Q_1 B_1'$ be another path with Q_1 to the right of Q, as before. Then $F(x) > 0$ on these paths, and we write

$$V_{B_1 B_1'} = -V_{B_1' B_1} - \oint_{B_1' B_1} F(x) \, \mathrm{d}y \leqslant - \oint_{H_1' H_1} F(x) \, \mathrm{d}y$$

(where BH_1, $B'H_1'$ are parallel to the x axis)

$$\leqslant - \oint_{B'B} F(x) \, \mathrm{d}y = V_{BB'} \tag{11.24}$$

(since, for the same values of y, $F(x)$ on $B'B$ is less than or equal to $F(x)$ on $H_1'H_1$).

(C) *From (A) and (B) we deduce that $V_{YY'}$ is monotonic decreasing to the right of A.*

(D) *$V_{BB'}$ tends to $-\infty$ as the paths recede to infinity.*

Let S be a point on the line $y = F(x)$, to the right of A, and let BQB' be an arbitrary path with Q to the right of S. The line $PSNP'$ is parallel to the y axis. Then, as before,

$$V_{BB'} = -\oint_{B'B} F(x)\,dy \leqslant \oint_{P'P} F(x)\,dy.$$

Also

$$F(x) \geqslant NS \quad \text{on} \quad PP'$$

since $F(x)$ is monotonic increasing by (iii); therefore

$$V_{BB'} \leqslant -NS \oint_{P'P} dy = -NS \cdot PP' \leqslant -NS \cdot NP. \qquad (11.25)$$

But as Q goes to infinity towards the right, $NP \to \infty$.

(E) *From (C) and (D), $V_{YY'}$ is monotonic decreasing to $-\infty$, to the right of A.*

(F) *$V_{YY} > 0$ when Q is at A or to the left of A.* (For then $F(x) < 0$ and $dy < 0$.)

From (E) and (F), there is one and only one path for which $V_{YY'} = 0$, and this, by eqn (11.17) and the symmetry of the paths, is closed.

Example 11.6 *The van der Pol equation $\ddot{x} + e(x^2 - 1)\dot{x} + x = 0$, $e > 0$, has a unique limit cycle. (The case $e < 0$ is exactly similar (put $-t$ for t). Also note that this proof holds even when e is not small.)*

Here, $f(x) = e(x^2 - 1)$, $F(x) = e(\tfrac{1}{3}x^3 - x)$. The conditions of the theorem are satisfied, with $a = \sqrt{3}$. The x-extremities of the limit cycle must be beyond $x = \pm\sqrt{3}$.

Example 11.7 *Show that the equation*

$$\ddot{x} + \frac{x^2 + |x| - 1}{x^2 - |x| + 1}\dot{x} + x^3 = 0$$

has a unique periodic solution.

$f(x)$ has the shape shown in Fig.. 11.15. Since $f(x) \to 1$ as $x \to \infty$, (iii) is

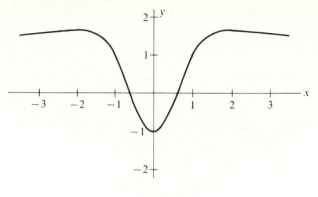

Fig. 11.15.

satisfied, and (i), (ii), and (iv) are obviously satisfied. Therefore a limit cycle exists, its extreme x-values being beyond the non-zero roots of $F(x) = 0$.

11.4. Van der Pol's equation with large parameter

Van der Pol's equation with large parameter e

$$\ddot{x} + e(x^2 - 1)\dot{x} + x = 0$$

is equivalent (by putting $\delta = 1/e$) to the equation

$$\delta\ddot{x} + (x^2 - 1)\dot{x} + \delta x = 0 \tag{11.26}$$

with δ small. This can be regarded as a singular perturbation problem in terms of Chapter 6, in which the 'weight' of the various terms is differently disposed on different parts of the solution curves, or on the phase paths. The existence of a limit cycle is confirmed in Example 11.6. Here we shall look briefly at the analytical construction of the limit cycle of (11.26) for small δ in the usual phase plane with $\dot{x} = y$, and not the Liénard plane. It is necessary to piece together the limit cycle from several approximate solutions corresponding to different balances between the terms. The calculations are made for $y > 0$; the phase path is then completed by invoking its symmetry in the origin.

The equation for the phase path is

$$\frac{dy}{dx} = -\frac{x^2 - 1}{\delta} - \frac{x}{y}, \tag{11.27}$$

and the isocline for zero slope is the curve $y = \delta x/(1 - x^2)$.

In eqn (11.26) put $t = \mu\tau$, so that

$$\frac{\delta}{\mu^2}x'' + \frac{1}{\mu}(x^2 - 1)x' + \delta x = 0. \qquad (11.28)$$

When $\mu = \delta$ the first two terms predominate, and to the first order

$$x'' + (x^2 - 1)x' = 0. \qquad (11.29)$$

If $x = a$ (the amplitude) when $y = 0$, then integration of (11.29) gives

$$x' = x - \tfrac{1}{3}x^3 - a + \tfrac{1}{3}a^3, \qquad (11.30)$$

or

$$x' = \tfrac{1}{3}(x - a)(3 - x^2 - ax - a^2).$$

If $a > 1$, then $x' > 0$ (in the upper half of the phase plane) on

$$\tfrac{1}{2}\{-a + \sqrt{(12 - 3a^2)}\} < x < a,$$

and (11.30) represents the phase path in this interval: the amplitude a will be determined later.

When we put $\mu = \delta^{-1}$ in (11.28), the second and third terms predominate and are of the same order, so that

$$x' = x/(1 - x^2), \qquad (11.31)$$

which is essentially the zero-slope isocline of (11.27). This equation can only be valid for $y > 0$ and $-a < x < -1$ because of the singularity at $x = -1$.

The two approximate solutions given by (11.30) and (11.31) must be connected by a third equation in the neighbourhood of $x = -1$. To the required order the solution given by (11.30) must pass through $y = 0$, $x = -1$: whence

$$\tfrac{1}{2}\{-a + \sqrt{(12 - 3a^2)}\} = -1, \quad \text{or} \quad a = 2.$$

For the transition, put $z = x + 1$; then (11.28) becomes

$$\frac{\delta}{\mu^2}z'' + \frac{1}{\mu}\{(z - 1)^2 - 1\}z' + \delta z - \delta = 0.$$

Now put $\mu = \delta^\alpha$, and let $z = \delta^\beta u$; then a balance is achieved between the terms containing z'', z', z and the constant if

$$1 - 2\alpha + \beta = -\alpha + 2\beta = 1;$$

that is, if $\alpha = \tfrac{1}{3}$ and $\beta = \tfrac{2}{3}$. The differential equation for the transition region is

$$u'' - 2uu' - 1 = 0, \tag{11.32}$$

an equation without an elementary solution, relating u' and u.

There remains the solution around $x = -2$ to be constructed. Put $z = x + 2$ in (11.28):

$$\frac{\delta}{\mu^2} z'' + \frac{1}{\mu} \{(z-2)^2 - 1\} z' + \delta z - 2\delta = 0.$$

Again, put $\mu = \delta^\alpha$ and $z = \delta^\beta u$. The equation

$$u'' + 3u' = 2 \tag{11.33}$$

follows if $1 - 2\alpha + \beta = \beta - \alpha = 1$, or if $\alpha = 1$, $\beta = 2$. From (11.33)

$$\tfrac{1}{3} u' + \tfrac{2}{9} \log(2 - 3u') + u = \text{constant}.$$

When $u = 0$, $u' = 0$, so finally

$$\tfrac{1}{3} u' + \tfrac{2}{9} \log(2 - 3u') + u = \tfrac{2}{9} \log 2.$$

After restoring the original variables, x and t, the sections of the limit cycle on the phase plane for $y > 0$ are as follows:

(i) $-2 \leqslant x < -2 + O(\delta^2)$:

$$\tfrac{1}{3}\delta y + \tfrac{2}{9}\delta^2 \log(1 - \tfrac{3}{2}\delta^{-1} y) = -(x + 2);$$

(ii) $-2 + O(\delta^2) < x < -1 - O(\delta^{2/3})$:

$$y = \delta x/(1 - x^2);$$

(iii) $-1 - O(\delta^{2/3}) < x < -1 + O(\delta^{2/3})$:
the appropriate solution of

$$\delta \ddot{x} - 2(x + 1)\dot{x} = \delta;$$

(iv) $-1 + O(\delta^{2/3}) < x \leqslant 2$:

$$\delta y = x - \tfrac{1}{3}x^3 + \tfrac{2}{3}.$$

The complete limit cycle can now be constructed since it is symmetrical about the origin. Figure 11.16(*a*) shows the computed phase diagram for $\delta = 0.1$, and the corresponding solution is shown in Fig. 11.16(*b*).

The van der Pol equation for large parameter e (or small δ) is an example of a *relaxation oscillation*, in which typically, as Fig. 11.16(*b*) shows, the system displays a slow build-up followed by a sudden discharge, repeated periodically.

(a)

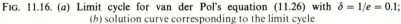

(b)

FIG. 11.16. (a) Limit cycle for van der Pol's equation (11.26) with $\delta = 1/e = 0.1$;
(b) solution curve corresponding to the limit cycle

Exercises

1. Consider the system

$$\dot{x} = y + P(x, y), \quad \dot{y} = -x + Q(x, y),$$

where P and Q are of order $x^2 + y^2$. Show that the origin is a centre if either
(i) $|P$ is odd in x, Q is even in x; or
(ii) P is even in y, Q is odd in y.

Deduce that the origin is a centre for the system

$$\dot{x} = y + cx(x^2 + y^2), \quad \dot{y} = -x + cy(x^2 + y^2).$$

2. A system has exactly one equilibrium point, n limit cycles, and no other periodic solutions. Explain why a stable limit cycle must be adjacent to unstable limit cycles, but an unstable limit cycle may have stable or unstable cycles adjacent to it.

Let c_n be the number of possible configurations, with respect to stability, of n nested limit cycles. Show that $c_1 = 2$, $c_2 = 3$, $c_3 = 5$, and that in general

$$c_n = c_{n-1} + c_{n-2}.$$

(This recurrence relation generates the Fibonacci sequence.) Deduce that

$$c_n = \{(2+\sqrt{5})(1+\sqrt{5})^{n-1} + (-2+\sqrt{5})(1-\sqrt{5})^{n-1}\}/(2^{n-1}\sqrt{5}).$$

3. By considering the path directions across each of the suggested topographic systems show that in each of the cases given there exists a limit cycle. Locate the region in which a limit cycle might exist as closely as possible. Show that in each case only one limit cycle exists:

(i) $\dot{x} = 2x + 2y - x(2x^2 + y^2)$, $\dot{y} = -2x + y - y(2x^2 + y^2)$

(topographic system $x^2 + y^2 = $ constant);

(ii) $\dot{x} = -x - y + x(x^2 + 2y^2)$, $\dot{y} = x - y + y(x^2 + 2y^2)$

(topographic system $x^2 + y^2 = $ constant);

(iii) $\dot{x} = x + y - x^3 - 6xy^2$, $\dot{y} = -\frac{1}{2}x + 2y - 8y^3 - x^2y$

(topographic system $x^2 + 2y^2 = $ constant);

(iv) $\dot{x} = 2x + y - 2x^3 - 3xy^2$, $\dot{y} = -2x + 4y - 4y^3 - 2x^2y$

(topographic system $2x^2 + y^2 = $ constant).

4. Show that the equation $\ddot{x} + e(x^2 + \dot{x}^2 - 1)\dot{x} + x^3 = 0$ has at least one periodic solution.

5. Show that the origin is a centre for the equations
(i) $\ddot{x} - x\dot{x} + x = 0$;
(ii) $\ddot{x} + x\dot{x} + \sin x = 0$.

6. Suppose that $f(x)$ in the equation $\ddot{x} + f(x)\dot{x} + x = 0$ satisfies the conditions of Theorem 11.3. Show that the equivalent linear equation (Section 4.5) with respect to $x = a\cos t$ is $\ddot{x} + x = 0$.

7. Show that the equation

$$\ddot{x} + e(x^2 - 1)\dot{x} + \tanh kx = 0$$

has exactly one periodic solution when $k > 0$. Decide on its stability.

The 'restoring force' resembles a step-function when k is large. Is the conclusion the same when it is exactly a step-function?

8. Show that

$$\ddot{x} + e(x^2 - 1)\dot{x} + x^3 = 0$$

has exactly one periodic solution.

9. Show that

$$\ddot{x} + (|x| + |\dot{x}| - 1)\dot{x} + x|x| = 0$$

has at least one periodic solution.

10. Show that the origin is a centre for the equation

$$\ddot{x} + (k\dot{x} + 1)\sin x = 0.$$

11. Analyse the equation

$$\varepsilon\ddot{x} + (x^4 - 1)\dot{x} + \varepsilon x = 0$$

for small ε by the singular perturbation technique outlined in Section 11.4.

12. Let F and g be functions satisfying the conditions of Theorem 11.4. Show that the equation

$$\ddot{u} + F(\dot{u}) + u = 0$$

has a unique periodic solution (put $\dot{u} = -z$). Deduce that Rayleigh's equation $\ddot{u} + e(\frac{1}{3}\dot{u}^3 - \dot{u}) + u = 0$ has a unique limit cycle.

13. Show that the equation

$$\ddot{x} + e(x^2 + \dot{x}^2 - 1)\dot{x} + x = 0,$$

unlike the van der Pol equation, does not have a relaxation oscillation for large positive e.

14. For the van der Pol oscillator

$$\delta\ddot{x} + (x^2 - 1)\dot{x} + \delta x = 0$$

for small positive δ, use the formula for the period, eqn (1.11), to show that the period of the limit cycle is approximately $(3 - 2\ln 2)\delta^{-1}$. (Hint: the principal contribution arises from that part of the limit cycle given by (ii) in Section 11.4.)

15. In the van der Pol plane a, b the equations for the response amplitudes $a(t)$, $b(t)$ of the forced van der Pol equation are (see eqn (7.35))

$$\dot{a} = \tfrac{1}{2}\varepsilon(1 - \tfrac{1}{4}r^2)a - \frac{\omega^2 - 1}{2\omega}b,$$

$$\dot{b} = \frac{\omega^2 - 1}{2\omega}a + \tfrac{1}{2}\varepsilon(1 - \tfrac{1}{4}r^2)b + \frac{\Gamma}{2\omega}$$

where $\varepsilon > 0$. Show that in Case III of Section 7.5, where this system has a single unstable node or spiral, there is a closed path.

12 Bifurcations, structural stability, and chaos

A CHARACTERISTIC OF nonlinear oscillating systems which has become a subject of considerable recent interest is the great variety of types of response of which they are capable as initial conditions or parameter values change. The passage from one set of responses to another often occurs very suddenly, or 'catastrophically': instances of this were shown in Section 1.7 in connection with a parameter-dependent conservative system, and in Section 7.3 on the jump phenomenon of Duffing's equation. Further examples of such changes are given in the present chapter, the linking concept being the idea of bifurcation, where the sudden change in behaviour occurs as a parameter passes through a critical value called a bifurcation point. A system may contain more than one parameter each with its own bifurcation points so that it can display extremely complex behaviour, and computer studies play an important part in providing a taxonomy for the behaviour of such systems. We shall look at some of the elementary characteristics of bifurcations as they arise in the fold and cusp catastrophes, the Hopf bifurcation, structural instability, Poincaré maps, and chaotic solutions and illustrate ways of exploring the field of responses by computer experimentation.

12.1. Examples of bifurcations

In this section we take an intuitive view of some simple bifurcations. Consider the system

$$\dot{x} = y, \qquad \dot{y} = -\lambda x$$

containing a parameter λ with values in $(-\infty, \infty)$. The phase diagram is a centre for $\lambda > 0$ and a saddle for $\lambda < 0$, these classifications representing radically different types of system behaviour. The change in nature occurs as λ passes through $\lambda = 0$: a bifurcation is said to occur at $\lambda = 0$, and $\lambda = 0$ is called the *bifurcation point*.

The behaviour of the damped system

$$\dot{x} = y, \quad \dot{y} = -ky - \omega^2 x, \tag{12.1}$$

with $\omega > 0$ given and k the bifurcation parameter, depends on the roots of the characteristic equation

$$m^2 + km + \omega^2 = 0.$$

The phase diagram is an unstable node for $k < -2\omega$, an unstable spiral for $-2\omega < k < 0$, a stable spiral for $0 < k < 2\omega$, and a stable node for $k > 2\omega$. It might appear at first sight that, for example, the transition from stable spiral to stable node should indicate a bifurcation, but an important common feature is that both are asymptotically stable, so that even though the stable spiral looks different from a stable node we do not call the value $k = 2\omega$ a bifurcation point. However, the transition of k from negative k to positive k through $k = 0$ is accompanied by a change of stability. Hence we might expect the system to have a bifurcation point at $k = 0$. We shall say more about the general linear problem in Section 12.3.

The equation

$$\ddot{x} - k(x^2 + \dot{x}^2 - 1)\dot{x} + x = 0$$

(see Example 1.6) has a limit cycle $x^2 + y^2 = 1$ in the phase plane with $\dot{x} = y$. Let k be the bifurcation parameter. The behaviour close to the origin is indicated by the linearized equation

$$\ddot{x} + k\dot{x} + x = 0$$

which is essentially the same as (12.1) with $\omega = 1$. The equilibrium point at $(0, 0)$ passes through the same bifurcation as that of (12.1). The limit cycle $x^2 + y^2 = 1$ is always present but its stability changes from stable to unstable as k increases through zero, so in this example the change in the phase diagram is not limited just to the equilibrium point but also to other features.

Example 12.1   *Find the bifurcation points of the system*

$$\dot{x} = -\lambda x + y,$$
$$\dot{y} = -\lambda x - 3y.$$

Let

$$\mathbf{x} = \begin{pmatrix} x \\ y \end{pmatrix}, \qquad A(\lambda) = \begin{pmatrix} -\lambda & 1 \\ -\lambda & -3 \end{pmatrix}$$

so that, in matrix form, the system is equivalent to

$$\dot{\mathbf{x}} = A(\lambda)\mathbf{x}.$$

Following the method of Chapter 2, we find the eigenvalues m from

$$|A(\lambda) - mI| = 0,$$

or

$$\begin{vmatrix} -\lambda - m & 1 \\ -\lambda & -3 - m \end{vmatrix} = 0.$$

Hence m satisfies

$$m^2 + (3 + \lambda)m + 4\lambda = 0,$$

which has the roots

$$m_1, m_2 = \tfrac{1}{2}[-\lambda - 3 \pm \sqrt{\{(\lambda - 1)(\lambda - 9)\}}].$$

We look for bifurcations where the character of the roots changes—having regard to whether they are real or complex, and to the sign when they are real or the sign of the real part when they are complex. The transitions are shown in Fig. 12.1. The system has a single bifurcation point $\lambda = 0$, where there is a change from a stable node to a saddle.

FIG. 12.1.

12.2. The fold and the cusp

Bifurcations of systems are closely linked to *catastrophes*. In general terms a system experiences a catastrophe when a a smooth change in the values of a parameter results in a sudden change in the response of the system. A comprehensive account of catastrophe theory is given in *Catastrophe theory and its applications* by Poston and Stewart (1978). Here we shall in a general way relate the two simplest catastrophes and their associated bifurcation sets of parameters as they arise in the 'conservative' systems of Section 1.3, which take the standard form

$$\ddot{x} = f(x) = -\mathcal{V}'(x)$$

where $\mathcal{V}$ is called the potential energy, or potential, of the system.

Suppose we have a potential containing a parameter λ, defined by

$$\mathcal{V}(\lambda, x) = \tfrac{1}{3}x^3 + \lambda x. \tag{12.2}$$

The equilibrium points occur where $(\partial \mathscr{V}/\partial x) = 0$, or where

$$x^2 + \lambda = 0. \tag{12.3}$$

If $\lambda < 0$ there are two equilibrium points, one if $\lambda = 0$, and none if $\lambda > 0$. The phase diagrams may be obtained by using the method of Section 1.3: a plot of $\mathscr{V}$ against x for constant λ is made, and must take one of the forms shown in Fig. 12.2, depending on the value of λ. If $\lambda < 0$ then $\mathscr{V}$ has a minimum at $x = \sqrt{\lambda}$ indicating a centre at this point, and a maximum at $x = -\sqrt{\lambda}$ indicating a saddle there. The case $\lambda = 0$ and $\lambda > 0$ may be discussed similarly.

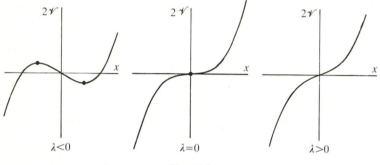

FIG. 12.2.

Equation (12.3) defines the so-called 'catastrophe manifold' for this problem: this is the curve shown in Fig. 12.3. In effect it shows the positions of the equilibrium points in terms of the values of λ, but its real significance is to display the fact that $\lambda = 0$ is a bifurcation point for the problem, where the nature of the solutions suddenly changes.

The stability of the equilibrium points to which the two branches of the curve relate can be deduced from the method of Section 1.7. If the region where $-(\partial \mathscr{V}/\partial x) > 0$ is shaded, then equilibrium points which form the boundary above this region are stable and the rest are unstable.

Suppose now that the potential consists of a fourth-order polynomial of the form

$$\mathscr{V}(\lambda, \mu, x) = \tfrac{1}{4}x^4 - \tfrac{1}{2}\lambda x^2 + \mu x \tag{12.4}$$

where λ and μ are two parameters. Equilibrium points occur where $(\partial \mathscr{V}/\partial x) = 0$, or where

FIG. 12.3. Catastrophe manifold for the fold catastrophe

$$M(\lambda, \mu, x) \equiv x^3 - \lambda x + \mu = 0. \tag{12.5}$$

Regarding (12.5) as a cubic equation for x, the nature of its solutions depends on the sign of the discriminant $D = -4\lambda^3 + 27\mu^2$ (see, e.g., Ferrar, 1950). If $D > 0$ there is one real solution; if $D = 0$ there are two real roots one of which is a repeated root; and if $D < 0$ there are three real roots, all different. The potential $\mathscr{V}$ then takes one of the forms shown in Fig. 12.4.

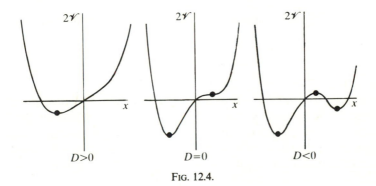

FIG. 12.4.

The catastrophe manifold defined by (12.5) is this time a surface in λ, μ, x space (Fig. 12.5). A fold grows from the origin in the general direction of the λ axis. The two curves radiating from the origin along which the fold turns over are of particular interest, since they form a

FIG. 12.5. Catastrophe manifold for the cusp catastrophe

demarcation between the parameter region where there is a single equilibrium point and the region in which there are three. We therefore project on to the λ, μ plane. At the edge of the fold

$$\frac{\partial M}{\partial x} = 3x^2 - \lambda = 0. \qquad (12.6)$$

The projection of the boundary of the fold is obtained by eliminating x between (12.5) and (12.6), and the resulting curves in the λ, μ-plane are given by

$$\mu = \pm \frac{2}{3\sqrt{3}} \lambda^{\frac{3}{2}}$$

as shown in Fig. 12.6. (This expression is equivalent to $D(\lambda, \mu) = 0$.)

Figure 12.6 is called a 'catastrophe map' (of a 'cusp catastrophe'): it indicates the critical values of the parameters at which sudden change in the nature of the equation's output can be expected. The *bifurcation set* consists of the cusp itself: in the shaded region between the two branches of the curve there are three equilibrium points, and outside there is one.

The technique of Section 1.7 can again be used to establish which parts of the surface of Fig. 12.5 represent stable equilibrium points, and which unstable, by noticing that the surface divides the space into two parts, and that in the 'lower' half $(\partial\mathscr{V}/\partial x)$ is negative and in the 'upper'

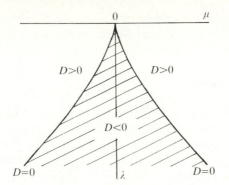

FIG. 12.6. Catastrophe map and bifurcation set for the cusp catastrophe

half positive (the separation occurring when $(-\partial \mathscr{V}/\partial x) = 0$, which is the equation of the surface itself). The connecting portion of the fold between the upper and lower parts therefore represents unstable equilibrium points, and the rest are stable. The unstable set lies between the branches of the cusp in Fig. 12.4. Thus if the parameter values undergo a change in which they cross the cusp a marked change in response of the system is to be expected, this change taking place 'catastrophically'.

It is shown in Section 7.3 how Duffing's equation generates a similar cusp catastrophe when the parameters are taken to be the forcing amplitude and the forcing frequency: the observed variables in that case are not equilibrium points and their stability, but amplitudes of steady forced oscillations and their stability, as parametrized in the van der Pol plane.

12.3. Structural stability and bifurcations

A formal definition of structural stability is beyond the scope and intention of this book, but some of the underlying ideas can be sketched in fairly broad terms. We shall consider only the autonomous system $\dot{\mathbf{x}} = \mathbf{X}(\mathbf{x})$, where it is assumed that the system is regular (see Appendix A) for $\mathbf{x} \in \mathscr{R}^n$ and $t \in [0, \infty)$. The vector field $\mathbf{X}(\mathbf{x})$ is said to define a *dynamical system* in $\mathscr{R}^n$. In simple terms, we wish to be able to decide whether a phase diagram will remain qualitatively unchanged when the *functional form* of the right-hand side of the system is slightly changed in any way. If the output characteristics are liable to change radically, for example by the appearance of a new equilibrium point, then the system is said to be structurally unstable. The special case where a change in a parameter has such an effect, as at a bifurcation point, involves similar considerations. We are not concerned with the stability as such of equilibrium points and the like, but only with changes in such characteristics.

The types of change, or *perturbation*, which may be used to test the system's response, and also the criteria by which the property of structural stability is assessed under the influence of such changes, may be severe or relaxed depending, for example, on the practical demands that are to be made on the hardware which the system represents. Generally the perturbations considered take the form of smooth, additive, arbitrarily small modifications of the vector field X, which may be subject to various further restrictions. In characterizing the effect upon the system questions arise which are analogous to those which made the distinction between Poincaré and Liapunov stability necessary.

Here we will consider a simple type of perturbation sufficient to illustrate the principles involved. For the n-dimensional system $\dot{x} = X(x)$ let $\mathscr{S}$ be the collection of vector functions, $s(x) \in \mathscr{S}$, which are *uniformly* bounded and have continuous derivatives which are also uniformly bounded: that is to say for some number $M > 0$

$$\|s(x)\| < M, \qquad \left\|\frac{\partial(s(x))}{\partial x_i}\right\| < M$$

for all $s \in \mathscr{S}$, $x \in \mathscr{R}^n$, and $i = 1, 2, \ldots, n$. If, for all the eligible s, there exists a value for ε such that the phase diagrams of all the systems

$$\dot{x} = X(x), \quad \dot{x} = X(x) + \varepsilon s(x) \tag{12.7}$$

are 'similar', in a sense which we shall define, we say that the system is *structurally stable*. It can be seen that the value of M chosen is immaterial, and that once a value for ε is established any smaller value will do also.

The meaning we shall take for 'similar' is *topological orbital equivalence*—that there should exist a *homeomorphism*, meaning a continuously invertible mapping, connecting every pair of diagrams, on which the *pattern of directions* (the arrows on the diagrams) is preserved by the differential equations.

It is often easier to establish structural instability rather than the alternative, since a single instance of an $s(x)$ which produces fresh equilibrium points will settle that the system is unstable, as in the following example.

Example 12.2 *Show that (a) the one-dimensional system $\dot{x} = x^2$ is structurally unstable; (b) the system $\dot{x} = x$ is structurally stable.*

(a) There is a single, unstable equilibrium point, at $x = 0$. Now consider the same system perturbed by $\varepsilon s(x)$ where $s(x) \equiv -1$, so that $\dot{x} = x^2 + \varepsilon$. This has two equilibrium points, at $x = \varepsilon^{\frac{1}{2}}$ and $x = -\varepsilon^{\frac{1}{2}}$. In the case of one-dimensional problems the homeomorphism always exists—the real line can be mapped into the real line in any number of ways. But there is no way to make the directions of the 'arrows' on the two diagrams correspond. This system is therefore structurally unstable.

(b) Consider the systems

$$\dot{x} = x + \varepsilon s(x) \tag{12.8}$$

where the s are continuously differentiable and a value of M exists such that $|s(x)| < M$, $|s'(x)| < M$, for all x and for all s.

The unperturbed system has a single unstable equilibrium point, at $x = 0$. The equilibrium points of the perturbed system are obtained by solving

$$X(x) = x + \varepsilon s(x) = 0. \tag{12.9}$$

Choose ε so that $\varepsilon < 1/2M$. Then $|\varepsilon s'(x)| < \varepsilon M < \frac{1}{2}$ for all x, so that

$$X'(x) = 1 + \varepsilon s'(x) > \frac{1}{2}$$

for all x. From this it is clear that the graph of $X(x)$ cuts the x axis at exactly one point, so that eqn (12.9) has exactly one solution. Since also $|s(x)| < M$ for all x it is clear from (12.9) that this solution approaches the origin as $\varepsilon \to 0$. The perturbed equilibrium point therefore approaches the unperturbed one. Since the M is the same for all the $s(x)$ to be considered, the same ε will do for all, and the first part of the definition of structural stability is satisfied.

If we can establish that the perturbed equilibrium points are always unstable, as is the original equilibrium point, then the second requirement referring to directions, or arrows, will also be satisfied. The linearized version of (12.8) is

$$\dot{x} = \{1 + \varepsilon s'(x_s)\}x$$

where x_s represents the new equilibrium point found above. By using the same value of ε as before the coefficient can be made positive, and the new system is therefore also unstable.

The system is therefore structurally *stable*—all its essential properties are unaffected when it suffers a small disturbance.

Systems of dimension 2 are far more difficult. For the system $\dot{\mathbf{x}} = \mathbf{X}(\mathbf{x})$, $\mathbf{x} \in \mathcal{R}^2$, a perturbed system of the form

$$\dot{\mathbf{x}} = \mathbf{X}(\mathbf{x}) + \varepsilon \begin{pmatrix} u(x) \\ v(x) \end{pmatrix}$$

has to be investigated, where the permitted classes u, v are uniformly bounded with uniformly bounded, continuous, first partial derivatives in $\mathscr{R}^2$. It is not possible to consider such systems in detail, but it is possible to illustrate the *kind of transition* between perturbed and unperturbed phase diagrams permitted by the criterion that they should be linked by a homeomorphism. For example, and loosely speaking, spirals and nodes are topologically orbitally equivalent, and we show this in the particular case of such features arising from *linear* systems:

Example 12.3 *Find a homeomorphism between the (linear) node with paths given by $y = A|x|^\gamma$, A arbitrary, $\gamma > 0$ a given constant, and the (linear) spiral with paths given in polar coordinates by $r = B\,e^{\mu\theta}$, $\beta > 0$ arbitrary, μ a given constant. (For the derivation of these features, see Chapter 2.)*

The map $(x, y) \to (x_1, y_1)$ which sends the origin into the origin and is defined elsewhere by

$$(x_1, y_1) = (x^{1/\gamma}, y) \qquad \text{for } x > 0 \text{ and all } y,$$

$$(x_1, y_1) = (-(-x)^{1/\gamma}, y) \quad \text{for } x < 0 \text{ and all } y,$$

takes the family $y = A|x|^\gamma$ into the 'star' $y_1 = A_1 x_1$ (A_1 arbitrary), and is a homeomorphism.

It is easy to show that the map which takes the star in (x_1, y_1) into the spiral family in (x_2, y_2) defined in polars by $r_2 = B\,e^{\mu\theta_2}$ is given by

$$\begin{pmatrix} x_2 \\ y_2 \end{pmatrix} = \begin{pmatrix} \cos\left(\dfrac{1}{\mu}\ln r_1\right) & -\sin\left(\dfrac{1}{\mu}\ln r_1\right) \\[2ex] \sin\left(\dfrac{1}{\mu}\ln r_1\right) & \cos\left(\dfrac{1}{\mu}\ln r_1\right) \end{pmatrix} \begin{pmatrix} x_1 \\ y_1 \end{pmatrix}$$

where $r_1 = \sqrt{(x_1^2 + y_1^2)}$ and $r_1 \neq 0$; and the origin is mapped into the origin. This is also a homeomorphism, and when the homeomorphisms are successively applied the node is mapped into the spiral.

There are no implications in this example for the structural stability of any system—it is intended simply to illustrate the possibility that phase diagrams which are distinctly different in general appearance may still be topologically orbitally equivalent and are not to be counted as different for the purpose of determining structural stability.

As regards linear systems, the structural stability boundaries occur for the degenerate cases when one eigenvalue is zero or both are pure imaginary; these are the demarcations $p = 0$ ($q > 0$) and $q = 0$ on Fig. 2.8. There is no homeomorphism taking a phase diagram into another which lies across these boundaries. Thus a centre is not a structurally stable feature of a linear system: it is readily persuaded to change into a spiral through a small modification of the right-hand side of its equation, and there is no homeomorphism connecting these.

Finally we will point out a relation between structural stability and bifurcation points of a parameter. Earlier examples illustrated how it comes about that at such a value of a parameter or vector of parameters the character of the solution structure may change radically. This can be linked to the definition of structural instability by considering that a small change in the value of the parameter, λ say, is equivalent to an additive modification of the right-hand side of the equations. This can be further modified, if necessary, to fit the boundedness requirements for a permissible $s(x)$, and this shows that the system concerned is structurally unstable at any bifurcation value of λ (see, for example, Exercises 6 and 8).

12.4. Hopf bifurcations

Some bifurcations generate limit cycles or other periodic solutions, and one such type is known as a Hopf bifurcation. Consider the system

$$\dot{x} = \mu x + y - x(x^2 + y^2), \tag{12.10}$$

$$\dot{y} = -x + \mu y - y(x^2 + y^2), \tag{12.11}$$

where μ is the bifurcation parameter. The system has a single equilibrium point, at the origin.

In polar coordinates the equations become

$$\dot{r} = r(\mu - r^2), \qquad \dot{\theta} = -1.$$

If $\mu \leqslant 0$ then the entire diagram consists of a stable spiral. If $\mu > 0$ then there is an unstable spiral at the origin surrounded by a stable limit

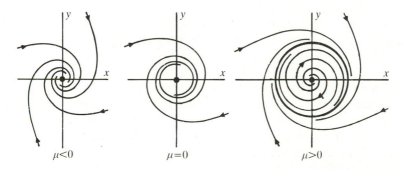

$\mu<0$ $\mu=0$ $\mu>0$

FIG. 12.7. Development of a limit cycle in a Hopf bifurcation cycle which grows out of the origin—the steps in its development are shown in Fig. 12.7. This is an example of a Hopf bifurcation. Typically for such cases the linearization of (12.10) and (12.11) predicts a centre at the origin, which proves to be incorrect: the origin changes from being asymptotically stable to being unstable without passing through the stage of being a centre.

The following is a simple version, restricted to polar-type equations, of a more general result.

THEOREM 12.1   *Given the equations*

$$\dot{x} = \mu x + y - x f(r),$$

$$\dot{y} = -x + \mu y - y f(r),$$

where $r = \sqrt{(x^2 + y^2)}$, $f(r)$ *is continuous for* $r \geqslant 0$, $f(0) = 0$, *and* $f(r) > 0$ *for* $r > 0$. *Then*

(i) *for* $\mu < 0$ *the origin is a stable spiral covering the whole plane;*
(ii) *for* $\mu = 0$ *the origin is a stable spiral;*
(iii) *for* $\mu > 0$ *there is a stable limit cycle whose radius increases from zero as* μ *increases from zero.*

Proof   Choose the Liapunov function $V(x, y) = \frac{1}{2}(x^2 + y^2)$. Then

$$\dot{V}(x, y) = r^2(\mu - f(r)) < 0$$

for $\mu \leqslant 0$ and $r > 0$. Hence by Theorem 10.2 the origin is asymptotically stable and its domain of attraction is the entire plane. That it is a spiral can be seen from the polar form of the equations

$$\dot{r} = r(\mu - f(r)), \qquad \dot{\theta} = -1.$$

Note that the case $\mu = 0$ is included.

Imagine now that time is reversed and t is replaced by $(-t)$ in the equations. Select the same Liapunov function. Then

$$\dot{V}(x, y) = r^2(f(r) - \mu) < 0$$

for $\mu > 0$ in some interval $0 < r < r_1$ by continuity of $f(r)$ and the inequality $f(r) > 0$ for $r > 0$. This proves that the origin is unstable, and the polar equation confirms that it is a spiral in $0 < r < r_1$.

Finally in some interval $0 < \mu < \mu_1$, $\mu - f(r) = 0$ must have exactly one solution for μ_1 sufficiently small. Again the polar equations confirm that this limit cycle must be stable.

Example 12.4 *Show that the equation*

$$\ddot{x} + (x^2 + \dot{x}^2 - \mu)\dot{x} + x = 0$$

exhibits a Hopf bifurcation as μ increases through zero.

We can apply Theorem 3.5 to show that the system has no periodic solutions for $\mu < 0$. Write the equation as

$$\dot{x} = X(x, y) = y,$$

$$\dot{y} = Y(x, y) = -(x^2 + y^2 - \mu)y - x.$$

Then

$$\frac{\partial X}{\partial x} + \frac{\partial Y}{\partial y} = -x^2 - 3y^2 + \mu > 0$$

for all (x, y) and all $\mu < 0$. Hence by Theorem 3.5, there can be no closed paths in the phase plane.

For the critical case $\mu = 0$, we can use the Liapunov function

$$V(x, y) = \tfrac{1}{2}(x^2 + y^2)$$

which results in

$$\dot{V}(x, y) = -(x^2 + y^2)y^2.$$

Since $\dot{V}(x, y)$ vanishes at the isolated value $y = 0$, but is otherwise strictly negative, the origin is an asymptotically stable equilibrium point (see the remark after Example 10.3). For $\mu > 0$ the system has a stable limit cycle with path $x^2 + y^2 = \mu$ (cf. Example 1.6) which evidently emerges from the origin at $\mu = 0$ with a radius which increases with μ.

12.5. Poincaré maps

Consider the autonomous system

$$\dot{x} = X(x, y), \qquad \dot{y} = Y(x, y)$$

and its phase diagram in the (x, y)-plane. Let Σ be a curve or *cross-section* of the plane with the property that it cuts each phase path transversely, that is, it is nowhere tangential to a phase path. Consider a point $A_0: (x_0, y_0)$ on the cross-section Σ shown in Fig. 12.8. If we follow the phase path through A_0 in its direction of flow then it next cuts Σ at $A_1: (x_1, y_1)$. This point is the *first return* or *Poincaré map* of the point A_0. We are not implying that such a point must exist, but if it does then it is called a first return. If we continue on the phase path, then the first return of A_1 is $A_2: (x_2, y_2)$ (see Fig. 12.8). We can represent the mapping by an operator P_Σ which is such that for the particular Σ chosen, and for

every (x, y) on Σ,

$$(x', y') = P_\Sigma(x, y)$$

where (x', y') is the point of first return of the path from (x, y). For successive returns starting from (x_0, y_0) we can use the notation

$$(x_2, y_2) = P_\Sigma(P_\Sigma(x_0, y_0)) = P_\Sigma^2(x_0, y_0); \qquad (x_n, y_n) = P_\Sigma^n(x_0, y_0).$$

FIG. 12.8. First returns associated with a spiral and a limit cycle and section Σ; $A_1, A_2, \ldots$ are first returns of $A_0, A_1, \ldots$, and B is a fixed point

The starting time for a sequence of first returns is immaterial since the system is autonomous. Note that the time lapse between returns is not in general constant.

Figure 12.8 also shows a stable limit cycle, and in this case successive first returns approach the limit cycle. From point B on the limit cycle, the first returns map on to themselves; in other words, the point B is a fixed point of the map P_Σ, and the sequence $A_0, A_1, A_2, \ldots$ approaches the *fixed point*. The same is true for a sequence of first returns on the inside of the limit cycle. We would expect the behaviour of these first returns to imply stability for the limit cycle.

Example 12.5 *Obtain the map of first returns for the differential equations*

$$\dot{x} = \mu x + y - x\sqrt{(x^2 + y^2)},$$
$$\dot{y} = -x + \mu y - y\sqrt{(x^2 + y^2)},$$

for the section Σ given by $y = 0$, $x > 0$.

Since the equations are autonomous the starting time t_0 is immaterial—assume that $t_0 = 0$. In polar coordinates the equations become

$$\dot{r} = r(\mu - r), \qquad \dot{\theta} = -1$$

with solutions

$$r = \mu r_0 / \{r_0 + (\mu - r_0) e^{-\mu t}\}, \qquad \theta = -t + \theta_0$$

where $r(0) = r_0$ and $\theta(0) = \theta_0$. $r = \mu$ is a limit cycle, and the approach directions show that it is stable. By eliminating t the paths are given by

$$r = \mu r_0 / \{r_0 + (\mu - r_0) e^{\mu(\theta - \theta_0)}\}.$$

The section given corresponds to $\theta_0 = 0$, and the required successive returns occur for $\theta = -2\pi$, -4π,... (bearing in mind that the representative points move round clockwise); with initial point $(r_0, 0)$, so

$$r_n = \mu r_0 / \{r_0 + (\mu - r_0) e^{-\mu 2n\pi}\}, \qquad \theta_n = 0 \qquad (n = 1, 2, \ldots).$$

As $n \to \infty$ the sequence of points approaches the fixed point $(\mu, 0)$, as shown on Fig. 12.9, corresponding to the intersection with the limit cycle.

In this particular case the time lapses between returns are all equal, the interval being 2π, but this will not usually be true.

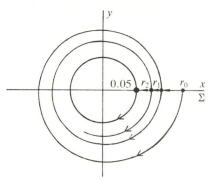

FIG. 12.9. Poincaré map of Example 12.5 with $\mu = 0.05$, $r_0 = 0.12$; the fixed point is at $r = 0.05$, $\theta = 0$

Important features of a system can be revealed by Poincaré maps. For example, as we have seen, in autonomous systems periodic solutions exist when fixed points appear, so some useful information is retained if the phase paths of Figs. 12.8 or 12.9 are deleted and only the dots of the Poincaré map retained. But for autonomous systems there is no particular advantage in using the map of first returns since the paths can be plotted (numerically if necessarily) and the phase diagrams interpreted without difficulty. For *nonautonomous systems*, on the other hand, the diagram of solution curves projected on to the (x, y)-plane which corresponds to the phase diagram in the autonomous case appears as a tangle of intersecting and self-intersecting curves, since each initial state or point in the plane generates an infinite number of curves corresponding to the various initial times t_0: see, for example, Fig. 7.11(a). Important features can be totally obscured in such a diagram, but Poincaré maps can be used to detect underlying structure, such as periodic solutions having the forcing or a subharmonic frequency.

In this context the investigation of periodic solutions, nearly-periodic solutions, and similar phenomena is to a considerable extent an exploratory matter in which computation plays at the present time a very significant part. A search for hidden periodicities, such as subharmonic periods, is best carried out by starting with a period in mind and then looking for solutions with this period. In that case a variant of the Poincaré map might seem to be more profitable—if the solutions sought are to have period T, then we should plot on the (x, y)-plane (if two dimensions are being considered) a sequence of points calculated at times T, $2T$, $3T$, ... starting from various states, and see whether any of these sequences indicate that we are approaching a periodic solution. However, this does not quite fit the definition of a Poincaré map, which does not involve any mention of time intervals but picks out intersections of a phase path under investigation with another given curve (the 'section'). The two procedures can with advantage for the analysis (especially in multidimensional cases) be brought together in a manner suggested by the following example.

Consider the first-order system

$$\dot{x} = -\tfrac{1}{8}x + \cos t \qquad (12.12)$$

The right-hand side has period 2π. In fact if t is replaced by $t - 2\pi$ throughout, the equation remains unchanged; but note that this does not mean that all the solutions have period 2π, because if we start with a given initial state and proceed through an interval of length 2π, then the state arrived at is the initial state for the next interval but generally this second state will be different from the first.

By the following artifice (12.12) can be written as an *autonomous* system of one dimension higher: we nominate a second variable, y, to take the place of t, and rewrite the relations as

$$\dot{x} = -\tfrac{1}{8}x + \cos y, \tag{12.13}$$

$$\dot{y} = 1 \tag{12.14}$$

(the independent variable is still t). It is also convenient to complete the identification of y with t by requiring

$$y(0) = 0. \tag{12.15}$$

The system (12.13) and (12.14) now has the property that if we substitute $y + 2n\pi$ for y, n being any integer, then the system is unchanged. By taking advantage of this property we can relate a plot of calculated values of x at equal time steps 2π to a Poincaré map associated with (12.13) and (12.14)—not in a plane space (x, y), but on a cyclindrical surface. The space is constructed by picking out the strip

$$-\infty < x < \infty, \qquad 0 < y \leqslant 2\pi$$

from the (x, y)-plane and wrapping it round a circular cylinder of circumference 2π, so that the side $y = 0$ becomes attached to the side

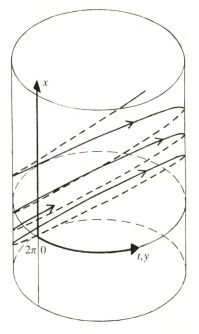

Fig. 12.10.

$y = 2\pi$. On this space, and for all times, eqs (12.13) and (12.14) hold good, and the solutions are represented by curves which wrap round the cylinder and, in the case of periodic solutions, wrap round it perhaps more than once and join up smoothly (see Fig. 12.10).

Suppose now that we are interested in marking the isolated values of $x(t)$ at times $0, 2\pi, \ldots$ Then these are simply the intersections of the solution curves in the new space with the 'section' $y = 0$, and thus constitute a Poincaré map in this space.

In a similar way the two-dimensional system

$$\dot{x} = X(x, y, t), \qquad \dot{y} = Y(x, y, t)$$

where X and Y are periodic with period 2π in t, is equivalent to the system

$$\dot{x} = X(x, y, z), \qquad \dot{y} = Y(x, y, z), \qquad \dot{z} = 1, \qquad z(0) = 0,$$

where X, Y are 2π-periodic in z. This may be visualized in a 'toroidal' space for which the plane $z = 2\pi$ is bent round to coincide with $z = 0$, and this serves as a section for a Poincaré map.

Example 12.6 *Find the map of 2π first returns on Σ: $t = 0$ for arbitrary initial states, and locate the fixed point, of the equation*

$$\ddot{x} + 3\dot{x} + 2x = 10\cos t$$

in the phase plane for which $\dot{x} = y$.

As an autonomous first-order system the equation breaks into the form

$$\dot{x} = y, \qquad \dot{y} = -2x - 3y + 10\cos z, \qquad \dot{z} = 1$$

$$z(0) = 0,$$

and the sequences of first returns appear as sequences of points in the $z = 0$, or the x, y plane. They can be found by referring to the solution with initial condition $x = x_0$, $y = y_0$, at $t = 0$

$$x = (-5 + 2x_0 + y_0)e^{-t} + (4 - x_0 - y_0)e^{-2t} + \cos t + 3\sin t.$$

The sequence starting at (x_0, y_0) in the plane $z = 0$ is given by

$$x_n = (-5 + 2x_0 + y_0)e^{-2n\pi} + (4 - x_0 - y_0)e^{-4n\pi} + 1,$$

$$y_n = -(-5 + 2x_0 + y_0)e^{-2n\pi} - 2(4 - x_0 - y_0)e^{-4n\pi} + 3$$

for $n = 1, 2, \ldots$. As $n \to \infty$, $(x_n, y_n) \to (1, 3)$ irrespective of the initial coordinates (x_0, y_0). It would not turn out to be the same point, $(1, 3)$, if we considered an initial time $t = t_0$ having a value different from zero or a multiple of 2π, but it might not be necessary to explore all these possibilities since the existence of a single fixed point must indicate either an equilibrium point or a periodic solution of period 2π or an integral multiple of 2π. In the latter case it provides an initial condition for calculating the whole cycle. Some typical first returns which approach $(1, 3)$ on radial paths are shown in Fig. 12.11.

FIG. 12.11. Poincaré map of $\ddot{x} + 3\dot{x} + 2x = 10 \cos t$ showing the fixed point at $(1, 3)$

When the exact solution to the differential equation is not obtainable, as will be the case in all instances of significance, computation must take the place of an algebraic derivation of the Poincaré map. This is fairly easy to compute and display graphically—a step-by-step solution of the equations is organized to print out a result at the end of every period, and the points are normally displayed on continuous linking curves (whose significance will be explained in Section 12.8) with the direction of movement indicated. The technique is sometimes referred to as the stroboscopic method since the representative point is, as it were, illuminated once in every cycle (see Minorsky 1962).

In seeking stable periodic solutions via the approach to fixed points of a Poincaré map it is advantageous to be able to start the calculation at an initial point as close as possible to the closed cycle in the (x, y)-plane which is sought. This is especially true if the evolution of periodic solutions as a parameter varies is being traced, as in Section 12.6, since otherwise spurious connections may be identified. For this purpose any of the approximate methods for finding periodic solutions described in the earlier chapters may be used. In the following example we indicate the use of the van der Pol plane for this purpose.

Example 12.7 *Find approximate positions for the fixed points of the Poincaré map of $2\pi/\omega$ first returns for the Duffing equation*

$$\ddot{x} + k\dot{x} + x + \beta x^3 = \Gamma \cos \omega t \qquad (k > 0, \ \beta < 0). \tag{12.16}$$

In Section 7.2 it is shown that there exists a stable oscillation of the approximate form

$$x(t) = a_0 \cos \omega t + b_0 \sin \omega t \tag{12.17}$$

where a_0 and b_0 are obtained by solving the equations

$$b\{\omega^2 - 1 - \tfrac{3}{4}\beta(a^2 + b^2)\} + k\omega a = 0,$$

$$a\{\omega^2 - 1 - \tfrac{3}{4}\beta(a^2 + b^2)\} - k\omega b = \Gamma. \tag{12.18}$$

On the 'phase plane' with $y = \dot{x}$ the approximate solution (12.17) traces out an ellipse with its centre at the origin. The true path should be a closed curve close to this ellipse. If we solve (12.18) for a_0 and b_0 numerically and substitute the values into (12.17), we may then use any point on the ellipse as a starting-point for an exact Poincaré map—say the point where $t = 0$:

$$x_0 = a_0, \qquad y_0 = \omega b_0.$$

This should lead comparatively quickly to a good estimate for the position of a fixed point which will lie somewhere on the true oscillation and serve as an initial condition for plotting the complete cycle in the (x, y)-plane. Similar techniques were used to assist in obtaining the later figures in this chapter.

12.6. Chaos and strange attractors

Recent studies of computer-generated solutions of forced second-order equations such as Duffing's equation and third-order autonomous systems have revealed unexpectedly complex solution structures arising from what might appear to be relatively simple nonlinear differential equations. The system

$$\dot{x} = a(y - x), \qquad \dot{y} = bx - y - xz, \qquad \dot{z} = xy - cz \tag{12.19}$$

(a, b, c constant) was first investigated by Lorenz (1963). The Lorenz equations, as they are now known, arose in a model for convective motion in the atmosphere. The solutions calculated by computer display very complex behaviour for a wide range of parameter values. By complex behaviour we include seemingly random or 'chaotic' output from the system although the solutions remain deterministic in terms of their initial values and there is no 'random' input. These phenomena arise specifically with third- and higher-order equations, or in forced second-order systems (they have no obvious counterpart in autonomous second-order systems). The Lorenz equations and similar models have been put forward as explanations of turbulent flow in fluid mechanics. However, certain aspects of the long-run behaviour of such systems show systematic characteristics with a degree of independence of initial conditions and parameter values, which is opening the way to a rational study even of their 'chaotic' behaviour.

An equation exhibiting some of these interesting features is the Duffing equation

$$\ddot{x} + k\dot{x} - x + x^3 = \Gamma \cos \omega t, \qquad (12.20)$$

(expressed in a convenient dimensionless form): note that this version has negative linear and positive cubic restoring terms. The corresponding autonomous system ($\Gamma = 0$) has equilibrium points at $x = \pm 1$, (stable spirals when $0 < k < 2\sqrt{2}$), and at $x = 0$ (a saddle point). As Γ increases from zero we might expect stable forced periodic solutions to develop from $x = \pm 1$ rather than from $x = 0$ as was the case in Section 7.2.

With this in view we shall look for shifted $2\pi/\omega$-periodic solutions of the form

$$x = c(t) + a(t) \cos \omega t + b(t) \sin \omega t. \qquad (12.21)$$

As with earlier applications of harmonic balance we assume that the amplitudes $a(t)$ and $b(t)$ are 'slowly-varying', so that in the subsequent working their second derivatives can be neglected (it is found in practice that this approximation would not be appropriate for $c(t)$). Also, harmonics of order higher than the first will be neglected in the approximations as before. With these assumptions, when (12.21) is substituted into (12.20) and the coefficients of $\cos \omega t$ and $\sin \omega t$, and the constant term, are matched on either side of (12.20) we find that

$$\ddot{c} + k\dot{c} = -c(c^2 - 1 + \tfrac{3}{2}r^2), \qquad (12.22)$$

$$k\dot{a} + 2\omega\dot{b} = -a(-1 - \omega^2 + 3c^2 + \tfrac{3}{4}r^2) - k\omega b + \Gamma, \qquad (12.23)$$

$$-2\omega\dot{a} + k\dot{b} = -b(-1 - \omega^2 + 3c^2 + \tfrac{3}{4}r^2) + k\omega a, \qquad (12.24)$$

where $r^2 = a^2 + b^2$. If we further put

$$\dot{c} = d, \qquad (12.25)$$

we have a van der Pol 4-space (see Chapter 7 for the van der Pol 2-space), (a, b, c, d), with paths defined by (12.23), (12.24), (12.25), and the new version of (12.22):

$$\dot{d} = -c(c^2 - 1 + \tfrac{3}{2}r^2) - kd. \tag{12.26}$$

The equilibrium points of the system, which correspond to steady oscillations, are obtained by equating the right-hand sides of (12.23)–(12.26) to zero. From (12.25), $d = 0$, and the rest become

$$c(c^2 - 1 + \tfrac{3}{2}r^2) = 0, \tag{12.27}$$

$$a(-1 - \omega^2 + 3c^2 + \tfrac{3}{4}r^2) + k\omega b = \Gamma, \tag{12.28}$$

$$b(-1 - \omega^2 + 3c^2 + \tfrac{3}{4}r^2) - k\omega a = 0. \tag{12.29}$$

By squaring and adding (12.28) and (12.29) we find that r satisfies

$$r^2[(-1 - \omega^2 + 3c^2 + \tfrac{3}{4}r^2)^2 + k^2\omega^2] = \Gamma^2. \tag{12.30}$$

There are two sets of solutions of (12.30)

TYPE I $c = 0$, $r^2[(-1 - \omega^2 + \tfrac{3}{4}r^2)^2 + k^2\omega^2] = \Gamma^2$; \tag{12.31}

TYPE II For $r \leqslant \sqrt{(2/3)}$ *only*:

$$c^2 = 1 - \tfrac{3}{2}r^2, \; r^2[(2 - \omega^2 - \tfrac{15}{4}r^2)^2 + k^2\omega^2] = \Gamma^2. \tag{12.32}$$

Figure 12.12 shows plots of r against $|\Gamma|$ using (12.30) (the curve $\mathscr{C}_1$) and (12.31) (the curve $\mathscr{C}_2$). The results are plotted for the case $k = 0.3$, $\omega = 1.2$, these values being chosen to display particularly interesting properties of the solutions. The best choice for the parameters is a matter of experiment with computed solutions and graphic displays,

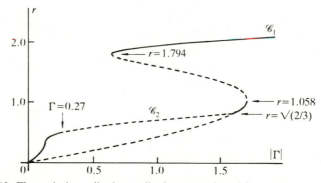

FIG. 12.12. Theoretical amplitude–amplitude curves plotted from (12.31) and (12.32); —— indicates experimentally stable period $2\pi/\omega$ solutions and – – – indicates their absence. $\mathscr{C}_1$ is the Type I solution for which a theoretical analysis of stability is given in the text. Type II solutions are shown as curve $\mathscr{C}_2$. The value $\Gamma \approx 0.27$ was found by numerical investigation. For $\Gamma > \sim 0.27$ the complex behaviour referred to in the text occurs

but for this value of k, the damping coefficient, the phenomena to be discussed occur over a wide range of frequency ω and amplitude Γ.

We need to know which of the solutions of (12.30)–(12.32) are stable. Consider Type I. Let $a = a_0$, $b = b_0$, $c = 0$ be an equilibrium point of (12.22), (12.23), and (12.24) (reverting to the form not involving d), and introduce local coordinates (ξ, η) by writing $a = a_0 + \xi$, $b = b_0 + \eta$. The linearized form of the equations is

$$\ddot{c} + k\dot{c} + c(\tfrac{3}{2}r_0^2 - 1) = 0, \tag{12.33}$$

$$k\dot{\xi} + 2\omega\dot{\eta} + A\xi + B\eta = 0, \tag{12.34}$$

$$-2\omega\dot{\xi} + k\dot{\eta} + C\xi + D\eta = 0; \tag{12.35}$$

where $r_0^2 = a_0^2 + b_0^2$ and

$$A = -1 - \omega^2 + \tfrac{9}{4}a_0^2 + \tfrac{3}{4}b_0^2, \qquad B = k\omega + \tfrac{3}{2}a_0b_0,$$
$$C = -k\omega + \tfrac{3}{2}a_0b_0, \qquad D = -1 - \omega^2 + \tfrac{3}{4}a_0^2 + \tfrac{9}{4}b_0^2.$$

Given that $k > 0$, (12.33) has asymptotically stable solutions if and only if

$$r_0^2 > \tfrac{2}{3}. \tag{12.36}$$

(12.34) and (12.35) can be written in the form

$$L\dot{\boldsymbol{\xi}} + M\boldsymbol{\xi} = 0 \tag{12.37}$$

where $\boldsymbol{\xi} = (\xi, \eta)^t$ and

$$L = \begin{pmatrix} k & 2\omega \\ -2\omega & k \end{pmatrix}, \qquad M = \begin{pmatrix} A & B \\ C & D \end{pmatrix}. \tag{12.38}$$

The eigenvalues of the system, m_1, m_2, can be obtained by solving the quadratic equation

$$|L^{-1}M - mI| = |L^{-1}||M - mL| = 0. \tag{12.39}$$

The conditions for negative eigenvalues, which implies asymptotic stability, are (see Fig. 2.8) that the coefficient of m in the quadratic form should be positive, which reduces to

$$r_0^2 > \tfrac{2}{3}(1 - \omega^2), \tag{12.40}$$

and that the constant term should be positive, or

$$|L^{-1}||M| > 0,$$

which reduces to

$$27r_0^4 - 48(1 + \omega^2)r_0^2 + 16(1 + \omega^2)^2 + 16k^2\omega^2 > 0. \tag{12.41}$$

These conditions are necessary and sufficient for asymptotic stability. For the case of Fig. 12.12, when $k = 0.3$, $\omega = 1.2$, conditions (12.40) and (12.41) are satisfied if and only if

$$r_0 > 1.794 \quad \text{or} \quad 0.816 < r_0 < 1.058.$$

The boundaries are shown in Fig. 12.12. It can also be shown that these critical values occur where the Γ, r_0-curve turns over—this is a catastrophe situation similar to that discussed in Section 7.3.

Type II, in which the centre of the oscillation is not at the origin, is somewhat more complicated to analyse, but it predicts that all solutions corresponding to $\mathscr{C}_2$ are stable for the given values of k and ω. The true behaviour is, however, far more complex than this, and we shall exhibit the complications by describing a computer exploration of solutions calculated directly from (12.20) for a range of values of Γ (in the case $k = 0.3$, $\omega = 1.2$ under discussion). We focus attention on the solutions which emerge, for small and increasing values of Γ, from the equilibrium point at $x = 1$.

Between $\Gamma = 0$ and about 0.27, stable harmonic solutions of period $2\pi/\omega$ are found, as predicted by the theory. A typical solution from this range has been depicted in two ways: for $\Gamma = 0.2$ the solution curve is plotted directly in Fig. 12.13(*a*); and in Fig. 12.15(*a*) it is shown as a closed curve on the 'phase plane' with coordinates $(\dot{x}, x)$. Figure 12.16 is designed to exhibit the later occurrence of amplitude bifurcations: on this are plotted only the 'peaks' and 'troughs' (local maxima and minima), of the simple oscillations occurring in $0 \leqslant \Gamma < 0.27$. Also

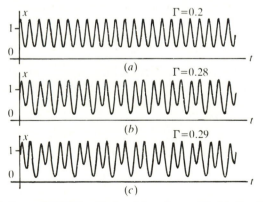

Fig. 12.13. Solutions of (12.20) with $k = 0.3$ and $\omega = 1.2$. (*a*) $\Gamma = 0.20$ (period 1); (*b*) $\Gamma = 0.28$ (period 2); (*c*) $\Gamma = 0.29$ (period 4)

indicated for comparison are the quantities $c(\Gamma) \pm r(\Gamma)$ calculated from (12.32), showing good agreement with the solutions computed directly from the differential equation.

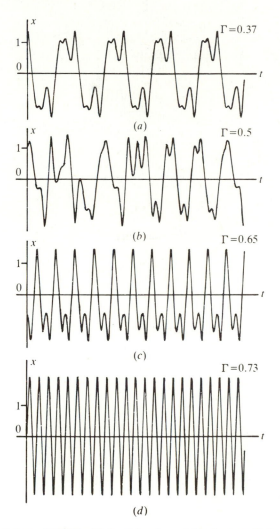

FIG. 12.14. Solutions of (12.20) with $k = 0.3$ and $\omega = 1.2$ in the cases (a) $\Gamma = 0.37$ (period 5); (b) $\Gamma = 0.5$ (chaotic response); (c) $\Gamma = 0.65$ (period 2); (d) $\Gamma = 0.73$ (period 1). (See also Figs 12.15 and 12.16)

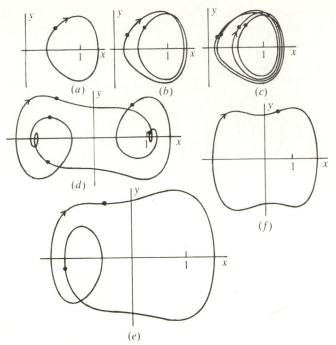

FIG. 12.15. Some typical periodic solutions of (12.20) projected on to the (x, y)-plane for $k = 0.3$ and $\omega = 1.2$ in the cases (a) $\Gamma = 0.2$ (forced period); (b) $\Gamma = 0.28$ (period 2); (c) $\Gamma = 0.29$ (period 4); (d) $\Gamma = 0.37$ (period 5); (e) $\Gamma = 0.65$ (period 2); (f) $\Gamma = 0.73$ (forced period and drawn half-size). (See also Figs. 12.12–12.14)

For $\Gamma > 0.27$ this agreement breaks down completely. Instead of a more or less harmonic solution of continuously changing amplitude a variety of complex behaviour commences. By $\Gamma = 0.28$ the harmonic solution becomes increasingly disturbed through a regularly occurring difference between adjacent cycles (see Fig. 12.13(b)). The solution is still periodic, but with period $4\pi/\omega$—this is a subharmonic of order $\frac{1}{2}$, or a *period 2 oscillation*. Figure 12.15(b) shows the double loop on the phase plane, and in Fig. 12.16 the upper and lower pairs of local maxima and minima are displayed so as to indicate amplitude bifurcation. At around $\Gamma = 0.29$ the amplitudes bifurcate again to produce a period 4 solution having period $8\pi/\omega$ (see Figs 12.13(c), 12.15(c), and 12.16). This period doubling continues over a very short range of Γ (these are not displayed), up to about $\Gamma = 0.3$.

Between $\Gamma = 0.3$ and 0.36 no obvious regularity is observable: the solution plots reveal wandering solutions of an irregularly oscillating type without any uniform pattern. Such solutions are said to display chaotic behaviour. Close to $\Gamma = 0.37$, however, solutions of period 5 appear (see Figs. 12.14(*a*), 12.15, and 12.16)—these are centred on $x = 0$. Another chaotic regime follows between $\Gamma = 0.37$ and about 0.65 (a segment of a solution plot is shown in Fig. 12.14(*b*) for $\Gamma = 0.5$). In an interval around $\Gamma = 0.65$ more period 2 solutions appear, centred somewhere close to $x = 0$ but of a very unsymmetrical type.

FIG. 12.16. $\mathscr{C}_1$ corresponds to Type I solutions and $\mathscr{C}_2$ to Type II. The curves and dots show: (–·–·–) predicted stable amplitude; (– – –) predicted unstable amplitude; (————) directly computed amplitude; (●) local max./min. of subharmonics

This goes some way to show the wide variety of solution types which can be obtained from Duffing's equation over a comparatively narrow band of forcing amplitude. As Γ increases, the production of responses at the forcing frequency seems to fail after the predicted maximum displacement $c+r$ reaches its maximum value, and do not reappear until the stable Type I oscillation is picked up at about $\Gamma = 0.66$. Figure 12.16 does not exhaust the possibilities even at the forcing frequency

$\omega = 1.2$: there may well be other response frequencies undetected. (It should be noted in particular that the 'dots' on this figure are only representative indications—isolated sections of bifurcations which in fact arise gradually, but over short intervals in Γ.)

The search for periodicities can be assisted by plotting Poincaré maps (Section 12.5) appropriate to the period being looked for: $2\pi/\omega$ for the forcing frequency, $4\pi/\omega$ for period doubling, and so on. Periodic solutions of period $2\pi/\omega$ produce maps in the (x, y)-plane having a single fixed point, period doubling two fixed points, etc. (see Fig. 12.15). If we apply the same procedure to the 'chaotic' output, say for $\Gamma = 0.5$, we find that after all there does appear to be some degree of underlying structure. Figure 12.17 shows a Poincaré map for this case. The section shown is for $t = 2n\pi$ $(n = 0, 1, 2, \ldots)$, and nearly 4000 points are displayed. Not all the points calculated are plotted—working from a given initial condition the earlier points are subject to a form of transient behaviour, and not until they find their way into the set displayed are they recorded. There are no fixed points observed in the chaotic regime: instead the set plays a similar role. The set is to a large extent independent of the initial values, and has the property that any point once in the set generates a sequence of first returns all of which lie in the set: it is described as a *strange attractor*. Computer studies point also to the fact that, given any point in the attracting set, some future first return will come arbitrarily close to it. The shape of the

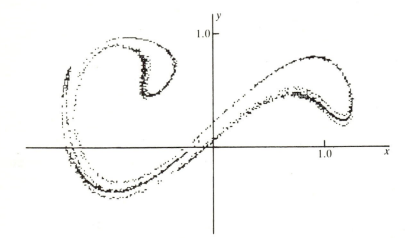

FIG. 12.17. Poincaré map of eqn (12.20) for sections $t = 2n\pi/\omega$ $(n = 0, 1, \ldots)$ showing the strange attractor for a single orbit

strange attractor depends upon which section is chosen: for example the first returns for the section $t = (2n+1)\pi$ $(n = 0, 1, 2, \ldots)$ will fill a differently-shaped region in the plane. In x, y, t-space the solutions (after the disappearance of 'transients' as above) will remain in a periodically repeated tube (or in a torus if the tube is imagined to be bent round until its ends meet), whose cross-section at $t = 0$ is of the shape shown in Fig. 12.17.

There have been theories proposed recently to explain why these strange attractors appear as the forcing amplitude increases. In the example we have been discussing, as Γ increases from zero to about 0.27 the amplitude of the response increases, tending to carry the representative point into the neighbourhood of the centre of the unstable mode of oscillation which exists close to $x = 0$ (Type I above). It is the interaction between the growing incursion of the stable periodic solution into the region of unstable oscillation which progressively destroys its periodicity (see Fig. 12.16). This subject will be discussed further in Section 12.8.

12.7. Perturbation analysis of an amplitude bifurcation

The previous Section describes the break-up, as Γ increases, of the stable Type II response at the forcing frequency; first into subharmonics of period 2, then into a cascade of bifurcating subharmonics of periods 4, 8, and so on, then into a chaotic region punctuated by subharmonics of other orders, until finally another stable response at the forcing frequency is encountered which is of Type I. The value of r, the amplitude, which signals the onset of this behaviour by the appearance of a period 2 subharmonic beginning as a small disturbance of the main harmonic, can be estimated as follows by a simple perturbation procedure.

The equation for $x(t)$ is

$$\ddot{x} + k\dot{x} - x + x^3 = \Gamma \cos \omega t. \tag{12.42}$$

In the neighbourhood of the critical point, which occurs at an unknown value of r, $r = r_0$, where $\Gamma = \Gamma_0$ say (calculated from (12.32)), we put

$$x(t) = x_1(t) + \varepsilon x_{\frac{1}{2}}(t), \tag{12.43}$$

$$\Gamma = \Gamma_0 + \varepsilon \gamma.$$

Here

$$x_1(t) = c + a \cos \omega t + b \sin \omega t \tag{12.44}$$

where c, a, b are given through eqns (12.32), (12.28), and (12.29), and x_1 represents the previously stable solution of period $2\pi/\omega$.

$$x_{\frac{1}{2}}(t) = A \cos \tfrac{1}{2}\omega t + B \sin \tfrac{1}{2}\omega t, \tag{12.45}$$

represents the gradually superimposed subharmonic starting with zero amplitude at r_0, Γ_0. The ε is really a dummy symbol enabling a reasonable separation into orders of magnitude to be achieved.

By substituting (12.44) and (12.45) into (12.42) we obtain

$$\ddot{x}_1 + k\dot{x}_1 - x_1 + x_1^3 + \varepsilon(\ddot{x}_{\frac{1}{2}} + k\dot{x}_{\frac{1}{2}} - x_{\frac{1}{2}} + 3x_1^2 x_{\frac{1}{2}} + O(x_{\frac{1}{2}})^2)$$

$$= (\Gamma_0 + \varepsilon\gamma)\cos\omega t.$$

From a match between the terms independent of ε

$$\ddot{x}_1 + k\dot{x}_1 - x_1 + x_1^3 = \Gamma \cos\omega t,$$

which is the same as (12.42) and so has the solution approximately of the form (12.44), with a, b, and c determined from Γ_0 via (12.32), (12.28), and (12.29). This is the generating solution from which the subharmonic starts to grow. By matching the remaining terms, neglecting the square and cube of $x_{\frac{1}{2}}$, we find

$$\ddot{x}_{\frac{1}{2}} + k\dot{x}_{\frac{1}{2}} + (3x_1^2 - 1)x_{\frac{1}{2}} = \gamma \cos\omega t. \tag{12.46}$$

When powers of the trigonometric functions are reduced to sums of sines and cosines at $\tfrac{1}{2}\omega t$, ωt, and $\tfrac{3}{2}\omega t$, and only the dominant frequency $\tfrac{1}{2}\omega t$ is retained, (12.46) reduces to terms of r^2 only, becoming

$$\{A(2 - \tfrac{1}{4}\omega^2 + 3ac - 3r^2) + B(\tfrac{1}{2}\omega k + 3bc)\} \cos \tfrac{1}{2}\omega t$$

$$+ \{A(-\tfrac{1}{2}\omega k + 3bc) + B(2 - \tfrac{1}{4}\omega^2 - 3ac - 3r^2)\} \sin \tfrac{1}{2}\omega t = 0.$$

The coefficients of $\cos \tfrac{1}{2}\omega t$, $\sin \tfrac{1}{2}\omega t$ must vanish, giving a pair of linear, homogeneous equations for A and B. For nonzero solutions the appropriate determinant must be zero, and by using eqns (12.32) the condition becomes

$$(2 - \tfrac{1}{4}\omega^2 - 3r^2)^2 - 9r^2(1 - \tfrac{3}{2}r^2) + \tfrac{1}{4}\omega^2 k^2 = 0.$$

Using $\omega = 1.2$, $k = 0.3$, the smallest solution is $r = 0.43$, which may be compared with the corresponding value found experimentally (Fig. 12.16) of about 0.52. (The second solution, in the neighbourhood of $r = 0.8$, does not correspond to the requirement that $x_{\frac{1}{2}}$ should grow out of a perturbation of a harmonic solution at the forcing frequency). The actual agreement may be better than the value 0.52 suggests, since it is in principle impossible to decide numerically the exact point at which a smoothly-growing disturbance just emerges from the value zero.

12.8. Homoclinic bifurcation

The subharmonic analysis described above does not assess the stability of the harmonic and subharmonic responses, nor does it supply the mechanism which eventually triggers the period doubling and the chaotic behaviour. The theoretical discussion of the previous section does indicate that subharmonics of order $\frac{1}{2}$ are available whilst the numerical evidence (Fig. 12.16) shows that period doubling occurs for $\Gamma > 0.27$. The cause of the progressive bifurcations is a complex and less obvious phenomenon.

For $\Gamma < 0.27$, approximately, Duffing's eqn (12.42) has three periodic solutions, two stable ones symmetrically disposed about the origin and one unstable one centred at the origin. It is the unstable periodic solution which causes bifurcations in the harmonic solutions. This process is best viewed by constructing maps of Poincaré first returns. For first returns with initial time $t = 0$ and period $2\pi/\omega$, the unstable periodic solution has a fixed point just as the stable ones have, but its fixed point is unstable. Consider for a moment the linear system

$$\ddot{x} + \dot{x} - 2x = 10 \cos t.$$

As in Example 12.6 the first returns from (x_0, y_0) are given by

$$x_n = \tfrac{1}{3}(2x_0 + y_0 + 5)\, e^{2n\pi} + \tfrac{1}{3}(x_0 - y_0 + 4)\, e^{-4n\pi} - 3,$$

$$y_n = \tfrac{1}{3}(2x_0 + y_0 + 5)\, e^{2n\pi} - \tfrac{2}{3}(x_0 - y_0 + 4)\, e^{-4n\pi} + 1,$$

but in this case the fixed point $(-3, 1)$ is unstable, in the sense that there are points in its neighbourhood from which the set of successive returns do not approach the fixed point—this arises from the instability of the oscillation corresponding to the fixed point. We shall identify two special sets of points associated with the fixed point. One is *the set such that the first returns from all points in it approach the fixed point* as $t \to \infty$—this is called the *stable manifold*, and in our case it consists of the straight line‘

$$2x_0 + y_0 + 5 = 0,$$

The other is *the set from which the first returns approach the fixed point when time is reversed*, that is, as $t \to -\infty$—this is called the *unstable manifold*, and is given in our case by

$$x_0 - y_0 + 4 = 0,$$

(a starting time $t = 0$ always being assumed). These curves are analogous to the separatrices of a saddle point in the plane for an autonomous system.

Construct now the same manifolds for the unstable fixed point of Duffing's equation. For Γ small we can expect stable and unstable manifolds since, in the unforced case $\Gamma = 0$, these curves reduce to the separatrices of the saddle point in the phase diagram. They have been computed and are shown in Fig. 12.18. If A denotes the fixed point in the figure, then DA and EA are the stable manifolds and AB and AC the unstable manifolds.

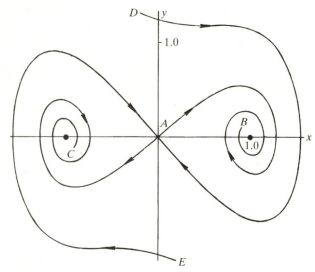

FIG. 12.18. Stable and unstable manifolds for the unforced Duffing equation (12.42) with $\Gamma = 0$, $k = 0.3$, $\omega = 1.2$

These manifolds have also been computed for the cases $\Gamma = 0.20$, $\Gamma = 0.28$, and $\Gamma = 0.50$ (again with $k = 0.3$, $\omega = 1.2$ as in the cases discussed in Section 12.6), and their graphs are displayed in Fig. 12.19. In the unforced case (Fig. 12.18) the unstable manifolds approach *equilibrium points*. In Fig. 12.19(a), the unstable manifolds AB and AC approach fixed points on the *stable periodic solutions*: the right-hand fixed point S corresponds to the oscillation in Fig. 12.15(a). In Fig. 12.19(b), the unstable manifolds go to the stable *subharmonic* represented by S', S'', which corresponds to the oscillation shown in Fig. 12.15(*b*).

The unstable manifolds AB and AC are plotted by first locating the fixed point A and the directions of the unstable manifolds at A, and then by iterating first returns from a succession of initial values over a short segment of the manifolds near A. The stable manifolds can be computed in a similar manner by reversing time.

Manifolds from the same fixed point are said to be *homoclinic*. For $\Gamma = 0.2$ in Fig. 12.19(*a*), the homoclinic manifolds do not intersect although they appear to be closer at their closest approach than they were in the unforced case. For $\Gamma = 0.28$ in Fig. 12.19(*b*) the paths AB and DA, and the paths AC and EA have now intersected. These homoclinic intersections have important implications for the solutions of Duffing's equation. Consider the case of a homoclinic crossing as represented in Fig. 12.20, and let the first intersection occur at M_0. Let the first return of M_0 on AD be M_1. By the uniqueness of solutions of this differential equation the first return of M_0 on AB must also be M_1. If this were not the case there would exist two distinct solutions starting from M_0 at time $t = 0$. The intersections are 'orientation-preserving' in the sense that AB crosses DA in the same direction at M_0 and M_1 implying that there must be at least one further intersection between them. This matter will not be discussed here. By repetition of the argument there must be further crossings of the manifolds at the first returns M_2, M_3,.... Since the approach to the fixed point is asymptotic there will be an infinity of intersections of AB with DA. Since the unstable manifold must continually intersect the stable manifold it can no longer approach any other fixed point, and a bifurcation must therefore take place for some value of Γ. It occurs when *homoclinic tangency* first appears between the manifolds at about $\Gamma = 0.26$. Even though the unstable manifold breaks away from the fixed point at this tangency, the fixed point appears to remain for a short interval of Γ-values above $\Gamma = 0.26$. This phenomenon is known as *homoclinic bifurcation*. A similar homoclinic bifurcation takes place at what appears to be the same value of Γ to the left of A between EA and AC.

There must also be a further sequence of intersections between AD and AB arising from prior returns. This can be visualized by reversing time in the problem and considering first returns which arise from M_0. We can denote the first return by M_{-1} which must lie on both AM_0 and AD. There must exist a sequence of returns M_{-r} which approach A along M_0A. A loop arising from such a return is shown in Fig. 12.19(*b*).

(a)

(b)

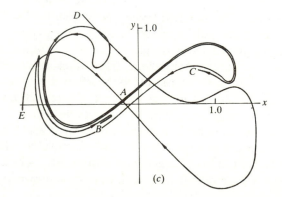

(c)

Any contact between the stable and unstable manifolds implies the existence of a solution of the differential equation, through this point of contact, which approaches the unstable periodic solution both as $t \to \infty$ and as $t \to -\infty$.

The manifolds for $\Gamma = 0.5$ are shown in Fig. 12.19(c). In this case the unstable manifold AC crosses both DA and EA, which increases the complexity of the first returns at points of intersection of the manifolds. Fig. 12.20 only represents a simplified view of the oscillations of the unstable manifold. As we shall see, it is possible for the manifold to backtrack on itself and intersect the stable manifold between the two previous intersections.

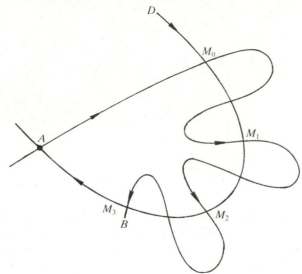

FIG. 12.20. Behaviour of the unstable manifold in a homoclinic bifurcation; there will be prior intersections between AM_0 and AD which are not shown in the figure

Some idea of what happens to these curves in the large can be given by the following descriptive line of reasoning. Choose a segment of the first loop of the unstable manifold between two points L and M on AB as shown in Fig. 12.21. (Homoclinic intersections for Duffing's equation rapidly become compressed against the unstable manifold near A which

FIG. 12.19. Stable and unstable manifolds for the forced Duffing equation (12.42) for $k = 0.3$; $\omega = 1.2$ in the cases (a) $\Gamma = 0.20$; (b) $\Gamma = 0.28$; (c) $\Gamma = 0.50$. A prior 'return loop' from the stable manifold AD is also shown in (b); the dotted link on this manifold includes a large loop which crosses into the half-plane $x > 0$. In (a) and (b) the stable fixed points for $x > 0$ are shown. There is some exaggeration in curve separation in (b) and (c)

FIG. 12.21. Horseshoe domain in a simple homoclinic bifurcation

makes it very difficult to illustrate realistic detail—Fig. 12.21 is an expanded version of a possible developed homoclinic bifurcation.) L and M must be chosen to be on the same side of AD. Let $\mathcal{H}_0$ consist of a set of initial points lying close to the segment LM in the manner shown in Fig. 12.21, that is to say in the form of a horseshoe (which is the conventional name) crossed twice by the stable manifold AD.

   Now consider mappings of the horseshoe generated by P: let $\mathcal{H}_0$ be mapped by P, P^2, P^3,... into $\mathcal{H}_1$, $\mathcal{H}_2$, $\mathcal{H}_3$,...; or backwards by P^{-1}, P^{-2},... into $\mathcal{H}_{-1}$, $\mathcal{H}_{-2}$,.... We expect, from consideration of the saddle-like character of the fixed point A, that, as we repeatedly map backwards once, twice,..., r times, the sets $P^{-r}(\mathcal{H}_0)$, or $\mathcal{H}_{-r}$, will progressively form collections of points well spread out but falling eventually into a narrow strip which overlaps the horseshoe as shown schematically on Fig. 12.21 for a particular value of r (Note that LM maps into $L'M'$.) The intersections are shown shaded. Choose any point Q from the shaded region, which belongs both to $\mathcal{H}_{-r}$ and $\mathcal{H}_0$: then

$P^r(Q) \in \mathscr{H}_0$ since $Q \in \mathscr{H}_{-r}$, and $P^r(Q) \in \mathscr{H}_r$ since $Q \in \mathscr{H}_0$ and the definition of $\mathscr{H}_r$. Therefore $\mathscr{H}_0$ and $\mathscr{H}_r$ have a common point, $P^r(Q)$, so that $\mathscr{H}_0$ and $\mathscr{H}_r$ intersect, implying a distortion of the simple 'horseshoe' shape attributable to $\mathscr{H}_r$ in the simplest picture. The loop of the unstable manifold which is contained in $\mathscr{H}_r$ is carried along with it into the neighbourhood of LM: the type of picture which emerges in practice is illustrated in Fig. 12.21 for the case $r = 1$. If $P^r(Q)$ also lies in $\mathscr{H}_{-r}$, then $P^{2r}(Q)$ lies in $\mathscr{H}_0$ with the result that another loop has to interlace between the existing paths. There also exists the possibility of period r solutions through the possible relation $P^r(Q) = Q$.

As Γ increases beyond about 0.28, the fixed points of the period-doubled solutions are limit points for initial values which are not on either the stable or the unstable manifolds. The phenomenon of period doubling will be explained later in terms of difference equations which correspond to the map P. The increasing complexity of both branches of the unstable manifold, which can intersect both stable manifolds as in Fig. 12.19(c), means that the first returns of solutions of the differential equation rapidly become entrapped by the convolutions of the unstable manifold, and thereafter approach a limit set which lies on the unstable manifold of A as indicated by Fig. 12.17 for $\Gamma = 0.5$. This is the situation in any 'chaotic' region where the limit set is the strange attractor. The first few returns of a solution starting at (0.5, 0.5) for $\Gamma = 0.5$ are shown in Fig. 12.22. Comparison with Fig. 12.17 shows that these returns are on the attracting set of points in that figure.

There are still intervals of Γ values where, within the chaotic region,

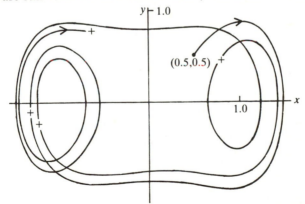

FIG. 12.22. Solution of (12.42) with initial value (0.5, 0.5) at $t = 0$ projected on to the (x, y)-plane in the case $k = 0.3$, $\omega = 1.2$, $\Gamma = 0.5$. The crosses ($+$) mark successive first returns

order seems to reappear: for example Fig. 12.15(*d*) confirms that a period 5 solution occurs in an interval around $\Gamma = 0.37$. The precise details of the mechanism which creates the attracting set of the strange attractor are difficult to put together for Duffing's equation, but the phenomenon can be accounted for in a general way by analysing the iterated effect of the horseshoe map described earlier—by using assumptions similar to those involved in discussing the convolutions of the unstable manifold it can be shown that there does exist a set of points having the attracting property, as follows.

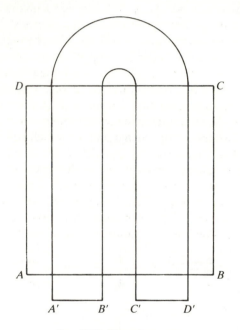

FIG. 12.23. Horseshoe map

Assume for simplicity that homoclinic bifurcation causes a square *ABCD* (Fig. 12.23) to be mapped in a particular manner, which will be specified, into a horseshoe *A'B'C'D'*, by the operator *P* of first returns (again for simplicity we take $r = 1$). The mapping is assumed to be carried out in this way: the square is stretched in the direction *AD*, compressed in the direction *AB*, bent through 180°, and placed back over the square (Fig. 12.23). Suppose the mapping is repeated for such points as still lie in the square, and that this process is iterated. Figure

FIG. 12.24. Successive horseshoe maps showing generation of the Cantor set. Sections similarly shaded correspond under the mapping

12.24 shows the first two iterations. The horizontal shaded strips in the square are *chosen* so as to map onto the vertical parts of the horseshoe. These two vertical strips now map onto the pair of thinner horseshoes. We are interested, at each stage, in the parts of the mapped regions which are shared by the regions from which they are mapped—for the first two stages, shown in Fig. 12.24; these are shown as the hatched regions. After two iterations there are points remaining in 16 'squares'; after the third iteration there will be 64 'squares', and so on. The limit set as the number of iterations tends to infinity turns out to be very much like a *Cantor set*. In one-dimensional terms a Cantor set can be formed as follows. Consider the interval $[0, 1]$. Delete the open interval $(\frac{1}{3}, \frac{2}{3})$. Now delete the middle thirds $(\frac{1}{9}, \frac{2}{9})$ and $(\frac{7}{9}, \frac{8}{9})$ of the remaining intervals, and carry on deleting middle thirds. The limit set of this process is a Cantor set, and the set is uncountable. The limit set of the horseshoe map has a similar, but two-dimensional, structure. The implication is that there exists an uncountable number of points in the initial square which, when treated as initial states at $t = 0$ for iterated first returns, lead ultimately to endlessly repeated scans of a certain set of points—the limit set—which constitutes the strange attractor. The associated oscillations will include periodic motions and bounded non-periodic motions, as indicated by Fig. 12.17. The elements of this set are distributed on the unstable manifold since horseshoes can be constructed for each loop across the stable manifold. An extensive account of homoclinic bifurcations and horseshoe maps can be found in Guckenheimer and Holmes (1983).

The properties of Poincaré maps are closely connected with those of nonlinear difference equations. For any differential equation it is impossible to form the difference equation for the first returns without first knowing the exact solution with the same initial conditions. This is certainly the case with Duffing's equation. It is possible to obtain an approximation to the difference equation (Holmes 1979), and many numerical and analytical studies have been made of difference equations which have similar properties. The Duffing map is cubic as we might expect. A cubic map with properties similar to this is the map

$$u_{i+1} = v_i,$$

$$v_{i+1} = \mu u_i + v v_i - v_i^3, \quad (\mu > 0)$$

where μ and v are parameters. Here (u_{i+1}, v_{i+1}) are the first returns of (u_i, v_i) and the elapse time is one. Periodic solutions occur if $u_{i+1} = u_i$ and $v_{i+1} = v_i$ for all i; period 2 solutions occur if $u_{i+2} = u_i$ and $v_{i+2} = v_i$ $(u_{i+1} \neq u_i, v_{i+1} \neq v_i)$ for all i; and so on. We shall simply illustrate here that a period 2 bifurcation can take place for the case $v = 0$. The general case really requires computational assistance to find the bifurcation cascade which leads to period doubling. The difference equations have periodic solutions at $(0, 0)$ and also at $\pm(\sqrt{(\mu-1)}, \sqrt{(\mu-1)})$ provided $\mu > 1$. The solution at the origin is stable if $\mu < 1$, and unstable if $\mu > 1$.

We now search for period 2 solutions for $\mu > 1$, by putting

$$u_{i+2} = u_i = u, \qquad v_{i+2} = v_i = v \text{ (say)}.$$

Then

$$u_{i+2} = v_{i+1}, \quad v_{i+2} = \mu u_{i+1} - v_{i+1}^3$$

Hence

$$u = \mu u - v^3, \quad v = \mu v - (\mu v - v^3)^3$$

or

$$u(\mu-1) = v^3, \quad v(\mu-1) = (\mu u - v^3)^3.$$

Elimination of u gives

$$v = 0 \quad \text{or} \quad v^8 = (\mu-1)^4.$$

Hence there always exists a period 2 solution. Although we must exclude period one solutions from this set, genuine period 2 solutions exist for $\mu > 1$. In this simple cubic map the fixed point bifurcates into a period 2 map at $\mu = 1$. No further bifurcations take place in this example, unlike the Duffing model where further bifurcations take place to produce period 4, 8, 16 solutions and ultimately the strange attractor.

Further details of cubic maps associated with the Duffing oscillator can be found in the paper by Holmes (1979).

Exercises

1. Find the bifurcation points of the linear system $\dot{\mathbf{x}} = A(\lambda)\mathbf{x}$ with $\mathbf{x} = (x_1, x_2)^t$ and $A(\lambda)$ given by

(i) $A(\lambda) = \begin{pmatrix} -2 & \frac{1}{4} \\ -1 & \lambda \end{pmatrix}$;

(ii) $A(\lambda) = \begin{pmatrix} \lambda & \lambda-1 \\ 1 & \lambda \end{pmatrix}$.

2. In a conservative system, the potential is given by

$$\mathscr{V}(x, \lambda) = \tfrac{1}{3}x^3 + \lambda x^2 + \lambda x$$

(cf. eqn (12.2)). Find the equilibrium points of the system, and show that it has bifurcation points at $\lambda = 0$ and at $\lambda = 1$. What type of catastrophes occur for $\lambda < 0$ and $\lambda > 1$?

3. Let

$$\mathscr{V}(x, \lambda, \mu) = \tfrac{1}{4}x^4 - \tfrac{1}{2}\lambda x^2 + \mu x$$

as in eqn (12.4). Draw projections of the bifurcations given by the 'vertical' tangents to the cusp catastrophe surface $x^3 - \lambda x + \mu = 0$ on to both the (x, λ)-plane and the (x, μ)-plane. Draw rough sketches of the sections (i) $\lambda = 0$; (ii) $\lambda = 1$; (iii) $\mu = 0$; (iv) $\mu = 1$ of the surface. (This should help to visualize the shape of the surface.)

4. Discuss the stability of the equilibrium points of the parameter-dependent conservative system

$$\ddot{x} = -\mathscr{V}_x(x, \lambda)$$

where $\mathscr{V}(x, \lambda) = \tfrac{1}{4}x^4 - \tfrac{1}{2}\lambda x^2 + \lambda x$. Show that the equation exhibits a fold catastrophe for the equilibrium states for $\lambda \geqslant 27/4$.

5. Discuss the bifurcations of the following one-dimensional systems:
   (i) $\dot{x} = x^2 - \lambda$;
   (ii) $\dot{x} = x^2 - \lambda x$,
   (iii) $\dot{x} = \lambda x - x^3$.

6. Use the result in Exercise 5 (iii) as a guide to finding an $s(x)$ (Section 12.3) which will serve to show that the equation $\dot{x} + x^3 = 0$ is structurally unstable.

7. Consider the system $\dot{x} = y$, $\dot{y} = x(\lambda - x^2)$, $-\infty < \lambda < \infty$. Investigate the phase diagrams for $\lambda < 0$, $\lambda = 0$, and $\lambda > 0$. Describe the bifurcation of the system as λ increases through zero.

8. Use Question 7 as a guide to an $\mathbf{s}(x)$ (section 12.3) which will serve to show that the system $\dot{x} = y$, $\dot{y} = -x^3$ is structurally unstable.

9. Investigate the bifurcation of the system

$$\dot{x} = x, \quad \dot{y} = y^2 - \lambda,$$

at $\lambda = 0$. Show that, for $\lambda > 0$, the system has an unstable node at $(0, \sqrt{\lambda})$ and a saddle point at $(0, -\sqrt{\lambda})$. Sketch the phase diagrams for $\lambda < 0$, $\lambda = 0$, and $\lambda > 0$. (This is an example of a *saddle-node* bifurcation.)

10. A *homoclinic path* is a phase path which joins an equilibrium point to itself in an autonomous system. Show that

$$\dot{x} = y, \quad \dot{y} = x - x^2$$

has such a path and find its equation. (In this case the equilibrium point is a saddle and the path is called a *homoclinic saddle connection*.) Sketch the phase paths for the perturbed system

$$\dot{x} = y + \lambda x, \quad \dot{y} = x - x^2,$$

for both $\lambda > 0$ and $\lambda < 0$. (The homoclinic saddle connection is destroyed by the perturbation; the system undergoes what is known as a *homoclinic bifurcation* at $\lambda = 0$, and is structurally unstable.)

11. A *heteroclinic path* is a phase path which joins two different equilibrium points. Find the heteroclinic saddle connection for the system

$$\dot{x} = xy, \quad \dot{y} = 1 - y^2.$$

Sketch the phase paths of the perturbed system

$$\dot{x} = xy + \lambda, \quad \dot{y} = 1 - y^2$$

for both $\lambda > 0$ and $\lambda < 0$. Show that the perturbed system has no saddle connections.

12. Let

$$\dot{x} = -\mu x + y + x/(1 + x^2 + y^2),$$
$$\dot{y} = x - \mu y + y/(1 + x^2 + y^2).$$

Show that the equations display a Hopf bifurcation as $\mu > 0$ decreases through $\mu = 1$. Find the radius of periodic solution for $0 < \mu < 1$.

13. Show that the system

$$\dot{x} = x - \gamma y - x(x^2 + y^2), \quad \dot{y} = \gamma x + y - y(x^2 + y^2) - \gamma \quad (\gamma > 0),$$

has a bifurcation point at $\gamma = \frac{1}{2}$, by investigating the numbers of equilibrium points for $\gamma > 0$.

14. Obtain the solutions for the usual polar coordinates r and θ in terms of t, for the system

$$\dot{x} = x + y - x(x^2 + y^2), \quad \dot{y} = -x + y - y(x^2 + y^2).$$

Let Σ be the section $\theta = 0$, $r > 0$. Find any fixed points for first returns on this section, and obtain a Poincaré map for an arbitrary initial point in this section.

15. Find the map of 2π first returns on the section Σ: $t = 0$ for

$$\ddot{x} + 2\dot{x} + 2x = 2 \sin t$$

in the usual phase plane. Find also the coordinates of the fixed point of the map and discuss its stability. Where is the fixed point of the map if the section is $t = \frac{1}{2}\pi$?

16. Using averaging methods show that the van der Pol equation

$$\ddot{x} + \varepsilon(x^2 - \mu)\dot{x} + x = 0$$

for ε positive and small undergoes a Hopf bifurcation as μ increases through zero.

17. Let x_1 satisfy

$$\ddot{x}_1 + \tfrac{1}{4}\omega^2 x_1 = \Gamma \cos \omega t$$

Obtain the solutions for x_1 and $x_2 = \dot{x}_1$ given that $x_1(0) = x_{10}$ and $x_2(0) = x_{20}$. Let Σ be the section $t = 0$ and find the first returns of period $2\pi/\omega$. Show that the mapping is

$$P_\Sigma(x_{10}, x_{20}) = \left(-x_{10} - \frac{8\Gamma}{3\omega^2}, x_{20} \right),$$

and that

$$P_\Sigma^2(x_{10}, x_{20}) = (x_{10}, x_{20}).$$

Deduce that the system exhibits period doubling for all initial values except one. Find the coordinates of this fixed point.

18. For the autonomous system

$$\dot{x} = X(x, y), \quad \dot{y} = Y(x, y),$$

it is possible to plot T-period first returns by introducing the section z = constant in the x, y, z space with z defined by $\dot{z} = 1$ with $z(0) = 0$, and $z \in \mathbb{R}$ (modulo T). In fact the introduction of z is merely a device to make the first returns conform to the definition. The Poincaré map appears as a pattern of dots along the phase paths with successive elapse times of T.

With $T = 1$, sketch some typical Poincaré maps for

$$\dot{x} = \lambda x, \quad \dot{y} = \lambda y$$

for each of the cases $\lambda < 0$, $\lambda = 0$, $\lambda > 0$. Discuss the nature of any fixed points in each case. Assume that $x(0) = x_0$, $y(0) = y_0$ and show that

$$x_{n+1} = e^\lambda x_n, \quad y_{n+1} = e^\lambda y_n, \quad n = 0, 1, 2, \ldots.$$

19. Two rings can slide on two fixed horizontal wires which lie in the same vertical plane with separation a. The two rings are connected by a spring of unstretched length l and stiffness μ. The upper ring is forced to move with displacement $\phi(t)$ from a fixed point O as shown in Fig. 12.25. The resistance on the lower ring which has mass m is assumed to be $mk \times (\text{speed})$. Let y be the relative displacement between the rings. Show that the equation of motion of the lower ring is given by

$$\ddot{y} + k\dot{y} - \frac{\mu}{ma}(l-a)y + \frac{\mu l}{2ma^3}y^3 = -\ddot{\phi} - k\dot{\phi}$$

to order y^3. (If $l > a$ then a Duffing equation of the type discussed in Section 12.6 can be obtained. This could be a model for a strange attractor for the appropriate parameter values. Cosine forcing can be reproduced by putting $-\ddot{\phi} - k\dot{\phi} = \Gamma \cos \omega t$.)

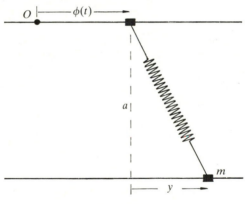

FIG. 12.25.

20. Search for period doubling in the undamped Duffing equation

$$\ddot{x} - x + x^3 = \Gamma \cos \omega t$$

using the form

$$x = c + a_1 \cos \omega t + a_2 \cos \tfrac{1}{2}\omega t,$$

where c, a_1, and a_2 are constants. If frequencies $3\omega/2$ and above are neglected, show that the shift and the amplitudes satisfy

$$c[-1+c^2+\tfrac{3}{2}(a_1^2+a_2^2)]+\tfrac{3}{4}a_2^2a_1 = 0,$$

$$a_1(-\omega^2-1+3c^2+\tfrac{3}{4}a_1^2+\tfrac{3}{2}a_2^2)+\tfrac{3}{2}a_2^2c = \Gamma,$$

$$a_2(-\tfrac{1}{4}\omega^2-1+3c^2+\tfrac{3}{2}a_1^2+3ca_1+\tfrac{3}{4}a_2^2) = 0.$$

Deduce that for harmonic solutions ($a_2 = 0$), c and a_1 are given by solutions of

(i) $c = 0$, $a_1(-\omega^2-1+\tfrac{3}{4}a_1^2) = \Gamma$,

or

(ii) $c^2 = 1-\tfrac{3}{2}a_1^2$, $a_1(-\omega^2+2-\tfrac{15}{4}a_1^2) = \Gamma$.

Sketch the amplitude $|a_1|$–amplitude $|\Gamma|$ curves corresponding to Fig. 12.12. Explain why subharmonic bifurcations (period doubling, that is) might start where c and a_1 satisfy

$$-\tfrac{1}{4}\omega^2-1+3c^2+\tfrac{3}{2}a_1^2+3ca_1 = 0$$

and either (i) or (ii) above (the question of stability is unresolved by this analysis). The amplitude $|a_1|$ at which period doubling emerges can be obtained by solving these equations for a_1.

21. Design a computer program to plot $2\pi/\omega$ first returns for the system

$$\dot{x} = X(x, y, t), \quad \dot{y} = Y(x, y, t)$$

where $X(x, y, t)$ and $Y(x, y, t)$ are $2\pi/\omega$-periodic functions of t. Apply the program to the following systems:

(i) $X(x, y, t) = y$, $Y(x, y, t) = -ky+x-x^3+\Gamma\cos\omega t$

for $k = 0.3$, $\omega = 1.2$, and Γ taking a selection of values between 0 and 0.8. Let the initial section be $t = 0$.

(ii) $X(x, y, t) = y$, $Y(x, y, t) = -ky-x^3+\Gamma\cos t$

for $k = 0.1$ and $\Gamma = 12$. Show that a strange attractor becomes visible on a graphics plot of the Poincaré map. Use the initial section $t = 0$.

22. Find the equations of the stable and unstable manifolds in the (x, y)-plane of

$$\ddot{x}+\dot{x}-2x = 10\cos t, \quad \dot{x} = y$$

for Poincaré maps of period 2π and initial time $t = 0$.

Appendix A: Existence and uniqueness theorems

We state, without proof, certain theorems of frequent application in the text.

THEOREM A1   *For the nth order system*

$$\dot{\mathbf{x}} = \mathbf{f}(\mathbf{x}, t) \tag{A1}$$

suppose that $\mathbf{f}$ is continuous and that $\partial f_j / \partial x_i$, $i, j = 1, 2, \ldots, n$ are continuous for $\mathbf{x} \in \mathscr{D}, t \in I$, where $\mathscr{D}$ is a domain and I is an open interval. Then if $\mathbf{x}_0 \in \mathscr{D}$ and $t_0 \in I$, there exists a solution $x^*(t)$, defined uniquely in some neighbourhood of $(\mathbf{x}_0, t_0)$, which satisfies $\mathbf{x}^*(t_0) = \mathbf{x}_0$.

We call such a system *regular* on $\mathscr{D} \times I$ (the set $(\mathbf{x}, t)$ where $\mathbf{x} \in \mathscr{D}$ and $t \in I$). For brevity, we refer in the text to a system regular on $-\infty < x_i < \infty, i = 1, 2, \ldots, n$, $-\infty < t < \infty$ as being a *regular system*. Points at which the conditions of the theorem apply are called *ordinary points*.

The theorem states only that the initial value problem has a unique solution on some sufficiently small interval about t_0, a limitation arising from the method of proof. A question of importance is how far the solution actually extends. It need not extend through the whole of I, and some possibilities are shown in the following examples.

Example A1 $\dot{x}_1 = x_2, \dot{x}_2 = -x_1, \quad x_1(0) = 0, x_2(0) = 1$.

$\mathscr{D}$ is the domain $-\infty < x_1 < \infty, -\infty < x_2 < \infty$, and I is $-\infty < t < \infty$. The solution is $x_1^*(t) = \sin t, x_2^*(t) = \cos t$. This solution is defined and is unique on $-\infty < t < \infty$.

Example A2   *The one-dimensional equation $\dot{x} = 3x^2 t^2, \quad x(0) = 1$.*

$\mathscr{D}$ is $-\infty < x < \infty$, and I is $-\infty < t < \infty$. But

$$x^*(t) = (1 - t^3)^{-1}, \; ($$

so the solution is only valid in $-\infty < t < 1$, since $x^*(t)$ approaches infinity as $t \to 1-$.

Example A3   *The one-dimensional equation $\dot{x} = 2|x|^{1/2}, x(0) = 1$.*

A family of solutions is $x(t) = (t + c)^2, t > -c$ (so that $\dot{x} > 0$); there also exists the solution $x(t) \equiv 0$. The required solution is shown in Fig. A1, made up from the appropriate parabolic arc together with part of the zero solution (the derivative is continuous at A). Here, $\mathscr{D}$ is the domain $x > 0$, I is $-\infty < t < \infty$. The solution in fact leaves $\mathscr{D} \times I$ at the point A.

FIG. A1

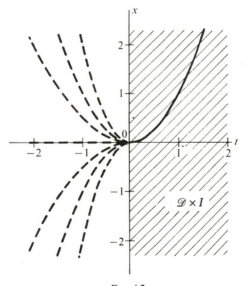

FIG. A2

Example A4   *The one-dimensional equation $\dot{x} = x/t$, $x(1) = 1$.*

$\mathcal{D}$ is $-\infty < x < \infty$, and I is $t > 0$. The general solution is $x = ct$, and the required solution is $x = t$. The solution leaves $\mathcal{D} \times I$ at $(0, 0)$. Despite the singularity at $t = 0$, the solution continues uniquely in this case into $-\infty < t < \infty$.

Example A5   *The one-dimensional equation $\dot{x} = 2x/t$, $x(1) = 1$.*

$\mathcal{D}$ is $-\infty < x < \infty$, and I is $t > 0$. The general solution is $x = ct^2$, and the required solution is $x = t^2$, $t > 0$ (t remaining within I). However, as Fig. A2 shows, there are many continuations of the solution (having a continuous derivative), but uniqueness breaks down at $t = 0$.

For a regular system we shall define I_{max} (the maximal interval of existence of the solution in Theorem A1) as the *largest interval of existence of a unique solution* x^*. I_{max} can be shown to be an open interval.

THEOREM A2   *Under the conditions of Theorem A1, either $(x^*(t), t)$ approaches the boundary of $\mathcal{D} \times I$, or $x^*(t)$ becomes unbounded as t varies within I.* (For example, the possibility that as $t \to \infty$, $x^*(t)$ approaches a value in the interior of $\mathcal{D}$ is excluded.)

In Example A2, I_{max} is $-\infty < t < 1$, and as $t \to 1-$, $x^*(t)$ becomes unbounded. In Example A3, $x^*(t) \in \mathcal{D}$ only for $-1 < t < \infty$, and the solution approaches the boundary of $\mathcal{D} \times I$ as $t \to -1+$, at the point $(-1, 0)$. In Example A5, the boundary at $(0, 0)$ is approached.

The following theorem shows that *linear systems* have particularly predictable behaviour.

THEOREM A3   *Let $\dot{x} = A(t)x + h(t)$, $x(t_0) = x_0$, where A and h are continuous on $I: t_1 < t < t_2$ (t_1 may be $-\infty$, and t_2 may be ∞). Then $I \subseteq I_{max}$ for all $t_0 \in I$, and for all x_0.*

In this case the unique continuation of the solutions is guaranteed at least throughout the greatest interval in which A, h are continuous.

Appendix B: Hints and answers to the exercises

Chapter 1

1.   (i) Paths $y = kx + C$.
    (ii) Paths $y = 4x^2 + C$; note that paths approach equilibrium points in finite times.
   (iii) Paths $y = C$, $(|x| \leqslant 1)$; $\frac{1}{2}y^2 = kx + C$, $(|x| > 1)$.
   (iv) Stable node.
    (v) Unstable spiral.
   (vi) $y > 0$: stable node; $y < 0$: unstable node; joined along $y = 0$.
  (viii) Paths $y^2 + x^2 \operatorname{sgn}(x) = C$.

2. $\alpha > 0$, centre at $(0, 0)$; $\alpha < 0$, centre at $(0, 0)$ and saddles at $(\pm\sqrt{(-\alpha^{-1})}, 0)$.

8. Paths $\frac{1}{2}y^2 - x^2 = C$. Step lengths of equal time intervals δt are $\delta s \simeq \delta t \sqrt{(6x^2 + 2C)}$.

9. Constant increments in the polar angle give constant time intervals on all paths (all periodic solutions have the same period).

11. Centre at $x = 0$, saddle points at $x = \pm\sqrt{6}$. For the exact equation, the saddle points are at $x = n\pi$.

12. An impulse-excited oscillation: the sudden injection of energy in each cycle offsets the damping. Note that the amplitude is independent of α.

13. $E/2F_0$.

14. $ma^2\ddot{\theta} = mga \sin\theta + F_0 \operatorname{sgn}(\dot{\theta} - \Omega)$. Two equilibrium states, at $\sin^{-1}(F_0/mga)$ and $\pi - \sin^{-1}(F_0/mga)$, when $F_0 < mga$; one, at $\frac{1}{2}\pi$, if $F_0 = mga$; none if $F_0 > mga$.

15. Centre at $x = -1$, centre/saddle point at $x = 0$, saddle point at $x = 1$.

17. Centre at $\theta = 0$, saddle points at $\theta = \pm\pi$, when $\lambda \geqslant 1$. Centres at $\theta = \cos^{-1}\lambda$, saddle points at $\theta = 0, \pm\pi$, when $0 < \lambda < 1$.

18. For $x = \lambda$, the equilibrium points are stable for $0 < \lambda \leqslant 1$, unstable for $\lambda \leqslant 0$ and $\lambda > 1$. For $x = \sqrt{\lambda}$ the equilibrium points are stable for $\lambda \geqslant 1$ and unstable for $0 \leqslant \lambda < 1$. For $x = -\sqrt{\lambda}$, all are unstable.

19. $\omega^2 < g$: origin is a centre. $\omega^2 = g$: every x is an equilibrium point. $\omega^2 > g$: the origin is a saddle point (note that $y = \pm\sqrt{(\omega^2 - g)}$ are phase paths).

20. $|x| \leqslant a$. Paths $y^2 + (x - a)^2 = C$ for $x > a$; $y = C$ for $|x| \leqslant a$; $y^2 + (x + a)^2 = C$ for $x < -a$.

23. $[2E e^{-2\pi/3\sqrt{3}}/(1 - e^{-4\pi/\sqrt{3}})]^{1/2}$.

24. Take initial conditions $x = 0$, $\dot{x} = v_0$, at $t = 0$. Cycle ends with $\dot{x} = V$, where $V^2 = v^2(1 - e^{-2\pi/\sqrt{3}}) + v_0^2 e^{-2\pi/\sqrt{3}}$. Then show $v_0 < V < v$ when $v_0 < v$, and $v < V < v_0$ when $v_0 > v$.

29. $-2\lambda x[\sqrt{(a^2 + x^2)} - a]/\sqrt{(a^2 + x^2)} = m\ddot{x}$.

32. $\theta = 0$: centre if $c < mg(h-a)^3/h$, saddle point if $c > mg(h-a)^3/h$. $\theta = \pi$: centre if $c > mg(h+a)^3/h$, saddle point if $c < mg(h+a)^3/h$. $\theta = \cos^{-1}[a^2+h^2 - (ch/mg)^{2/3}/2ah]$ is an equilibrium point if $mg(h+a)^3/h > c > mg(h-a)^3/h$, a centre.

35. $z = \int[\exp\int g(x)\,dx]\,dx$.

Chapter 2

1. (i) Centre; (ii) saddle; (iii) stable node; (iv) saddle; (v) unstable node; (vi) unstable spiral.

2. (i) Single eigenvalue: degenerate node, straight paths superposed.
    (ii) Line of e.p.'s on $y = x$; paths $y = 2x + constant$.
    (iii) Saddle (one of the straight paths is $x = 0$).
    (iv) Node (one of the straight line paths is $x = 0$).
    (v) Paths are all radial straight lines.
    (vi) Line of e.p.'s on the y-axis; paths $x = constant$.

3. (i) $(0,0)$ unstable spiral; $(1,1)$ saddle.
    (ii) $(-1,0)$ saddle; $(1,0)$ centre.
    (iii) $(1,1)$ stable spiral.
    (iv) $(-1,0)$ stable node; $(1,1)$ unstable spiral; $(0,0)$ saddle.
    (v) $(-1,-1)$ saddle; $(1,1)$ unstable spiral.
    (vi) $(-1,-1)$ stable spiral; $(4,4)$ unstable spiral.
    (vii) $(0,2)$ centre; $(0,-2)$ centre; $(2,0)$ saddle; $(-2,0)$ saddle.
    (viii) $(0,0)$ centre ($x = \pm 1$ is a line of e.p.'s).
    (ix) e.p.'s at $(n\pi, m\pi)$. $n+m$ even—saddle, $n+m$ odd—centre.

4. (i) $(-1,0)$ saddle; $(0,0)$ centre; $(1,0)$ saddle.
    (ii) $(0,0)$ centre.
    (iii) $(-1,0)$ saddle; $(0,0)$ centre; $(1,0)$ saddle.
    (iv) $(0,0)$ stable spiral.
    (v) Diagram has period 2π. Saddles at $x = 0, \pi$; centres at $x = \pm\frac{1}{3}\pi$; and at periodic extensions.

7. Use polar coordinates.

12. Let $D = a_1c_2 - a_2b_1$. There are 3 equilibrium points when $D > 0$: at $(H,P) = (0,0)$, saddle point; $(a_1/b_1,0)$, saddle point, and $(a_2/c_2, D/c_1c_2)$, stable node if $D > a_2b_1^2/4c_2$, stable spiral if $D < a_2b_1^2/4c_2$. If $D < 0$ there are two significant points: $(0,0)$, saddle point; $(a_1/b_1,0)$, stable node. P disappears if $D < 0$, and approaches D/c_1c_2 if $D > 0$.

13. $(0,0)$, $(a_1/b_1,0)$, $(a_1b_2 - c_1a_2, b_1a_2 + c_2a_1)/(b_1b_2 + c_1c_2)$, $(0, a_2/b_2)$.

14. One equilibrium point at $(a_1c_2/a_2c_1, a_1/c_1)$; a stable node if $a_2 > 4a_1$, a stable spiral if $a_2 < 4a_1$.

17. $v = 0$ and any θ; the roots of $(\gamma + 1)v - \gamma\sqrt{(2v)} + (\gamma - 1)c = 0$ and the corresponding θ.

18. $u = (h^2/\gamma)^{(\alpha-3)^{-1}}$, $\alpha \neq 3$; none if $\alpha = 3$, except the trivial case. A centre if

$\alpha < 3$, a saddle point if $\alpha > 3$. The gravitational orbit, $\alpha = 2$, is stable in this sense.

21. (i) $P = \frac{1}{2}mg$ (ii) $P(0) = \frac{1}{2}mg, P(\frac{1}{2}\pi) = mg$.

22. $\vartheta = 0, \pi, \pm\cos^{-1}(1/4)$.

30. (i) $\partial P/\partial y < 0, \partial Q/\partial x < 0$. (ii) $P(x,0) > 0$ for $x < a$; $P(x,0) < 0$ for $x > a$, $Q(0,y) > 0$ for $y < b$, $Q(0,y) < 0$ for $y > b$. (iii) There exists $K > 0$ such that $P(x,y) \leqslant 0$ and $Q(x,y) \leqslant 0$ if $x \geqslant K$ or $y \geqslant K$.

32. Equilibrium point at $x = a\sqrt{m_1}/(\sqrt{m_1} + \sqrt{m_2})$.

33. Write the system in the form $\dot{x} = -\sigma x + g(y), \dot{y} = g(x) - \sigma y$, and sketch the curves $y = g(x)/\sigma$, $x = g(y)/\sigma$, looking for the symmetry.

Chapter 3

1. (i) 0; (ii) 0; (iii) 1; (iv) 1; (v) -2.

2. Eq. pts at $\theta = n\pi$, index -1 if n odd, 1 if n even.

3. (i) -2; (ii) -2; (iii) 1, -1.

4. The sense in which Γ is described is the key to the result.

5. $a^2 < 4\lambda$, no eq. pts $a^2 = 4\lambda$, eq. pt at $x = \frac{1}{2}a$, index 0. $a^2 > 4\lambda$, eq. pt at $x = \frac{1}{2}a + \frac{1}{2}\sqrt{(a^2 - 4\lambda)}$, index -1, and at $x = \frac{1}{2}a - \frac{1}{2}\sqrt{(a^2 - 4\lambda)}$, index 1.

8. Write as $(X + iY)e^{-i\alpha}$.

12. Spiral at $(0,0)$, index 1; saddle point at $(1,1)$, index -1. Index at infinity must be 2.

13. The paths are almost parallel in a small region.

14. Let $\theta(s)$ be the angle between $\mathbf{X}_1$ and $\mathbf{X}_2$ on Γ, s being a parameter for the curve, $\alpha \leqslant s \leqslant \beta$. Since $|\theta(s)| < \pi$, continuity gives $\theta(\alpha) = \theta(\beta)$, which is essentially the result.

15. In the notation of Exercise 14 put $\mathbf{X}_1 = (y, -x), \mathbf{X}_2 = (y, -\sin x)$. On any closed curve containing $(0,0)$, $\mathbf{X}_1$ and $\mathbf{X}_2$ have the same first component.

17. See Theorem 3.3.

28. Put $\rho = x^3$. Also, $x(t) \equiv 0$ is a solution.

29. (i) No eq. pt; (ii) only eq. pt is on the path $y = 0$; (iii) no eq. pt; (iv) no eq. pt; (v) no eq. pt; (vi) only eq. pt is on path $y = 0$; (vii) Bendixson's test.

31. Yes.

32. $\operatorname{div}(\mathbf{X})$ is of one sign for $r < \sqrt{2}$.

34. $a \approx 2$.

37. For any periodic solution, $\displaystyle\int_{\mathscr{D}} \operatorname{div} \mathbf{X}_1 \, dx \, dy = 0$, where $\mathscr{D}$ is the interior of the corresponding path for the system $\dot{\mathbf{x}} = \mathbf{X}_0$. This may be satisfied for all such $\mathscr{D}$, in which case the origin remains a centre with paths close to those of the original system. There may be no such $\mathscr{D}$ giving the integral as zero, or there may be one or more corresponding to limit cycles.

For stability we require paths to approach the limit cycles from both sides. Since, for any $\mathscr{B}$, $\displaystyle\int_{\mathscr{B}} \operatorname{div} \mathbf{X} \, dx \, dy = \int_{\mathscr{C}} \mathbf{X} \cdot \mathbf{n} \, ds$, where $\mathbf{n}$ is the outward-pointing unit normal, intuition suggests that a sufficient condition for stability is that

$$\varepsilon \int_{\mathscr{B}} \operatorname{div} \mathbf{X}_1 \, dx \, dy > 0 \quad \text{or} \quad < 0$$

according as $\mathscr{D}$ is contained in $\mathscr{B}$, or $\mathscr{B}$ is contained in $\mathscr{D}$, respectively.

The example is a variant of van der Pol's equation. When $\varepsilon = 0$ the paths are $\frac{1}{2}x^2 + y^2 = c$. When $\varepsilon \neq 0$, the limit cycle is close to $c = \sqrt{2}$.

38. Use Bendixson's Negative Criterion.

39. $x_1^2 + x_2^2 = $ constant, $x_3 = $ constant are closed.

42. If $ax + by + c = 0$ contains the points it intersects each of the conics $X(x, y) = 0$, $Y(x, y) = 0$ at three points. Hence X and Y have the common factor $ax + by + c$.

43. Compare Exercise 42: since X and Y are quadratic and relatively prime there can be no more than two points of tangency of any straight line to the path.

Chapter 4

1. (i) $x^2 + y^2 = 1$, stable.   (ii) $x^2 + y^2 = 1$, unstable.

2. $r = r_k$ stable if $f'(r_k^2) < 0$, unstable if $f'(r_k^2) > 0$, either possible if $f'(r_k^2) = 0$.

3. (i) $a = 1$, stable.

   (ii) $a = 2$, stable.

   (iii) $a = 2^{3/4}$, stable.

   (iv) $a = \sqrt{(n\pi)}$, $n = 1, 2, \ldots$: n even, stable; n odd, unstable.

   (v) $a = 3\pi/4$, stable.

   (vi) $a = 2$, stable.

   (vii) $a = 2\sqrt{3}$, stable.

4. Stable. $a(\theta) = 2/\{1 - (1 - 4/a^2(0)) e^{2\varepsilon\theta}\}^{1/2}$.

5. Stable. $a(\theta) = \frac{3}{4}\pi/\{1 - (1 - 3\pi/4a(0)) e^{\frac{1}{2}\varepsilon\theta}\}$.

7. $a = 2$, $T = 2\pi(1 + \frac{3}{2}\varepsilon)$.

9. $a = 0.197$.

12. $\ddot{x} + \varepsilon(a^2 - 1)\dot{x} + x = 0$.

13. Amplitude 8/3, frequency $\sqrt{(19/3)}$.

17. Saddle at $x = 0$, centres at $\pm 1/\sqrt{\alpha}$: the separatrix cuts the x axis at $\sqrt{(2/\alpha)}$. The periodic solutions found embrace all these points.

18. The 'restoring force' is unsymmetrical: effectively, a bias is present, displacing the mean displacement. Centre at $x = 0$, saddle at $x = 1/\alpha$ with separatrix cutting the x axis at $x = -1/2\alpha$ (paths $x^2 - y^2 - \frac{2}{3}\alpha x^3 = C$): periodic solutions confined to $-1/2\alpha < x < 1/2\alpha$.

20. Compare answer to 18.

Chapter 5

1. $\Omega = \pm 1$: none. $\Omega \neq \pm 1$: $\Gamma \cos t/(\Omega^2 - 1)$. Also for $\Omega \neq \pm 1$ (p, q integers):
   $\Omega = p \neq \pm 1$, $A \cos pt + B \sin pt + \Gamma \cos t/(\Omega^2 - 1)$, period 2π;
   $\Omega = p/q \neq \pm 1$, $A \cos (pt/q) + B \sin(pt/q) + \Gamma \cos t/(\Omega^2 - 1)$, period $2\pi q$.

2. (i) $-0.15\cos t + 0.03\cos 3t$.
    (ii) $3.2\cos t - 0.027\cos 3t$.
    (iii) $2\cos t + 1.2\sin t - 0.024\sin 3t$.
4. (i) $x = a\cos t + \varepsilon a^2(\tfrac{1}{6}\sin 2t - \tfrac{1}{3}\sin t)$
    (ii) $x = a\cos t + \varepsilon a^2(\tfrac{1}{6}\sin 2t - \tfrac{1}{3}\sin t)$
7. $x = \dfrac{\Gamma + \varepsilon a_1}{\Omega^2 - 1}\cos t + \dfrac{\varepsilon a_3}{\Omega^2 - 9}\cos 3t$.
8. $x = 2\sqrt{(\beta/3)}\cos(3t - \alpha) + \tfrac{1}{8}\varepsilon\gamma\cos t$;
13. $|x|/a \ll 1$, $\lambda l/2m\omega^2 a \ll 1$. (Note: a dimensionless form is required.)
15. $x = -\dfrac{1}{8}\left\{\Gamma + \tfrac{1}{8}\varepsilon\left(\dfrac{3\Gamma^3}{256} - \beta\Gamma\right)\right\}\cos 3t$.
17. $x = -\dfrac{\varepsilon\Gamma^2}{2\Omega^2(\Omega^2 - 1)^2} + \dfrac{\Gamma}{\Omega^2 - 1}\cos t - \dfrac{\varepsilon\Gamma^2}{2(\Omega^2 - 1)^2(\Omega^2 - 4)}\cos 2t + \ldots$
18. $x = a_0\cos 3t + b_0\sin 3t + \tfrac{1}{8}\Gamma\cos t - \tfrac{1}{18}\varepsilon(a_0^2 + b_0^2) - \tfrac{\varepsilon}{40}\Gamma(a_0\cos 2t + b_0\sin 2t)$
    $-(\Gamma^2\varepsilon/1152) - (\Gamma^2\varepsilon\cos 2t/640) + \ldots$
21. $u = k(e\cos\theta + 1) + \varepsilon k^3\{e\theta\sin\theta + \tfrac{1}{2}e^2 + 1 - (\tfrac{1}{3}e^2 + 1)\cos\theta - \tfrac{1}{6}e^2\cos 2\theta\}$.

Chapter 6
2. The same to $O(a^6)$.
4. $x = au$, $\mu = \varepsilon a^2$.
9. Uniform solution $e^1(e^{-x} - e^{-x/\varepsilon})$.
10. $y_0 = e^{\frac{1}{2}(1 - x^2)}$; $y_1 = e^{1/2}(1 - e^{-x/\varepsilon})$; $y = e^{1/2}(e^{-\frac{1}{2}x^2} - e^{-x/\varepsilon})$ (uniform).
11. For the outer expansion the coefficient of y' vanishes when $y = 1$; that is, the outer problem is singular.
14. $x = a\,e^{-\varepsilon t}\cos(t + \alpha)$, a and α constants.
15. The exact solution is periodic.
20. $y_0 = 1$, $y_1 = 1 - e^{-x/\varepsilon}$.
21. Put $x = X_0(\tau) + \varepsilon X_1(\tau) + \ldots$, $t = \tau + \varepsilon T_1(\tau) + \ldots$, and derive the equations for X_0 and X_1. Assume $t = 1 \Leftrightarrow \tau = \tau^* = 1 + \varepsilon\tau_1^* + \ldots$. Confirm that $\tau_1^* = -T_1(1)$, $X_0(1) = 1$, $X_1(1) = T_1(1)X_0'(1)$. Finally, remove the worst singularity from the equation for X_1 by putting $\tau^2 T_1' - \tau T_1 = 1$. The answer (which is exact) is $x = \tau^{-1}$, $t = \tau + \tfrac{1}{2}\varepsilon(\tau - \tau^{-1})$.
22. Uniform approximation $y = x + e^{-x/\varepsilon}$; exact solution $y = x - \varepsilon + (1 + \varepsilon)e^{-x/\varepsilon}$. The outer approximation is $x + O(\varepsilon)$ when $y(0) = O(1)$.
23. Compare the result of Exercise 17.
27. $t = \tfrac{1}{3}\sqrt{2}z^{3/2} + \mu\sqrt{2}\{\tfrac{1}{3}z^{3/2} + \tfrac{1}{2}z^{1/2} - \tfrac{1}{4}\log[(1 + \sqrt{z})/(1 - \sqrt{z})]\} + O(\mu^2)$ except near $z = 1$.

Chapter 7
2. In polars $\dot r = \dfrac{\Gamma}{2\omega}\sin\theta$, $\dot\theta = \dfrac{1}{2\omega}(\omega^2 - 1 - \tfrac{1}{8}r^2) + \dfrac{\Gamma}{2\omega}\dfrac{\cos\theta}{r}$.

8. Equilibrium points at $x = \pm 2.5$. Linearized equations $\ddot{\xi} \pm 0.8\xi = 0.2\cos t$.
18. The equations for the equilibrium points (see Exercise 19) in fact have only the zero solution except in the case given; however, it is easy to produce spurious solutions by 'squaring and adding'.

Chapter 8
3. $r = (2n-1)\pi$ (form $d(r^2)/dt$).
4. $y = e^{Cx}$, stable when $C < 0$, unstable when $C > 0$.
9. Use Theorem 8.8.
10. (i) Theorems 8.1 and 8.8. (ii) Substitute $u = e^t x$.
11. Solve and use Theorem 8.8.
12. Write $r(\theta) = a + \xi(\theta)$ and linearize.
13. (i) Lin. dep.; (ii) lin. dep.; (iii) lin. indep.; (iv) lin. dep.
19. Mean value of $\mu(t)$ is k. Unstable.

Chapter 9
1. $E = \Phi^{-1}(t)\Phi(t+2\pi) = \Phi^{-1}(0)\Phi(2\pi)$, $\mu = e^{\pm\pi}$, $\rho = \pm\pi$.
3. Use Theorem 9.4.
7. Use $E = \Phi^{-1}(t)\Phi(t+2\pi) = \Phi^{-1}(0)\Phi(2\pi)$.
11. The matching gives $\sigma^2 = -(\alpha+\frac{1}{4}) \pm \frac{1}{2}\sqrt{(4\alpha+\beta_1^2)}$. Reject negative sign since $\alpha = \frac{1}{4}$, $\beta = 0$ must give $\sigma = 0$. Unstable region is σ real.
12. Put $x = e^{-\frac{1}{2}\kappa t}\eta(t)$; then $\ddot{\eta} + (\nu + \frac{1}{2}\kappa^2 + \beta\cos t)\eta = 0$. Put $x = x_0 + \beta x_1 + \beta^2 x_2 + \dots$, $\nu = \nu_0 + \beta\nu_1 + \beta^2\nu_2 + \dots$, $\kappa = \beta^2\kappa_2$. Then $\nu_0 = 1$, $\nu_1 = 0$, $\nu_2 = \frac{1}{6} \pm \sqrt{(\frac{1}{16} - \kappa_2^2)}$.
18. Linearize and put $\tau = \frac{1}{2}\pi - \omega t$, giving $\theta'' + (1 + \alpha\cos\tau/a)\theta = 4\beta\sin 2\tau$. Then use Theorems 8.1 and 8.8 and the stability of Mathieu's equation near parameter value unity (Exercise 10). The solution is unstable.
20. Substitute $x = e^{-\sigma t}q(t)$; then $\ddot{q} + 2\sigma\dot{q} + (\frac{1}{4} + \beta c + \sigma^2 + \beta\cos t)q = 0$. Assume $\sigma = O(\beta)$, $\sigma = \beta s$, say, and use the perturbation method.
21. Put $x(t) = x^*(t) + \xi(t)$ and make suitable change of the time variable to give
$$\xi'' + \varepsilon(\tfrac{1}{4}r_0^2 - \tfrac{1}{2} - \tfrac{1}{4}r_0^2\sin\tau)\xi' + \{\tfrac{1}{4}(1-2\varepsilon\omega_1) - \tfrac{1}{4}\varepsilon r_0^2\cos\tau\}\xi = 0.$$
The substitution $\zeta(\tau) = \xi(\tau)\exp\{-\frac{1}{2}\int(\frac{1}{4}r_0^2 - \frac{1}{2} - \frac{1}{4}r_0^2\sin\tau)\,d\tau\}$ reduces this to Mathieu's equation near parameter value $\frac{1}{4}$. Exercise 20 gives the exponential behaviour of ζ, and, by the last equation, of ξ.
22. The method of Section 9.1 gives a Mathieu equation.

Chapter 10
1. (i) $V = x^2 + y^2$; (ii) $U = xy$; (iii) $U = xy$; (iv) $V = x^2 + y^2$; (v) $V = 15x^2 + 6xy + 3y^2$; (vi) $U = xy$; (vii) $U = x^2 + y^2$; (viii) $U = xy$; (ix) $U = xy$; (x) $V = 8x^2 + 11xy + 5y^2$; (xi) $U = 3x^2 + 14xy + 8y^2$.
4. $V = (x^2/a^2) + (y^2/b^2)$.
6. (i) $U = x^2 + y^2$; (ii) $U = x^2 + y^2$.
8. $V = x^2 + y^2$.

10. $V = x^2 + y^2$.

11. Let $V = W$ (or $U = W$ in the case where W takes positive values in every neighbourhood of the origin).

22. $V(x, y) = x^2 + y^2 < 1$ is the largest region enclosed by a level curve in which also $\{f(x)/x\} + \delta < 0$, so that $\dot{V} < 0$.

23. The interior of the greatest domain $V(x, y) = $ constant such that $x < 1$ and $x^2 + 2x^3 < 1$.

25. $V = x^2 + y^2$.

26. If the conditions of Exercise 22 hold for all x, and $\delta \int_0^x f(u)\,du \to \infty$ as $|x| \to \infty$.

27. If the conditions of Exercise 23 hold for all x, and $\int_0^x \{\delta f(u) - \beta g(u)\}\,du$ as $|x| \to \infty$.

Chapter 11

1. The origin is either a centre or a spiral. The symmetry requires a centre.

3. Show that all paths enter the annulus $\frac{1}{2} \leqslant x^2 + y^2 \leqslant 2$ from outside.

4. Theorem 11.2.

5. Theorem 11.3.

7. Theorem 11.4.

8. Theorem 11.4.

9. Theorem 11.2.

10. Theorem 11.3.

Chapter 12

1. (i) $\lambda = 1/8$; (ii) $\lambda = 0$.

2. Equilibrium points occur at $(-\lambda \pm \sqrt{(\lambda^2 - \lambda)}, 0)$ if $\lambda(\lambda - 1) \geqslant 0$; none otherwise.

5. They all have a bifurcation point at $\lambda = 0$. They are representative bifurcations in the sense that a single variable in a higher-order system may display them in projection. (i) is called a *saddle-node bifurcation*; (ii) a *transcritical bifurcation*; and (iii) a *pitchfork*.

6. $s(x) = x\,e^{-x^2}$ (to satisfy the uniform boundedness requirements on s).

7. The phase diagram bifurcates from a global centre to a saddle with symmetric centres.

8. $\mathbf{s}(x) = (0, x\,e^{-x^2})'$ (to satisfy the uniform boundedness requirements on $\mathbf{s}$).

12. The radius is $\sqrt{[(1 - \mu)/\mu]}$.

14. Fixed points occur at $(0, 0)$ and $(1, 0)$ in cartesian coordinates. The first returns are given by
$$r_1 = r_0\,e^{2\pi}/\sqrt{(1 - r_0^2 + r_0^2\,e^{4\pi})}.$$

16. For the section $t = 0$, $(-4/5, 2/5)$ is the fixed point; for the section $t = \frac{1}{2}\pi$, $(2/5, 4/5)$ is the fixed point.

17. The fixed point is $(-4\Gamma/3\omega^2, 0)$.

22. $2x_0 + y_0 - 5 = 0$, $x_0 - y_0 + 4 = 0$.

Bibliography

THE QUALITATIVE theory of differential equations was founded by H. Poincaré and I. O. Bendixson towards the end of the nineteenth century in the context of celestial mechanics. At about the same time A. M. Liapunov produced his formal definition of stability. Subsequent work has proceeded along two complementary lines: the abstract ideas of topological dynamics developed by G. D. Birkhoff and others in the 1920's, and the practical approach exploited particularly by Russian mathematicians in the 1930's, notably N. N. Bogoliubov, N. Krylov, and Y. A. Mitropolsky. Much of the earlier work is recorded in the first real expository text, by Andronov and Chaikin (1949). Since this time the subject has become generalized and diversified to the extent that specialized literature exists in all its areas. The short bibliography below contains the works referred to in the text and others providing introductory or specialist reading, often with extensive bibliographies.

ABRAMOWITZ, M. and STEGUN, I. A. (1965). *Handbook of mathematical functions.* Dover, London.

AGGARWAL, J. K. (1972). *Notes on nonlinear systems.* Van Nostrand, London.

ANDRONOV, A. A. and CHAIKIN, C. E. (1949). *Theory of oscillations.* University Press, Princeton.

ANDRONOV, A. A., LEONTOVICH, E. A., GORDON, I. I. and MAIER, A. G. (1973). *Qualitative theory of second-order dynamic systems.* Wiley, New York.

AMES, W. F. (1968). *Nonlinear ordinary differential equations in transport processes.* Academic Press, New York.

AYRES, F. (1962). *Matrices.* Schaum, New York.

BARBASHIN, E. A. (1970). *Introduction to the theory of stability* (trans. ed. T. Lukes). Wolters-Nordhoff, The Netherlands.

BARNETT, S. (1975). *Introduction to mathematical control theory.* Clarendon Press, Oxford.

BOGOLIUBOV, N. N. and MITROPOLKY, Y. A. (1961). *Asymptotic methods in the theory of oscillations.* Hindustan Publishing Company, Delhi.

CESARI, L. (1971). *Asymptotic behaviour and stability problems in ordinary differential equations* (3rd ed.). Academic Press, New York.

CODDINGTON, E. A. and LEVINSON, L. (1955). *Theory of ordinary differential equations.* McGraw-Hill, New York.

COHEN, A. M. (1973). *Numerical analysis.* McGraw-Hill, London.

COPSON, E. T. (1965). *Asymptotic expansions.* University Press, Cambridge.

CROCCO, L. (1972). Coordinate perturbations and multiple scales in gas-dynamics. *Phil. Trans. R. Soc.* **A272**, 275–301.

CRONIN, J. (1977). Some mathematics of biological oscillations. *SIAM Review* **19**, 100–138.

DAVIES, T. V. and JAMES, E. M. (1966). *Nonlinear differential equations.* Addison-Wesley, Reading, Mass.

FERRAR, W. L. (1950). *Higher algebra*. Clarendon Press, Oxford.

FERRAR, W. L. (1951). *Finite matrices*. Clarendon Press, Oxford.

FRAENKEL, L. E. (1969). On the method of matched asymptotic expansions. *Proc. Camb. Phil. Soc. Math. Phys. Sci.* **65**, 209–31.

GANDOLFO, G. (1971). *Mathematical methods and models in economic dynamics*. North-Holland, London.

GUCKENHEIMER, J. and HOLMES, P. (1983). *Nonlinear oscillations, dynamical systems, and bifurcations of vector fields*. Springer-Verlag, New York.

HALE, J. (1969). *Ordinary differential equations*. Wiley-Interscience, London.

HAYASHI, C. (1964). *Nonlinear oscillations in physical systems*. McGraw-Hill, New York.

HOLMES, P. (1979). A nonlinear oscillator with a strange attractor. *Phil. Trans. R. Soc.* **A292**, 419–448.

JONES, D. S. (1961). *Electrical and mechanical oscillations*. Routledge and Kegan Paul, London.

KRYLOV, N. and BOGOLIUBOV, N. (1949). *Introduction to nonlinear mechanics*. University Press, Princeton.

LA SALLE, J. and LEFSHETZ, S. (1961). *Stability by Liapunov's direct method*. Academic Press, New York.

LEIPHOLZ, H. (1970). *Stability theory*. Academic Press, New York.

LORENZ, E. N. (1963). Deterministic nonperiodic flow. *J. Atmospheric Sci.* **20**, 130–41.

MCLACHLAN, N. W. (1956). *Ordinary nonlinear differential equations in engineering and physical sciences*. Clarendon Press, Oxford.

MAGNUS, K. (1965). *Vibrations*. Blackie, London.

MINORSKY, N. (1962). *Nonlinear oscillations*. Van Nostrand, New York.

NAYFEH, A. H. (1973). *Perturbation methods*. Wiley, New York.

NEMYTSKII, V. V. and STEPANOV, V. V. (1960). *Qualitative theory of differential equations*. University Press, Princeton.

O'MALLEY, R. E. (1974). *Introduction to singular perturbations*. Academic Press, New York.

PAVLIDIS, T. (1973). *Biological oscillators: their mathematical analysis*. Academic Press, New York.

PIELOU, E. C. (1969). *An introduction to mathematical ecology*. University Press, Cambridge.

POSTON, T. and STEWART, I. (1978). *Catastrophe theory and its applications*. Pitman, London.

REISSIG, R., SANSONE, G. and CONTI, R. (1974). *Nonlinear differential equations of higher order*. Nordhoff, Leiden.

ROSEN, R. (ed.) (1973). *Foundations of mathematical biology, Volume III, supercellular systems*. Academic Press, New York.

SANCHEZ, D. A. (1968). *Ordinary differential equations and stability theory; an introduction*. Freeman, San Francisco.

SEWELL, M. J. (1976). Some mechanical examples of catastrophe theory. *Bull. Inst. Math. Applics.* **12**, No. 6.

STOKER, J. J. (1950). *Nonlinear vibrations*. Interscience, New York.

STRUBLE, R. A. (1962). *Nonlinear differential equations.* McGraw-Hill, New York.

URABE M. (1967). *Nonlinear autonomous oscillations.* Academic Press, New York.

VAN DYKE, M. (1964). *Perturbation methods in fluid mechanics.* Academic Press, New York.

WHITTAKER, E. T. and WATSON, G. N. (1962). *A course of modern analysis.* University Press, Cambridge.

WILLEMS, J. L. (1970). *Stability theory of dynamical systems.* Nelson, London.

WILSON, H. K. (1971). *Ordinary differential equations.* Addison-Wesley, Reading, Mass.

Index

The revised edition of this very successful textbook in the qualitative theory of nonlinear differential equations includes some significant changes to bring it into line with current thinking, including a completely new chapter on bifurcations, structural stability, and chaotic motion. The book is for students in mathematics or in mathematics-orientated combinations and is directed towards practical applications. There are over 400 examples and exercises in which a wide variety of mechanical, biological, and electrical applications appear. It treats recent developments and structures such as singular perturbations and the phenomena associated with forced nonlinear equations that have not been readily accessible before at this level. Topics include (as well as those listed above) the phase plane, linear approximation, nonlinear damping, geometrical and computational aspects, small-parameter expansions and singular perturbations, harmonics, sub-harmonics, entrainment, stability, and existence of limit cycles.

From reviews of the first edition (1977):
The book can profitably be used in a senior undergraduate course. It is well written, well motivated and contains some recent developments of interest which have not been readily accessible before at this level.

V. Lakshmikantham in *Mathematical Reviews*

Some other textbooks in this series

Partial Differential Equations
W. E. Williams

Numerical Solution of Partial Differential Equations (Third Edition)
G. D. Smith

Introduction to Mathematical Control Theory (Second Edition)
S. Barnett and R. G. Cameron

Mathematical Models in Applied Mechanics
A. B. Taylor

A First Course in Coding Theory
R. Hill

Vector Calculus
P. Baxandall and H. Liebeck

ISBN 0-19-859656-1

9 780198 596561

OXFORD UNIVERSITY PRESS
Also available in hardback